NANOMATERIALS AND DEVICES

NANOMATERIALS AND DEVICES

DONGLU SHI
ZIZHENG GUO
NICHOLAS BEDFORD

ELSEVIER

AMSTERDAM • BOSTON • HEIDELBERG • LONDON
NEW YORK • OXFORD • PARIS • SAN DIEGO
SAN FRANCISCO • SINGAPORE • SYDNEY • TOKYO
William Andrew is an imprint of Elsevier

William Andrew is an imprint of Elsevier
The Boulevard, Langford Lane, Kidlington, Oxford, OX5 1GB, UK
225 Wyman Street, Waltham, MA 02451, USA

Notices
Knowledge and best practice in this field are constantly changing. As new research and experience broaden our understanding, changes in research methods, professional practices, or medical treatment may become necessary.

Practitioners and researchers must always rely on their own experience and knowledge in evaluating and using any information, methods, compounds, or experiments described herein. In using such information or methods they should be mindful of their own safety and the safety of others, including parties for whom they have a professional responsibility.

To the fullest extent of the law, neither the Publisher nor the authors, contributors, or editors, assume any liability for any injury and/or damage to persons or property as a matter of products liability, negligence or otherwise, or from any use or operation of any methods, products, instructions, or ideas contained in the material herein.

ISBN: 978-1-4557-7754-9

British Library Cataloguing-in-Publication Data
A catalogue record for this book is available from the British Library

Library of Congress Cataloging-in-Publication Data
A catalog record for this book is available from the Library of Congress

For information on all William Andrew publications
visit our website at http://store.elsevier.com/

Typeset by MPS Limited, Chennai, India
www.adi-mps.com

Transferred to Digital Printing in 2014

CONTENTS

PREFACE

The advent of nanotechnology is becoming an ever-visible concept in various aspects of our lives, as evident by its popular (and often incorrect) usage in advertising/marketing and entertainment. Although pop-culture references to nanotechnology are often misused or are total science fiction, its origins are derived from a rapidly growing discipline of science and engineering. Nanotechnology can be defined as the fundamental study and application of materials displaying length scales of more than 100 nm. At this size, nanoscale materials exhibit physical and chemical properties that differ greatly from those of their bulk counterparts. These interesting properties can be advantageously exploited for a number of applications and have substantial real-world impacts in fields such as medicine and those that are energy-related. Although nanotechnology research is growing rapidly throughout the world, the teaching of this subject is lacking at the university level, particularly at the first-year and second-year levels. Introducing nanotechnology earlier during the college curriculum would be of immense benefit to students and to further progress in the field. As such, the motivation in preparing this book is to introduce the concept of nanotechnology in teaching while exposing students to current nanotechnology research. Given the ever-changing nature of nanotechnology, researchers in the field need to be constantly aware of new studies to update their knowledge and to keep current. With the advent of the Internet, research data and ideas are now readily accessible and communicated to the rest of the field and serve as a plentiful source for newcomers looking to learn more about this exciting field. This can be a double-edged sword, however, because the sheer amount of information can be difficult to organize and process. Furthermore, the Internet is a nonrefereed medium and, as such, information needs to be taken with a critical viewpoint. From an educational standpoint, a major challenge is teaching students how to recognize and collect useful online resources while simultaneously instilling an intuition in the students about potentially less credible or incorrect online resources. Motivated by the facts stated here, the author felt the need for a new perspective on the progress of nanotechnology research.

Nanomaterials and related nanoscale devices constitute the core infrastructure of nanoscience and technology. With the development of nanomaterials synthesis and characterization techniques, the fundamental knowledge base has grown considerably, resulting in a thorough understanding of nanoscale properties that can be used to develop novel devices in various application areas. To this end, the major focus of this book is nanomaterials and devices. We hope this textbook will become a useful tool for students to bridge their acquired knowledge to their current or future research activities, because a major aim of this text is to prompt research into practical applications. This book references knowledge from three areas: the author's own research activities, the selected literature, and Internet resources. Regarding content selected from online resources, we have performed extensive background studies to verify that the information is correct. Additionally, we also have cited direct references to a few online resources without the original source being indicated in the reference, and for that we must apologize here and acknowledge the original authors. Our thanks are hereby extended to all the original authors who may be involved in the contents herein.

It is our desire to publish this textbook for many years to come, updating future versions with the newest trends in nanotechnology research. We thank Tsinghua University Press for their support throughout the process of writing this book.

Given the targeted readership level, the short period of preparation, and the inherent diversity within the field of nanotechnology, there may be shortcomings that are inevitable in the book. All colleagues and readers are encouraged to kindly contact the authors with your professional opinions and suggestions for new material.

Donglu Shi
Materials Science and Engineering,
College of Engineering and Applied Science,
University of Cincinnati, Cincinnati, OH

Zizheng Guo
College of Science, South China Agricultural University

Nicholas Bedford
Department of Chemistry, University of Miami,
Coral Gables, FL

March 2014

1

BASIC PROPERTIES OF NANOMATERIALS

CHAPTER OUTLINE

In 1959, US physicist Richard Feynman, the famous Nobel Prize winner, first proposed the concept of "nanomaterials." Since then, Feynman's prediction has gradually become a reality in the development of nanoscience and nanotechnology. Peculiar physical properties of nanomaterials have a significant impact on people's lives and social development. Nanomaterials began to exhibit an increasing number of applications in sectors

Nanomaterials and Devices. DOI: http://dx.doi.org/10.1016/B978-1-4557-7754-9.00001-9

such as medicine, home appliances, computers and electronics, environmental protection, textile industry, machinery industry, and others.

1.1 The Nanometer and Its Brief History, Nanoscience, and Nanotechnology

Nano is the metric unit of the nanometer (nm) via transliteration. Like the millimeter and micron, the nanometer is defined as a scale of length, having no special physical meaning. Specifically, it is equivalent to one billionth of one meter (i.e., $1\ nm = 10^{-9}\ m$). One nanometer introduces a length of approximately two to three metal atoms being arranged together, or a "width" of 10 hydrogen atoms being arranged alone. A typical virus has a diameter of approximately 60–250 nm, a red blood cell has a diameter of approximately 2,000 nm, and the diameter of a hair is 30,000–50,000 nm (Figure 1.1).

Materials prefixed with "nano" can be traced back to the 1980s; it was used to define particles within a range of 1–100 nm. In July 1990, the first session of the International Symposium on Nanoscience and Nanotechnology was held

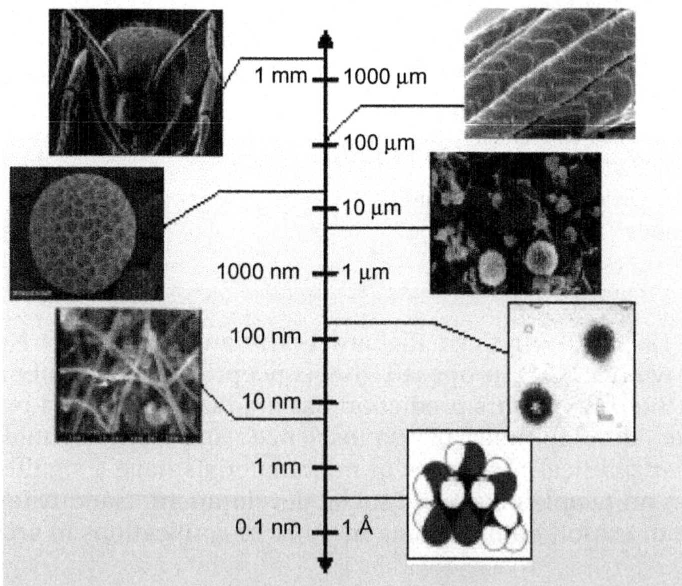

Figure 1.1 Comparison of physical scales.

in Baltimore, MD, and formally announced to the world the science of nanomaterials as a novel branch of materials science. Subsequently, a large number of scientific and technological personnel became engaged in the field of nanotechnology research, and this soon led to a "nano boom" worldwide.

In 1962, Kubo developed the quantum confinement theory on ultrafine particles, which promoted the exploration of nanoparticles in experimental physics. In 1984, the German Professor H. Gleiter and colleagues synthesized nanocrystals such as Pd, Fe, and others. In 1987, Dr. Siegel in the US-based Argonne National Laboratory prepared the nano-TiO_2 polycrystalline ceramics, which show good toughness, without any bending fracture under temperature conditions of 100°C or higher. This breakthrough brought about the first worldwide boom in nanotechnology, officially making it a branch of materials science.

As one of the most common elements in nature, the unique bonding orbital of carbon forms an abundant carbon family. People used to believe there were only three carbon allotropes in nature: diamond, graphite, and amorphous carbon. In 1985, Kroto and colleagues found the cage-like C_{60} molecules with a magic number of 60, in which 60 carbon atoms are respectively located at the top of football-shaped polyhedrons composed of 20 hexagons and 15 pentagons. By using the arc discharge of graphite electrodes, Kratschmer obtained a macro-amount of synthetic C_{60} for the very first time, triggering another wave of nanotechnology research. The later findings were a large family of spherical and spheroidal carbon allotropes.

In 1991, Professor Iljima from Japan's NEC Corporation found a hollow tube in the cathode rod with deposition of carbon black as a result of DC arc discharge in an Ar atmosphere. Under the transmission electron microscope, he found that this hollow tube had a diameter of one nanometer to tens of nanometers, and a length of tens of nanometers to one millimeter. Dozens of these tubes are structured together coaxially, leaving a radial spacing of approximately 0.34 nm between the adjacent hollow tubes, for example the plane spacing of graphite (002). This is what is now referred to as the carbon nanotube. Its unique molecular structure of a one-dimensional tube has opened a novel field in the study of one-dimensional nanomaterials. The discovery of carbon nanotubes led to another peak of nanotechnology study.

At present, nanoresearch involves three main areas: nanodevices, nanomaterials, and nanotechnology detection and characterization. The scientific significance of the research on nanostructure and nanomaterials is that it has opened a novel level of people's understanding of nature, and the subject itself has turned out to be a golden source of knowledge innovation. Nanoscale structural units (1–100 nm) are equivalent to many of the featured lengths in the substances, such as the de Broglie wavelength of electrons, the superconducting coherence length, the thickness of tunneling barriers, and the critical size of magnetic iron, thus making nanomaterials and nanostructures not only different from the microscopic atoms and molecules but also different from the macro-objects in terms of their physical and chemical properties. People's scope of exploring the nature and creating knowledge has been extended to a middle area between the macro- and micro-objects. In the field of nanotechnology, discovering novel phenomena, understanding novel laws, and developing novel concepts and theories, such activities will lay a foundation for building a scientific framework for nanomaterials. Furthermore, this also will greatly enrich the connotation of the study of nanophysics, nanochemistry, and other novel areas.

Nanotechnology renders human a mode of production and work on the nanometer scale, as well as novel tools and skills distinctively different from those in the traditional sense. For example, if we want to build robots that can enter the blood vessels, then we need to make them very small, so tools used by such robots must be made with nanomaterials. Recently, scientists have invented nanoshovels and nanospoons, which can be used by a vascular robot for operations in blood vessels. This is a typical example of nanotools.

Nanotechnology covers a wide range of contents, such as the following: the manufacturing technology of nanomaterials; the technologies of nanomaterials applied to various fields (including but not limited to high-tech fields); any device that is built in a nanoscope for double-cutting and operation of atoms and molecules; the understanding of new laws of the material transfer and energy transfer within the nanoscope; and others. So, we should not think that nanotechnology merely refers to nanomaterials, or that nanomaterials merely refer to nanopowder. Nanomaterials actually include nanofilaments, nanotubes, nanowires, nanocables, nanothin film, the three-dimensional nanoblock, composite materials, and other materials besides nanopowder. In addition, nanomaterials can be either solid or liquid. For example, there is such a thing called nanowater, which contains smaller clusters of water molecules after being processed by high-frequency ultrasound.

1.2 Characteristics of Nanomaterials

1.2.1 Perfect Law of Nanomaterials

In 1959, Feynman assumed the following: "Imagine that if one day, atoms and molecules could be arranged as what people want them to be, how different the world might be! There is no doubt that if we could control things on the very tiny scale, the scope of physical properties we can get can be greatly expanded." Now it is known that, in fact, people cannot organize atoms and molecules truly at will to form nanostructured materials, because their formation requires some special laws to be satisfied, such as the so-called perfect law of nanomaterials.

Atomic and electronic structures are commonly used to describe the structure of materials. The main parameters for atomic structure are the lattice constant, bond length, and bond angle, whereas the electronic structure has the energy band, quantum state, and distribution function as its main parameters. These parameters are constants determined for the macrosystem we are familiar with. But for the nanosystem, the majority of parameters may change as the atomic number changes. This is a typical characteristic found in materials and devices in nanotechnology that determines the diversity of nanomaterials. For the nanosystem there is an important law, and we call it the *perfect law of nanomaterials*. This can be expressed in simple language: "Existence is perfect, and only the perfect can be existent." It includes a magic number rule of nanocrystals; that is, atom clusters with the atomic numbers of 13, 55, 147, and others are considered to be stable. For example, carbon 60 and carbon 70 have the largest probability of existence in the fullerene structure, whereas structural systems such as carbon 59 or carbon 71 do not exist. That is why Smalley and colleagues [1] discovered that carbon 60 and carbon 70 exist in a number of fullerene structures and thus won the Nobel Prize. For one-dimensional nanostructures, including nanotubes and nanowires, similar rules also apply. The one-dimensional structure can be regarded as constituted by the shells, and each of the shells contains a more sophisticated structure known as a *unit*, and each unit is an atomic chain. The structure with the center containing one unit and the parcel layer containing seven units is expressed as the *7-1 structure*. If the structure also has a shell layer packaged with 11 units beyond it, then it is expressed as the *11-7-1 structure*, and so on. The 7-1 and 11-7-1 were only regarded as the most stable structures; this is called the magic number rule for the one-dimensional structure. A two-dimensional membrane is found to obey the defects melting rule; that is, it does not allow the existence of many defects. Once the defects reach a critical number, more defects will arise spontaneously and will completely destroy the two-dimensional crystalline structure. Such characteristics of low-dimensional structures are the specific interpretation of the Perfect Law.

1.2.2 Nano-Effect [2–4]

As materials are reduced to the nanometer scale—within the range of approximately 1–100 nm—the properties of the material may change abruptly so that the material may have some exceptional properties. Materials with such special properties that differ from both the original atomic or molecular components and the macroscopic material are called nanomaterials. Note that if the scale of the materials is within the nanometer range but they do not present special properties, then such materials cannot be called nanomaterials. People used to pay attention only to the microscopic objects like atoms or molecules, or to the macroscopic objects like the universe, and often overlooked this middle scope in between. Actually, a lot of materials exist within this scope in nature, except that we have never noticed the properties of physical objects of this scale before. It was Japanese scientists who took the initiative to gain a true reorganization of the performance of objects within this scale and introduced the concept of *nanotechnology*. In the 1970s, they successfully prepared advanced micro-ions by using the evaporation method and studied their performance. They found that metals such as copper and silver with electrical and thermal conductivities, after being reduced to the nanoscale, will lose their original nature and become nonconductive and nonthermal.

The same is true for magnetic materials, for example iron–cobalt alloy. When this alloy is prepared with a size of approximately 20–30 nm, the magnetic domain is changed to a single magnetic domain, exhibiting a coercivity 1,000 times higher than the original. In terms of magnetic susceptibility, nanomagnetic metal is 20 times more effective than that of ordinary metals. Its saturation magnetic moment is half that of ordinary metal. When a layer in multilayer film gets its thickness to nanosize, a giant magneto-resistive effect may occur.

Typically, $PbTiO_3$, $BaTiO_3$, and $SrTiO_3$ are ferroelectrics and can become paraelectrics when their size is reduced to nanoscale. Nanosilicon nitride ceramics are characterized by not having a typical covalent bond and being partially polarized on the interface bond with a small AC resistance. Nanoparticles prepared using inert platinum metals (platinum black) can become catalysts with excellent activity.

Changes in body surface area make the sensitivity of nanomaterials much higher than that of volume materials. Nano-optical material has unusual absorptive capacity, and nanometal presents a significantly decreased light reflection capacity: all these features are attributable to the small size and surface effects that give nanoparticles an extremely strong ability to absorb light.

Self-diffusion of Cu nanocrystals is 10^{16}- to 10^{19}-fold that of traditional crystals and is 10^3 times more than the spread of the crystal boundary. The specific heat of nano-Cu is twice that of traditional copper. The Pd thermal expansion rate of a nanosolid is double that of a normal solid. As a heat exchanger of dilution refrigeration fluid, Ag nanocrystals can be 30% more efficient than those of traditional materials.

When crystals are reduced to nanosize, the dislocation slip is limited to the border and shows hardness that is much higher than that of volume materials. As nanocrystals, copper may have five times stronger hardness than that of the micron-scale. The fracture strength of nano-Fe crystals (6 nm) can be increased 12 times more than polycrystalline Fe.

The following is a more detailed description of nanomaterials in regard to their special nature in the optical, thermal, magnetic, mechanical, and electrical aspects.

1.2.2.1 Exceptional Optical Properties

It has long been known that scale changes of materials come with changes in color. For example, the CaSe powder in larger particles presents a red color, whereas smaller particles of CaSe powder are yellow (Figure 1.2).

When gold was differentiated down to a size less than that of light wavelengths, it would lose its original rich luster and present a black color. In fact, all the metals in the state of being ultrafine particles are shown as black. The smaller the size, the darker the color. Silver–white platinum changes into a platinum black; the metal chrome changes into a chrome black.

Ultrafine particles of metal have a very low rate of light reflection, usually less than 1%. At a thickness of approximately a few microns, light can be completely eliminated. This feature can be used for highly efficient transformation of solar energy for heat and electricity and may also be used in infrared-sensitive devices or infrared stealth technology. The Gulf War

Figure 1.2 Color changes of CaSe particles of different sizes (upper: the powder in larger particles presents a red color; lower: the smaller particles are a yellow powder).

happened in the spring of 1991, and the coated materials used on the US F-117A stealth fighter contained a variety of nano-ultrafine particles, which have a strong ability to absorb electromagnetic waves of different bands. It can deceive radar to cloak the planes. The United States successfully used these fighter jets in striking important military targets in Iraq.

1.2.2.2 Exceptional Thermal Properties

A solid substance has a fixed melting point at its patterns in larger sizes, whereas this point would be significantly reduced in ultrafine forms. When particles are less than 10 nano-orders of magnitude, such decreases are particularly significant.

For example, the conventional melting point of gold is 1,064°C. When its particle size is reduced to 10 nm, the melting point will be 27°C; when the size is reduced to 2 nm, the melting point is only approximately 327°C. The conventional melting point of silver is 670°C, but that of its ultrafine particles can be less than 100°C (Figure 1.3). Therefore, the conductive paste prepared from ultrafine silver powder can be sintered at low temperatures. At this temperature level, the device substrate does not have to be high-temperature ceramic materials; instead, we can use more common materials or even plastic.

Atoms on the surface of metal nanoparticles are quite active. Powder of nanoparticles can be used as solid rocket fuel or catalyst. For example, adding 1% of aluminum or nickel ultrafine particles in rocket fuel (weight ratio) can double the combustion heat.

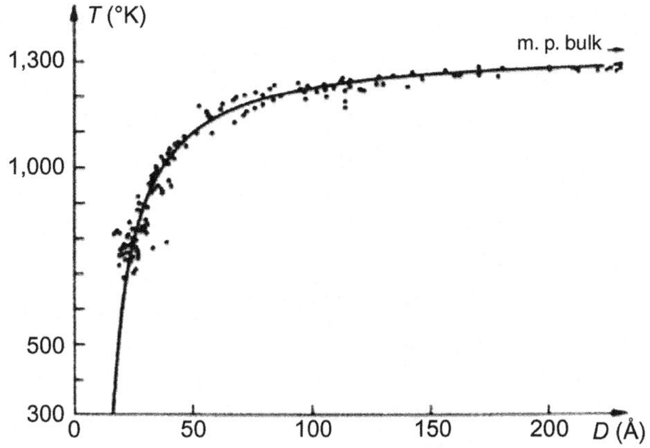

Figure 1.3 Changes of the Ag melting point at different sizes. From Ref. [3].

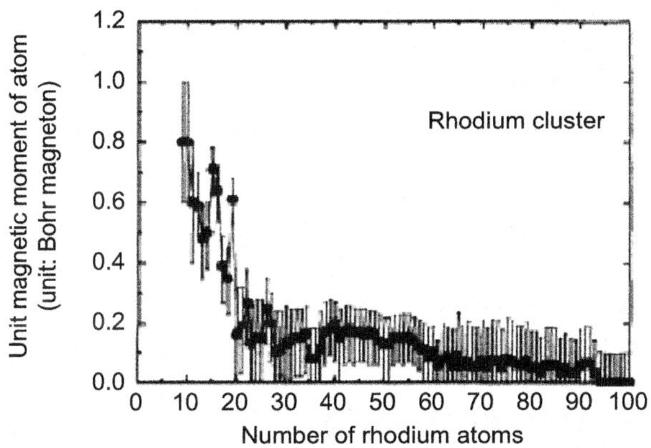

Figure 1.4 Changes in magnetic properties of rhodium clusters with different numbers of rhodium atoms. From Ref. [4].

1.2.2.3 Exceptional Magnetic Properties

When some materials are changed from bulk to nanoscale, great changes may take place regarding their magnetic properties. Rhodium is a typical example in this respect. We can verify this by increasing the number of rhodium atoms in Rh clusters (Figure 1.4).

When the particle size is reduced to 2×10^{-2} μm, its coercivity shows a 1,000-fold increase. However, if its size is further reduced to less than approximately 6×10^{-3} μm, the coercive force will conversely reduce to zero, presenting a surplus paramagnetic nature.

Magnetic properties of ultrafine particles with high coercivity can be used for the production of recording magnetic powder with high-density storage, which is currently used in the manufacture of tapes, disks, and magnetic cards in considerable volume. Based on the superparamagnetic feature, ultrafine magnetic particles can be used to prepare magnetic fluids that can be used for a wide range of purposes.

1.2.2.4 Exceptional Mechanical Properties

Nanomaterial particles have a very small size and have a good surface effect. One gram of nanomaterial can form a surface area of up to several hundred square meters. Products prepared using nanotechnology have superior strength, flexibility, and scalability. We can understand this by looking at a caterpillar with thousands of pairs of feet. When it is adsorbed on a smooth glass surface, its large contact area can even resist a level 12 typhoon.

Ceramic materials are usually very brittle, and a ceramic pot can be easily broken when something is thrown at it. However, nanoceramic material prepared by nano-ultrafine particles can have the same good tenacity as a spring.

Studies have shown that high intensity of people's teeth is due to their structure made of nanomaterials, such as calcium phosphate and others. Metal nanocrystals are as hard as three to five times that of the traditional metals of coarse grains. As for composite nanomaterials (such as metal–ceramic), their application is very broad.

1.2.2.5 Exceptional Electrical Properties

In the scope of nanomaterials, the electronic movement is restricted inside nanoparticles; as a result, quantization of electron energy occurs. Thus, we can prepare special metal particles that can be conductive under a certain voltage and nonconductive under other voltages. Metals used as conductors such as copper will lose their conductivity as their size is reduced to a few nanometers, whereas insulation materials like silicon dioxide will lose their insulating properties and become conductive.

There is also a strange phenomenon in that when a metal nanoparticle acquires an additional electron from the external circuit, it shows negative electricity. The Coulomb force of the electron is strong enough to exclude the next electron from having access to the metal particle from the external circuit, thus cutting off the continuity of current. This is the so-called *Coulomb blocking effect*. The Coulomb blocking effect inspires people to think about whether we can develop a master electronic device that is controlled by a single electron—the so-called single-electron device. Single-electron devices can be small enough to be integrated to make computer chips with capacity and computing speed many times that of the currently used chips.

1.2.3 Natural Nano-Effect

Nanomaterials are usually man-made synthetic materials, but there are many nanomaterials in nature that also demonstrate some peculiar properties. For example, note that some plant leaves, despite dirt in their environment, can remain clean. Over the years, people have found that such a feature involves hydrophobic materials on the rough surface and the outermost lower waxy surface arranged by nano-order (100–200 nm) rules. The nanorough surface is the primary factor in the mechanism that protects the leaves from dirt. Even when the leaves are covered by pollutants, their surface can be easily washed clean with water. We call this the self-cleaning effect, or *lotus effect*.

Nanometer-sized cilia exist on the surface of lotus leaves. Under an electron microscope, we can see that cilia form bulges on the surface of its leaves. This prevents the leaves from becoming easily adhesive to water and sludge particles, and that is why lotus leaves show an obvious effect of self-cleaning. Technically speaking, the surface of a lotus leaf is structured with villi that can strengthen its hydrophobic ability. The so-called hydrophobic effect refers to the fact that the surface of an object has a weak force with water molecules and it is difficult for it to absorb water molecules.

The self-cleaning effect of insects is similar to this. Insects need to maintain a balance of their bodies when they are flying. If their wings carry some dirt, then the weight can cause imbalances and incur flight problems. So, insects have to clean their wings from time to time. Insects with large-scale wings cannot clean their wings with legs, but because the surface of the wings of most insects has a nanostructure, they can be self-cleaning (Figures 1.5–1.7).

It was found that pigeons, dolphins, butterflies, bees, water magnetotactic bacteria, and other organisms have magnetic ultrafine particles within their bodies that enable this type of organism to navigate under a geomagnetic field. Magnetic ultrafine particles in the bodies of bees (G in Figure 1.8) are essentially a biological magnetic compass that can accurately position the geomagnetic field and determine the magnetic

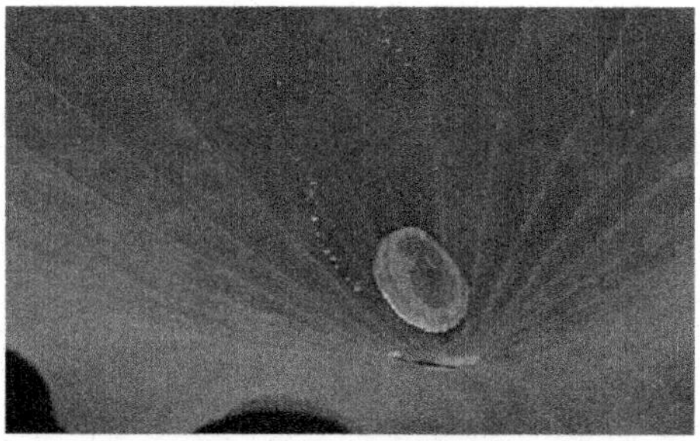

Figure 1.5 Lotus effect: plant leaves are usually able to remain clean without being polluted.

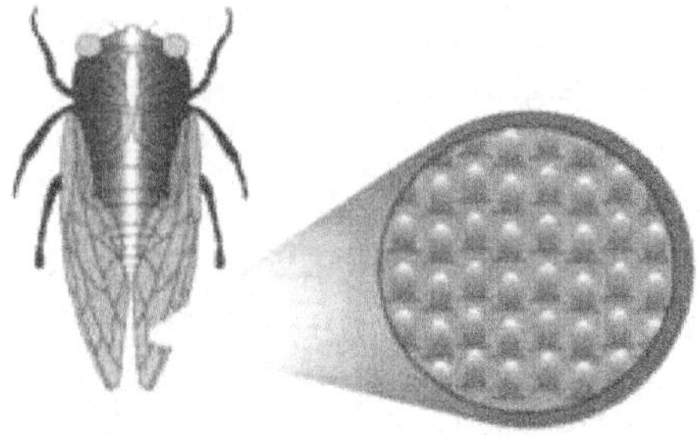

Figure 1.6 Self-cleaning effect of insect wings.

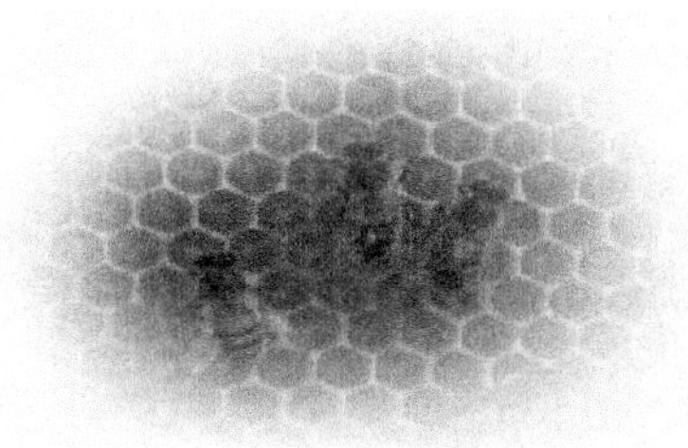

Figure 1.7 Bee wings.

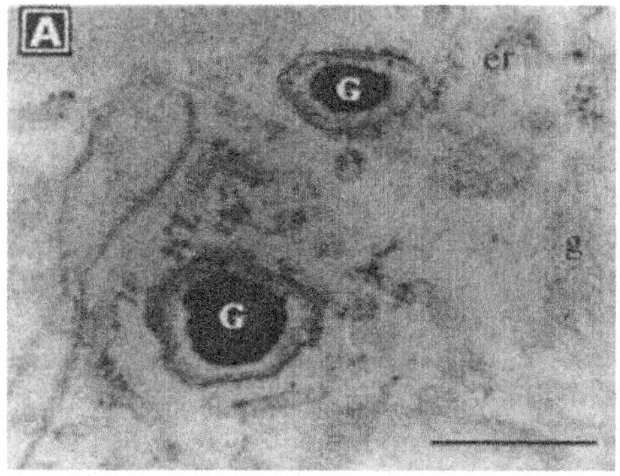

Figure 1.8 Magnetic nanoparticles in the body of bees.

declination and magnetic inclination; it is the flight navigation system of bees. Studies through electron microscopy have shown that magnetotactic bacteria living in water usually contain magnetic oxide particles with a diameter of approximately 2×10^{-2} μm. By virtue of these particles, magnetotactic bacteria can swim down to the nutrient-rich underwater world.

1.3 Physical Principles of the Nano-Effect

As mentioned, nanomaterials are different from the macro-objects in physical and chemical properties, as well as from the microscopic atoms and molecules. When material components reach nanoscale, nanomaterials will show a nature very different from that of volume materials. Nanomaterials can have unique characteristics that are brought about when the scale of a phase on a dimension in the composition of a material is reduced to nanoscale. At the nanometer scale, atomic and molecular interactions may strongly influence macroscopic properties of materials, resulting in changes in mechanical, electrical, and optical properties. Basically, the characteristics of nanomaterials are small size effect, surface interface effect, quantum size effect, quantum tunneling effect, and others [5–8]. These basic features can be explained qualitatively using the theory of Kubo and colleagues.

1.3.1 Discontinuity of Electron Levels

The band theory suggests that a single atom has discrete energy levels. With an increase in the number of atoms in solids (conduction electron number N), atomic levels may change into a discrete band (Figure 1.9). Macroscopic objects contain an unlimited number of atoms; that is, for the number of electrons N conductivity tends to infinity, and large particles or macroscopic objects may have a gap between the macrolevels that come close to zero. Therefore, at high temperature or with overall size, electrons near the Fermi level of metals are generally at a continuous level.

Regarding the nanoparticles containing a limited number of atoms, the value of N is quite small. This leads to a finite energy gap value, which means the energy levels are discrete at low temperatures. As a result of the energy level spacing being greater than thermal energy, magnetic energy, static magnetic energy, electrostatic energy, photon energy, or the concentration energy in the superconducting state, the nanoparticles will significantly vary with respect to their

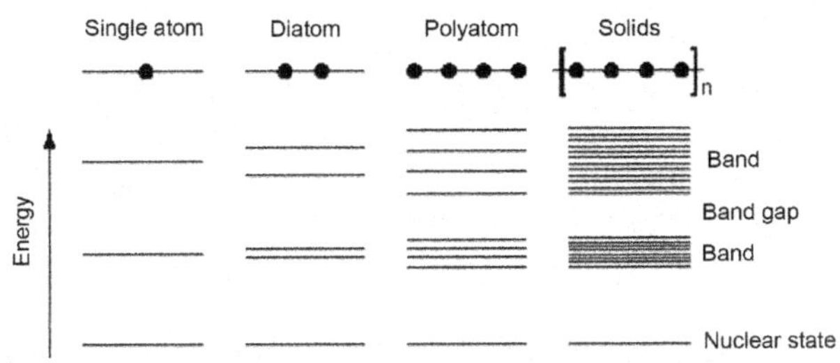

Figure 1.9 From the discrete atomic level to the level in solids band.

magnetic, optical, sound, heat, electricity, superconductivity, and macrocharacteristics.

1.3.2 Kubo Theory

Nanoparticle size is very small and contains very few atoms, thus its mass is minimal. Many phenomena cannot be explained with the nature of bulk matter that usually contains an unlimited number of atoms. This special phenomenon is usually referred to as a *volume effect*. One of the well-known theories is the Kubo theory, which is a typical example of the volume effect.

The Kubo theory refers to the state distribution of electron energy levels near the Fermi surface of metallic ultrafine particles. When particles enter the nanosize, the quantum size effect would trigger the quasi-discrete phenomena on continuous energy levels of the original bulk metal. At the beginning, electronic energy levels around the Fermi surface of a single small particle at low temperatures were regarded as the energy levels of the same spacing. A Fermi surface here refers to the equal energy surface with energy as ε_F in the space k at the temperature of absolute zero. The Fermi surface functions as a separator between the electronic orbits that are filled and not filled.

The specific heat of a single ultrafine particle can be expressed as the following equation:

$$C(T) = k_B \exp(-\delta/k_B T) \tag{1.1}$$

δ is level spacing, k_B is the Boltzmann constant, and T is absolute temperature.

At high temperatures, $k_B T \gg \delta$, $C(T) \to k_B$; that is, the specific heat is independent of temperature, which is consistent with that of the bulk metals. However, at low temperatures ($T \to 0$), $k_B T \ll \delta$, $C(T) \to 0$, which is totally different from bulk metals, because the specific heat and temperature are in a relationship as T^3.

Although the approximate model for the equal energy level is used to derive the specific heat formula of a single ultrafine particle at low temperatures, this practically cannot be confirmed by experiments because we can perform experiments only on a collection of ultrafine particles. Kubo made an outstanding contribution regarding the use of a novel theory with ultrafine particles to find a solution to the difficulties in consistency between theory and experiment. Kubo made two key assumptions of the electronic states of large aggregates of small particles.

1.3.2.1 Hypothesis Regarding Degenerate Fermi Liquid

Kubo took the electronic state of ultrafine particles near the Fermi surface as degenerate electronic gas under size restrictions and further assumed their energy levels as discrete energy levels of the quasi-particle state, while the interaction between quasi-particles can be neglected. As $k_B T \ll \delta$ (δ indicates the average energy level spacing between the two adjacent levels), the distribution of electronic energy levels in such a system near the Fermi surface follows a Poisson distribution.

Here, Δ is the spacing between two levels, $P_n(\Delta)$ is the probability density of the corresponding Δ, and n is the number of energy levels between these two energy states. If Δ is the spacing between adjacent energy levels, then $n = 0$. Kubo and colleagues pointed out that the probability $P_n(\Delta)$ of finding two energy states with a spacing of Δ is relevant to the nature of Hamiltonian transformation. For example, under the conditions of weak interaction of spinning and the orbital (note that the spinning–orbital interaction will enable the energy levels of all atoms with a valence electron to be split into two levels) and a small external magnetic field, the electronic Hamiltonian occurs with the space–time inversion invariance. In the case of Δ being relatively small, $P_n(\Delta)$ decreases as Δ decreases. Obviously, Kubo's model is superior to that of equal energy level spacing, because it reasonably explains physical characteristics of ultrafine particles at low temperatures.

1.3.2.2 Electrically Neutral Assumption of Ultrafine Particles

Kubo believed that it was rather difficult for an electron to be removed or placed in an ultrafine particle. He proposed the following well-known formula:

$$k_B T \ll W \approx e^2/d = 1.5 \times 10^5 k_B/dK\ (\text{Å}) \tag{1.2}$$

In this equation, W is the power made to overcome the Coulomb force for an electron removed from or placed in an ultrafine particle, d is the diameter of the ultrafine particle, and e is the electronic charge. This formula shows that, with the decrease of d values, W increases. Therefore, thermal changes at low temperatures make it very difficult to change the electron-neutrality of ultrafine particles. It is estimated that at temperatures that are low enough, a particle size of 1 nm might have W two orders of magnitude smaller than δ. According to Eq. (1.2), we know that $k_B T \ll \delta$. Small particles of 1 nm showed a highly obvious quantum size at low temperatures.

Kubo and colleagues proposed the relationship of the spacing between adjacent electronic levels and the particle diameter with the following well-known formula:

$$\delta = \frac{4}{3}\frac{E_F}{N} \propto V^{-1} \tag{1.3}$$

Here, N is the total number of conductive electrons in an ultrafine particle, V is the volume of the ultrafine particle, and E_F is Fermi level. It can be expressed as follows:

$$E_F = \frac{\hbar^2}{2m}(3\pi^2 n_1)^{2/3} \tag{1.4}$$

Here, n_1 represents the electron density and m represents equality. In Eq. (1.3), it can be seen that when the particle is spherical, if the particle diameter is expressed by d, then

$$\delta \propto \frac{1}{d^3} \tag{1.5}$$

that is, with the decrease of the particle size, the energy level spacing increases. Clearly, when the particles change into bulk pieces, the volume tends to infinity, and then $\delta \to 0$, in conformity with band theory.

1.3.3 Quantum Size Effect

Material is inversely proportional in level spacing to the atomic number N. Therefore, when the particle size is reduced to a certain extent, with the particle containing a finite number of atoms N, the electronic energy levels near the Fermi level of nanometal will change from the quasi-continuous state into a discrete state. For semiconductor nanoparticles, there are noncontinuous molecular orbitals that are occupied at the highest level and unoccupied at the lowest level, along with a widened energy gap. When the energy gap spacing is greater than material properties such as thermal energy, magnetic energy, electrostatic energy, photon energy, and others, it may lead to distinctively different properties than those of macromaterials. This phenomenon of a widened energy gap is known as the *quantum size effect*.

Nanoparticles contain a finite number of atoms, and the total number of electronic conductivity (N) has a very small value. This leads to a certain value of δ, which means a split-level space. When the energy level spacing is greater than thermal energy, magnetic energy, static energy, photon energy, or the pool of the superconducting state, the quantum size effect should be taken into consideration.

At 1 K, for example, different particle sizes may cause Ag to be either a conductor or an insulator. We can calculate the critical size for such a change as follows.

From $E_F = \hbar^2/2m(3\pi^2 n_1)^{2/3}$ and $\delta = 4/3\ E_F/N$, we get

$$\frac{\delta}{k_B} = (8.7 \times 10^{-18})/d^3 \ \ (\text{K cm}^3) \quad (1.6)$$

When $T = 1$ K, the minimum level spacing is $\delta/k_B = 1$, which when incorporated into the aforementioned formula gives $d = 20$ nm. According to the Kubo theory, only when $\delta > k_B T$ will the split level occur, thus giving rise to the quantum size effect; that is,

$$\frac{\delta}{k_B} = (8.7 \times 10^{-18})/d^3 > 1 \quad (1.7)$$

We can see that as the particle diameter $d_0 < 20$ nm, Ag of nanoparticles turns into a nonmetallic insulator. If the temperature is higher than 1 K, then the requirement $d_0 \ll 20$ nm must be satisfied to be able to become an insulator. It should be noted here that in the actual situation, the metal insulator change requires meeting the conditions of both $\delta > k_B T$ and e-life $\tau > \hbar/\delta$. Experiments found that nano-Ag does have very high resistance, similar to that of an insulator, which means that nano-Ag meets the preceding two conditions.

The problem of materials from the ultraviolet to the visible light range has always been a hot topic. The light emission we are referring to is approximately associated with the microprocess of radiative electron transition. Nanostructured materials have very small particles and that small size will lead to the quantum confinement effect. With the disorder of interfacial structure, it is easy to form excitons, particularly surface excitons. The interface may occupy a larger size, with the existence of many defects in the interface, such as hanging bonds, unsaturated bonds, and impurities. This may have many additional energy gaps in the energy gap. Because of the shifting cycle being damaged in nanostructured materials, rules for choice of electronic transitions in the conventional materials of momentum space are not likely to apply to nanomaterials. These will lead to nanostructured materials being different from conventional materials regarding light emission and having their own characteristics instead.

1.3.4 Small Size Effect

When the ultrafine particle size is smaller than the wavelength of light, the length of magnetic exchange, the width of magnetic domain wall, the de Broglie wavelength of conduction electrons as well as the correlation length or the perspective depth of the superconducting state and other physical characteristics or even less periodic boundary conditions of crystals will be destroyed. At the same time, near the particle surface layer of amorphous nanoparticles, the atomic density reduces and physical properties also show a novel effect, such as changes from magnetic order to magnetic disorder, changes in magnetic coercivity, and decrease in metal melting point. This has led to a novel small size effect that can be found in sound, light, electricity, magnetic, mechanical, and other characteristics, such as a significant increase in light absorption accompanying the plasmon resonance absorption peaks, changing from an ordered magnetic state to the disordered state, superconducting phase changing to the normal phase, changes in phonon spectrum, and so on.

Nanoparticles with small size effect have opened some novel areas of technology application. For example, strong magnetic particles in nanoscale (Fe–Co alloy, iron oxide, and others) can be used in the manufacture of magnetic cards, magnetic keys, magnetic tickets, and magnetic fluids. Among these, magnetic fluid has important applications in electro-acoustic devices, damping devices, rotary seals, and lubrication. Nanoparticles can have melting points well below those of bulk metals and can provide novel technology for the powder metallurgy industry. The nature of plasmon resonance frequency changing with the size can be used in combination with controlled displacement of the absorption edge for the manufacture of nanomaterials of microwave absorption of a certain bandwidth.

The optical properties of a material depend on its reflective properties or the performance of absorbing sunlight. For example, if a leaf is green, then it means that it absorbs other wavelengths of light while reflecting the wave of the green. Red pigment indicates that it absorbs other wavelengths of light while reflecting the red wave. Nanoparticle size can be as small as several nanometers or dozens of nanometers, showing a peculiar small size effect and surface interface effect. Thus, its optical properties are also different from those of the conventional block and coarse materials. Nanometal powders have a special role in the absorption of electromagnetic waves and can be used as high-performance military stealth materials in millimeter-wave infrared stealth materials and structural stealth materials, as well as radiation shielding materials for mobile phones. For example, glass is an insulator, and it cannot release the electromagnetic wave that it absorbs. However, heavy metals can be vaporized to generate nanomaterials with highly conductive properties. Therefore, the absorption of static electricity can be derived through the ground wire connected to the shield to eliminate static electricity that might cause harm to the human body. In addition, computer screens emit electromagnetic waves at non-uniform frequencies. Therefore, the glass surface is not evenly coated in nanomaterials. Instead, based on the changes of radio-frequency, the coating can offset the frequency changes of electromagnetic waves. Screens processed in this way can protect the eyes from being injured by the flashing lights and make the screen look clearer.

As nanoparticles are coupled with the surface effect and quantum size effect, the diameter of nanoparticles is equivalent to the superconducting coherence wavelength, Bohr radius, and the de Broglie wavelength of electrons. At the same time, atoms and electrons on the surface of a particle present a greater difference than those inside the particle. This feature has a significant impact on the optical properties of nanoparticles. For example, bulk metals have a luster of different colors, indicating that they are different regarding the reflection and absorption of various wavelengths in the visible range. When reduced to nanometer size, nanoparticles of almost all metals would turn to black, indicating that they have very low reflectance of visible light.

Nanoparticles have a stealth role against infrared and electromagnetic waves because of the following. First, nanoparticle size is much smaller than the infrared and radar wavelengths, with the transmission of such waves being much higher than that of conventional materials. This can greatly reduce wave reflectivity, so the infrared detectors and radar can receive very weak signals and thus achieve a stealth effect. Second, the surface area ratio of nanoparticle materials is much greater than that of conventional materials, making the reflection signal strength received by infrared detectors and radar greatly reduced. At present, the stealth coating has become useful in modern military confrontation. In developing the fourth generation of supersonic fighter aircraft, its body structure is made of composite materials, wing–body integration, and radio-absorbing coatings. Electromagnetic wave-absorbing coatings and shielding paint have started to be placed on the stealth aircraft. Because of the excellent absorption characteristics of nanomaterials, coupled with features such as broadband, compatibility, small quality, and slight thickness, the United States, Russia, France, Germany, Japan, and other countries have used nanomaterials as a new generation of stealth materials in their research. For metals, metal oxides, and some nonmetallic materials, an increasing number of atoms might deposit onto the surface of the nanoscale ultrafine powder during the refining process, thus increasing the activity of nanomaterials. Under the radiation in the microwave field, the atomic and electronic movements are intensified to promote magnetization. Electronic energy is transferred into heat energy, thereby increasing the absorption of waves. For example, it has been reported that there is nanometal automotive paint. It is a new generation of high-level automotive coating material that uses a variety of nanometal powder material combined with advanced foreign technology in the production of nanometal automotive paint. The product has extraordinary adhesion and high resistance against chemicals such as acids, alkalis, and antioxidants. The automotive paint also features a flip-flop effect and provides excellent protection against physical impact such as wear, scratches, and collisions. In addition, it can also absorb harmful radiation to protect the human body and extend the life of the paint.

1.3.5 Surface Effect

The surface effect of nanomaterials refers to the ratio of surface atoms and the total atomic number of nanoparticles, which has a sharp increase, resulting in a change of nature as the particle size becomes smaller. Nanoparticles feature a small size and high surface energy, with atoms at the surface accounting for a large proportion (Table 1.1).

As can be seen from Figure 1.10, the percentage of surface atoms will rapidly increase because particle size is less than 10 nm. As the particle size reduces to 1 nm, the surface atomic percentage will be approximately 90%, and the atoms are almost concentrated on the surface of nanoparticles. As the particle size becomes smaller, the number of atoms at the surface will comprise a larger proportion compared with the number of internal atoms. An increase in the number of surface atoms, the smaller atomic coordination, and the high surface energy combine to present high surface activity. Meanwhile, these atoms will become extremely unstable and easily combine with other atoms. For example, nanoparticles of metal can burn in the air. In addition, they will adsorb gases and react with them. Another example is copper. For copper particles of 100 nm in diameter, the surface area ratio is 6.6 m^2/g, and

Table 1.1 Relationship between the Size of Nanoparticles and the Number of Surface Atoms

Size of Nanoparticle, *d* (nm)	Number of Surface Atoms	Proportion of Surface Atoms (%)
10	3×10^4	20
4	4×10^3	40
2	2.5×10^2	80
1	30	90

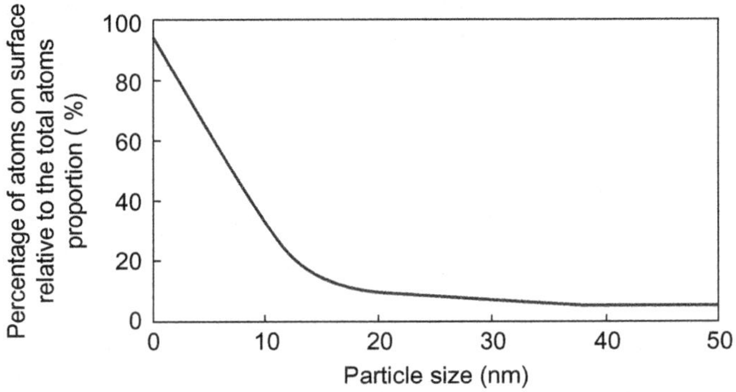

Figure 1.10 Percentage of surface atoms and the total number of atoms of nanoparticles of different sizes.

10-nm particles have a surface area ratio of 66 m^2/g. When the particle size is as small as 1 nm, the surface area ratio will increase to 660 m^2/g.

Activity of surface atoms can induce atomic transport and surface nanostructure changes, and also can lead to conformational changes of spin and electron spectroscopy. An extreme case is the carbon nanotube, which contains only surface atoms, without any internal ones.

1.3.6 Dielectric Confinement Effect

Materials essentially have dielectric properties in addition to the dielectric loss, which constitute the most important physical characteristics of dielectric materials. Polarization of conventional materials is associated with the orderly structure; however, in the structure there is a significant difference between nanomaterials and conventional coarse-grained materials. Nanomaterials have their own unique dielectric behavior (dielectric constant, dielectric loss) that is mainly found in the dielectric constant and dielectric loss, which are strongly dependent on particle size. Frequency of the electric field may have a strong effect on dielectric behavior.

When nanoparticles are dispersed in heterogeneous dielectric materials, the interface can produce dielectric enhancement of the system. This phenomenon is called the *dielectric confinement effect* [9]. It mainly comes from the surface of particles and the local area inside.

When there is a significant difference in the refractive index between medium and particles, a border of the refractive index will appear. This leads to a marked increase of field strength at the particle surface and inside, rather than of the incident field strength. This enhancement of the local area is known as dielectric confinement.

Dielectric confinement effects of nanoparticles are important for light absorption and photochemical and nonlinear optical properties. The effect on the absorption of light reflected in the absorption spectrum is shown with a clear red shift. It is easy to understand this phenomenon. Because of the fact that the size of nanoparticles is smaller than the free range of the carrier, the photovoltaic composite carrier can be lowered. As the particle size decreases, the nature of particles will be strongly influenced by the surface state. When materials with a smaller dielectric constant are modified at the surface of the semiconductor ultrafine particles, larger changes take place in their optical properties compared to those of bare ultrafine particles. This is because, in relation to the surrounding medium of the exposed particles, the power line of the charge carrier surrounded in ultrafine particles can easily pass through this layer of film. Therefore, the shielding effect is weakened while the Coulomb force between charged particles is enhanced, resulting in enhancement of excitons in binding energy and oscillator strength. In the absorption spectrum, this is reflected by a clear red shift.

In the Brus formula, we can give a quantitative or qualitative analysis [10] of dielectric confinement effects of optical absorption band edge shift (blue shift, red shift). The Brus formula indicates that:

$$E(r) = E_g(r = \infty) + \frac{h^2\pi^2}{2\mu r^2} - 1.786\frac{e^2}{\varepsilon r} - 0.248E_{Ry} \qquad (1.8)$$

Here, $E(r)$ is the nanoparticle band gap in absorption, $E_g(r = \infty)$ is the bulk band gap, r is the particle radius, and

$\mu = [1/m_e + 1/m_h]^{-1}$ is the equivalent quality of the particle, where m_e and m_h are the effective mass of the electron and the hole respectively. The second item is the quantum confinement item (blue shift). The fourth item is the effective Rydberg energy. The third item is dielectric confinement, and it is defined as negative. Therefore, it causes a red shift of the absorption spectrum.

Transition metal oxides, such as Fe_2O_3, Co_2O_3, Cr_2O_3, and Mn_2O_3, and other nanoparticles, while dispersed in sodium dodecylbenzenesulfonate, will present enhanced third-order nonlinear optical effects. Measurements of Fe_2O_3 nanoparticles found that the third-order nonlinear coefficient $\chi^{(3)}$ can be as high as 90 m^2/V^2, which is higher by two orders of magnitude in the water. This third-order nonlinear enhancement is also attributed to the phenomenon of the dielectric confinement effect.

References

[1] H.W. Kroto, J.R. Heath, S.C. O'Brien, R.F. Curland, R.E. Smalley, C60: buckminsterfullerene, Nature 318 (1985) 162.

[2] D.S. Yan, D. Feng, Nova of Materials: ¾ Nano Materials Science, Hunan Science and Technology Press, Changsha, 1998.

[3] P.h. Buffat, J.-P. Borel, Size effect on the melting temperature of gold particles, Phys. Rev. A 13 (1976) 2287.

[4] A.J. Cox, J.G. Louderback, S.E. Apsel, L.A. Bloomfield, Magnetism in 4d-transition-metal clusters, Phys. Rev. B 49 (1994) 12295.

[5] M.S. Cao, Introduction to Nano-Materials, first ed., Harbin Institute of Technology Press, Harbin, 2007.

[6] Z.K. Zhang, Z.L. Cui, Nano-Technology and Nano-Materials, National Defence Industry Press, Beijing, 2000, 147.

[7] L.D. Zhang, J.M. Mou, Nano-Materials and Nano-Structures, Science Press, Beijing, 2001, 144.

[8] R. Denton, B. Muhlschlegel, D.J. Scalapino, Electronic heat capacity and susceptibility of small metal particles, Phys. Rev. Lett. 26 (1971) 707.

[9] T. Takagahara, Effects of dielectric confinement and electron–hole exchange interaction on excitonic states in semiconductor quantum dots, Phys. Rev. B 47 (1993) 4569–4584.

[10] L. Brus, Diffusion controlled reactions: a variational formula for the optimum reaction coordinate, J. Chem. Phys. 79 (18) (1983) 5566.

2

CHARACTERIZATION AND ANALYSIS OF NANOMATERIALS

CHAPTER OUTLINE

Complete characterization and analysis of nanomaterials include particle composition, particle size distributions, morphology/shape, structural analysis, surface characterization, surface area analysis, optical properties, magnetic properties, and others [1].

Conventional characterization methods for nanomaterials can vary from system to system but commonly include transmission electron microscopy (TEM), X-ray diffraction (XRD), field-emission scanning electron microscopy (FE-SEM), energy dispersive X-ray spectroscopy (EDS), X-ray photoelectron spectroscopy (XPS), X-ray absorption fine structure (XAFS), inductively coupled plasma mass spectroscopy (ICP-MS), vibrating sample magnetometer (VSM), Auger electron spectroscopy (AES), Mossbauer spectroscopy, and differential scanning calorimetry (DSC), to

Nanomaterials and Devices. DOI: http://dx.doi.org/10.1016/B978-1-4557-7754-9.00002-0

Table 2.1 Comparison of the Performance of Different Testing and Analytical Instruments for Nanomaterials

Main Categories	Instrument	Main Functions	Resolution
Surface analysis	AES	Surface composition, chemical bonding	1 μm
	AFM	Surface structure	5 nm
	XPS	Surface composition, chemical bonding	5 nm
	SIMS	Surface composition, in-depth analysis	50 μm
Electron microscopy	SEM (×2)	Surface microstructure, composition analysis, material analysis	0.1 μm
	TEM/STEM	Internal microstructure, crystal structure, composition analysis	0.4/20 nm
	HREM	Crystal (atomic/molecular) structure, interface structure	0.18 nm
XRD	XRD	Crystal structure, phase identification	
	TFXRD	Phase identification of films, film thickness	
	TC	Hole image	

name a few. For nanoparticles with a size less than 10 nm, different techniques are required, such as high-resolution electron microscopy (HRTEM), Raman spectroscopy, nuclear magnetic resonance (NMR), ultraviolet photoemission spectroscopy (UPS), scanning tunneling electron microscopy (STEM), secondary ion mass spectroscopy, second neutral-atom mass spectroscopy (SNMS), and field-emission scanning transmission electron microscopy (FE-STEM). Table 2.1 highlights characterization techniques used for nanomaterials with the type of information obtained and the resolution of the instruments (TFXRD indicates thin-film XRD and TC indicates texture cradle).

2.1 Detection and Analysis of Particle Size

First, we need to find a way to define the size of nanomaterials. For spherical nanoparticles, the diameter is defined as the size of the nanoparticle. As for nanomaterials of asymmetric shapes, the following four definitions are usually used: *geometric diameter*, *equivalent diameter*, *SSA* (specific surface area) *diameter*, and *refraction diameter*.

Table 2.2 Determination of Particle Size

Size Definition	Determination	Range (μm)	Distribution Benchmark
Geometric diameter	Optical microscopy	500–0.2	Number distribution
	Electron microscopy	10–0.01	Number distribution
Equivalent diameter	Gravity sedimentation	50–1	Mass distribution
	Centrifugal sedimentation	10–0.01	Mass distribution
	Gas precipitation	50–1	Mass distribution
	Proliferation	0.5–0.001	Mass distribution
SSA size	Adsorption (gas)	20–0.001	Average SSA size
	Infiltration (gas)	50–0.2	Average SSA size
	Wetting heat	12–0.001	Average SSA size
Refraction diameter	Refraction	12–0.001	Volume distribution
	X-ray line width	0.05–0.0001	Volume distribution
	X-ray scattering at small-angle	0.1–0.001	Volume distribution

Geometric diameter: For particles of any geometric shape, the largest projected area can be converted into a circle of the same size. The diameter of this circle is the *geometric diameter* of particles.

Equivalent diameter: The size of powder particles can be measured by using the sedimentation method, centrifugal method, mechanical method, or hydraulic method. Homogeneous spherical particles, for example, have the same terminal settling velocity as nanoparticles; their diameter shall have an *equivalent diameter* of the nanoparticles.

SSA diameter: Using a variety of possible techniques, the SSA of the nanoparticle can be determined. From the surface area, a diameter can be calculated by one of the even spherical particles with the same formula:

$$d_s = \frac{6\left[\frac{\rho}{\rho_m}\right]V}{S}$$

Here, d_s is the SSA particle size, ρ is the sample density, V is volume of material tested, ρ_m is the density of the bulk materials, and S is the calculated surface area of the particles.

Refraction diameter: The diameter of the nanoparticle as determined using XRD techniques.

Typically, the calculated diameter will vary depending on the method used, as shown in Table 2.2.

Table 2.3 Comparison of Physical Adsorption and Chemical Adsorption

Properties	Category	
	Physical Absorption	*Chemical Absorption*
Adsorbability	van der Waals force	Atomic bonding force
Heat of adsorption	10 kcal/more	10–100 kcal/more
Optionality	None (applicable to any system at low temperatures)	Selectivity
Absorption rate	Quickly (unable to be determined)	Usually not too fast (able to be determined)
Adsorption layer	Adsorption on multi-molecular layer	Adsorption on single-molecule layer
Fixed temperature adsorption	Decreased at high temperatures (decreasing with temperature increase)	Increased at high temperature (increasing with temperature increase)
Reversibility	Easy to fall off	Not easy to fall off

SSA *diameter* can be measured by using chemical and physical adsorption methods. Physical adsorption and chemical adsorption are compared in Table 2.3.

The BET (Brunauer, Emmett, and Teller) method using multilayer gas adsorption is commonly used for the measurement of SSA materials in the solid phase. It is generally performed using two methods: the volumetric and the gravimetric methods. The volumetric method uses differences in sample volume of a known quantity of gas before and after adsorption to quantify the gas absorbed on the surface. The gravimetric method uses the weight difference of the sample before and after adsorption to determine the amount of adsorbed gas. The gravimetric method is generally considered to be more accurate. Both methods require a high vacuum and rigorous predegassing treatment. The accuracy of the BET method for SSA determination mainly depends on particle shape and defects, such as pores, cracks, and others. These factors may cause a negative deviation of the measured results.

2.2 Detection and Analysis of the Electrical Properties

The electrical properties of nanomaterials can be analyzed by the material's electric resistance and conductivity, dielectric properties, piezoelectric properties, and surface impedance/volume impedance measurements.

In general, nanomaterials have increased resistance compared to bulk materials; the size scale limits the motion of a large number of electrons through a small crystal size. The more disordered the crystal atoms are arranged, the larger the crystal boundary thickness becomes, thus inhibiting electron mobility. Such an energy barrier on the interface is responsible for the increased resistance. Meanwhile, the relationship between resistance and temperature significantly deviates from those of bulk materials, and some anomalies may even arise, such as in the case of Pd nanomaterials.

Nanomaterials can differ greatly in structure from conventional bulk materials. Their dielectric properties are unique and are mainly represented in the high dependence of the dielectric constant and dielectric loss on particle size. In addition, the electric field frequency may have a strong influence on dielectric behavior. In the materials, the electric displacement and the applied electric field strength E are proportional by a constant known as the dielectric constant ε, with the following relationship: $D = \varepsilon E$.

Because of the existence of the external electric field, the polarization process may result in electric energy being transferred into heat. Such energy loss is called *dielectric loss*. Dielectric loss is usually expressed as loss tangent or loss factor $\tan \delta$, which is defined as:

$$\tan \delta = \frac{\varepsilon'_r}{\varepsilon''_r}$$

Here, ε'_r and ε''_r are the respective real and imaginary parts of the complex dielectric constant ε_r.

Some materials can separate charge under mechanical action (stress or strain). Such a phenomenon is called the *piezoelectric effect*. The prefix *piezo-* comes from Greek, meaning "pressure." Regarding the mechanism of the piezoelectric effect, Voigt pointed out as early as 1894 that this effect can only be found in crystals formed by the noncentrosymmetric lattice. So far, research of micro-theory on the piezoelectric effect still faces many problems, mainly with inconsistencies with experimental results. Yet researchers have made it clear that the piezoelectric effect is essentially caused by the polarization of crystal medium. Piezoelectric material has electrical response to mechanical stress. Conversely, if this operation can be reversed, that is, by applying a voltage across the material, a mechanism stress of strain occurs. Objects with a piezoelectric effect are known as piezoelectric materials.

Detection and analysis of electrical properties of nanomaterials can be performed by using the following instruments:

1. Digital multimeter (with three functions), which tests AC and DC voltage and current, resistance, and diode.
2. High impedance meter, which tests the nature of electrical insulation.
3. Four-point probe, which tests the material surface resistance.
4. Metal insulator semiconductor component, which measures dielectric constant, threshold voltage, carrier concentration, the thickness of the insulating layer, insulating layer capacitors, and others.

In addition, scanning probe microscopy (SPM) can also be used. At present, SPM has developed into a major technique for the measurement of electrical properties of nanomaterials.

2.3 Detection and Analysis of Magnetic Properties

Magnetic properties of materials are related to the composition, structure, and status of the specific material. Some magnetic properties such as magnetization and susceptibility are closely related to the material's crystal size, shape, distribution, and defects. Other magnetic properties such as saturation magnetization and Curie temperature are related to the crystal phase and number of materials.

Nanomaterials differ greatly from conventional bulk materials in properties, especially in their magnetic properties. A majority of magnetic properties are different in nanomaterials when compared to the bulk nanoparticles, namely magnetic transformation and superparamagnetism, coercive force, Curie temperature, and susceptibility (see Chapter 5). Here, only a brief introduction is provided.

Néel temperature represents the characteristic temperature (critical temperature) of magnetic properties of certain metals, alloys, and salts. Below this temperature, the material has a spontaneous nonparallel magnetic ordering and displays antiferromagnetism; however, above this temperature, the material becomes paramagnetic. Néel temperature T_n is determined by the nearest coordination number of atoms, atomic spacing, as well as the types of nearest neighbor atoms.

Magnetic susceptibility is the ratio of the material magnetization and magnetic field strength. When the two physical quantities are not in the same direction, this ratio is a tensor; when they are in the same direction, the ratio is a simple scalar.

Materials tend to be magnetized under an external magnetic field. To reduce its intensity of magnetization to zero, the reverse magnetic field strength to be applied is called the *coercive force*. If the size of nanoparticles is larger than the critical size in the superparamagnetic state, then the coercive force will actually increase along with a reduction of the size. There are two interpretations of this phenomenon. One is that the small size makes each particle basically a single magnetic domain, in which nanoparticles behave like a permanent magnet. To reverse the magnetic moment formed by a combination of particle contributions, the reverse magnetic field required will naturally be great. Another interpretation states that due to the magnetostatic effects, spherical nanoparticles may form chains, resulting in an increased coercivity.

The Curie temperature indicates the characteristic temperature (critical temperature) at which a material loses strong magnetism or ferrimagnetism. Above this temperature, the material is paramagnetic. Curie temperature (T_C) is one of the major magnetic parameters of materials, and it is related to the atomic structure of a material and its spacing. These phenomena are also the main metrics in assessing magnetic properties of nanomaterials.

Magnetic properties of nanomaterials can be detected by using methods such as *magnetic force microscopy* (MFM) and NMR.

The MFM is a surface detection technology that involves using magnetic interactions between silicon probes coated with magnetic thin film and using the sample to obtain the surface magnetic structure. In 1987, two scientists, Martin and Wickramasinghe, developed the MFM technology [2]. Initially, only a magnetic image was obtained, whereas the appearance of the structure of materials was not available. At present, MFM is executed through two-step scanning in which intermittent contact scanning is first used to obtain the surface shape of a sample, and then the probe is raised to a certain height for scanning to capture magnetic images. The principle is the same as in noncontact atomic force microscopy (AFM). It has higher resolution ($\sim$50 nm) and is easy to use and applicable in various environments. MFM technology has gradually become an important detection technique in magnetic materials research and can be applied to magnetic film, magnetic memory devices, magnetic recording results, and others. Rugar and colleagues designed a magnetic force microscope that used a small bias voltage

(1–10 V) that was applied between the probe and the sample to increase the attraction between the tip to improve the stability.

In a strong magnetic field, some nuclei may have split spin energy levels that will be able to absorb electromagnetic waves (i.e., electromagnetic radiation) to produce a resonance phenomenon, called NMR. NMR can be generated under the following conditions: (1) when the nuclear magnetic moment is not zero; (2) when there is a strong and uniform external magnetic field; (3) when there is a specific frequency of electromagnetic waves to be applied to create nuclear resonance.

The NMR method uses nondestructive detection and can be applied to analyze the physical and chemical properties of materials. Compared with the same nondestructive inspection as X-ray detection and infrared detection, NMR has the following advantages: (1) quick detection rate; (2) a specific functional group that can be locked to detect whether there are precursors or byproducts through incomplete reaction, which facilitates the control of the purity of products and finds ways to enable purification and separation; and (3) detection is based on the electromagnetic wave, so it is not subject to the impact of the size, appearance, and color of material samples.

2.4 Detection and Analysis of the Mechanical Properties

Yield strength is defined as the stress at which a material irreversibly deforms. After the stress exceeds this elastic limit, deformation increases rapidly, including elastic and plastic deformation. When the stress reaches a particular value, plastic strain increases dramatically and a small region of fluctuation may occur on the stress–strain curve. This phenomenon is known as *yielding*. The maximum and minimum stress values at this stage are known as the upper and lower yield points respectively. Because the lower yield point value tends to be more stable, it is defined as the yield point or yield strength. Yield strength is the upper strength limit of a material at which point it can still be safely used. According to the empirical formula by Hall Petch, the yield strength of materials can be expressed as:

$$\sigma_{\mathrm{y}} = \sigma_0 + \frac{C_1}{\sqrt{d}}$$

Here, σ_0 is friction stress, d is the average size of crystals, and C_1 is an empirical constant.

TEM and AFM research has shown that carbon nanotubes have excellent mechanical properties, such as high elastic modulus, high elastic strain, and high rupture strain, which all match those in computational experimentation. Compared with the conventional bulk materials, nanostructured materials commonly double in strength and hardness. Furthermore, nanocomposites have shown dramatic improvements in strength, toughness, abrasion resistance, aging resistance, pressure resistance, compactness, and waterproof properties. Nanoceramic materials pressed from nanoparticles exhibit good toughness and ductility.

2.5 Detection and Analysis of Thermal Properties

Regarding the thermal properties of nanomaterials, physical quantities that require characterization include the thermal conductivity coefficient, specific heat, thermal expansion, thermal stability, and melting point.

As the thin film layer of material reaches a certain thickness, the grain boundary effect will have an increasingly significant impact on the thermal conductivity. Furthermore, the thermal conduction coefficient perpendicular to the film tends to decrease as the film thickness decreases.

Theoretical predictions and experimental results confirmed that nanostructured materials have specific heat values much higher than those of conventional bulk materials. Nanomaterials have a comparatively chaotic distribution of atoms on the structure, which has a higher volume as compared with bulk counterparts. Thus, entropic contributions due to this noncrystalline surface contribute much more to specific heat than the average coarse crystalline materials, leading to increased specific heat.

Nanocrystals are almost twice as large as the average crystals in the thermal expansion coefficient, with the increased t mainly contributed by the composition of the crystalline boundaries. The main instrument for measuring the thermal expansion coefficient of materials is known as a thermal expansion analyzer, but it is also known as a *thermal dilatometer analyzer* or a *thermomechanical analyzer*. The analysis of the thermal expansion coefficient of materials can provide an understanding of molecular motion, structural changes, and thermal expansion behavior. To solve problems such as heat bonding of different materials in the manufacture of semiconductor devices, a thermal expansion analyzer is the best tool for the analysis.

Melting point is the temperature at which a material is transferred from solid into liquid. For crystalline objects, there is a clear melting point; however, noncrystalline objects have an ill-defined melting point. The temperature may increase to a value at which a small number of atoms in an overall structure begin to move simultaneously with a liquid-like behavior. This temperature is known as the *glass transition temperature* (T_g). At temperatures below T_g, the glass material is in a solid state; at temperatures higher than T_g, it is a supercooled liquid. Expressed in mechanical terms, if the temperature is below T_g, then elastic deformation will occur; if the temperature is higher than the T_g, then viscosity (liquid type) deformation begins.

Thermal decomposition temperature is a value at which the material bonds can be heated to a broken state and dissociated into other substances.

For noncrystalline or amorphous nanomaterials that are plasticized, glass transition temperature and thermal dissociation temperature other than the melting point are also very critical thermal properties. Table 2.4 shows the melting point of several kinds of material at different scales.

Table 2.4 Melting Point of Several Materials at Different Scales

Material Type	Particle Size: Diameter (nm) or the Total Number of Atoms	Melting Point (K)
Au	Conventional bulk materials	1340
	300 nm	1336
	100 nm	1205
	20 nm	800
	2 nm	600
Sn	10–30	555
	500	480
Pb	Conventional bulk materials	600
	30–45	583
CdS	Conventional bulk materials	1678
	2 nm	≈910
	1.5 nm	≈600
Cu	Conventional bulk materials	1358
	20 nm	≈312

Thermal properties of nanomaterials are usually detected and analyzed by use of *thermogravity analysis* (TGA) and *derivative thermogravimetry* (DTG).

TGA can provide continuous measurement based on weight changes of materials in the process of heating during the measurement. Specifically, changes in mass are monitored as a function of temperature at a predetermined temperature rate and can be correlated to mass losses and thermal transitions in the material. Differential treatment can be completed at the same time. Namely, the recording of quality changes constitutes the DTG measurement technique.

By means of TGA (or DTG), a number of thermal properties of materials can be found, for example the aging temperature during pyrolysis and aging dynamics, aging behavior at different temperatures and in different gas environments, IC packaging materials used in the manufacturing process of executable semiconductor devices, flexible printed circuit boards, and glass substrates, ceramic substrates, and other components of analysis.

In the colloidal system, the related thermal properties of particles also include Brownian motion, diffusion and sedimentation balance, among others.

In Brownian motion, the average particle displacement $\overline{X}$ can be expressed as:

$$\overline{X} = \sqrt{\frac{RT}{N_0}\frac{Z}{3\pi\eta r}}$$

where R is the ideal gas constant, T is the absolute temperature, N_0 is the Avogadro constant, Z is the interval of observation time, η is the viscosity of the dispersion medium, and r is the particle radius.

Brownian motion has significant implications in the nature of colloidal particles. Brownian motion is an important factor that may affect the stability of a dispersion system of colloidal particles. Due to Brownian motion, the sedimentation of colloidal particles occurs not because of gravimetric forces but because of colloidal aggregation caused by interparticle collisions, resulting in sedimentation.

The phenomenon of diffusion is related to mass transfer, which arises from the Brownian motion of particles (Brownian motion) when there is a concentration gradient. The larger the particles and the smaller the thermal velocity, the less apparent the diffusion becomes. In general, the diffusion coefficient is used for measuring the diffusion velocity. It is a physical quantity of a material that indicates the diffusion capacity.

In the colloidal system, the diffusion coefficient D can be expressed as:

$$D = \frac{RT}{N_0} \cdot \frac{1}{6\pi\eta r}$$

Here, R is the ideal gas constant, T is the absolute temperature, N_0 is the Avogadro constant, η is the viscosity of the dispersion medium, and r is the particle radius.

Because diffusion coefficient is correlated with the average displacement, the derived diffusion coefficient D can also be expressed as:

$$D = \frac{\overline{X}^2}{2Z}$$

Here, Z is a specific interval of observation time and $\overline{X}$ is the average particle displacement in Brownian motion. Table 2.5 shows the diffusion coefficient of sol resulting from gold nanoparticles at 291 K.

When the particles suspended in fluid show a settling velocity equal to the diffusion speed, the system reaches a state of equilibrium, namely the sedimentation equilibrium. At the sedimentation equilibrium, the concentration of colloidal particles is subject to the Gaussian distribution law.

The Gaussian distribution law for colloidal particles can be expressed as:

$$n_2 = n_1\, e^{-\frac{N_0}{RT}\cdot\frac{4}{3}r^3(\rho_p - \rho_0)(x_2 - x_1)g}$$

Here, n_1 and n_2 are the concentration of particles at the cross-section at the height of x_1 and x_2 respectively, R is the ideal gas constant, T is absolute temperature, A is Avogadro's constant, r is the particle radius, ρ_0 is the density of colloidal particles, ρ_p is the density of the dispersion medium, and g is the acceleration of gravity.

Table 2.5 Diffusion Coefficient of Sol Resulting from Nano-Au Particles at 291 K

Size of Nano-Au Particles (nm)	Diffusion Coefficient (109 m^2/s)
1	0.213
10	0.0213
100	0.00213

2.6 Detection and Analysis of Optical Properties

The main differences in optical properties between nanomaterials and conventional bulk materials can be explained by the following: (1) strong broadband optical absorption of nanoparticles; (2) band shift of optical absorption; (3) the quantum confinement effect; (4) luminescence of nanoparticles; and (5) optical properties of the dispersion system.

1. *Strong broadband absorption characteristics of nanoparticles*

   Nanoparticles have an enhanced absorption of light at various wavelengths within the visible frequency range, whereas the reflectivity is decreased. Nanoparticles of silicon nitride (Si_3N_4), silicon carbide (SiC), alumina (Al_2O_3), and other powders have a strong broadband absorption spectrum range of infrared light. Zinc oxide (ZnO), iron oxide (Fe_2O_3), and titanium dioxide (TiO_2) have strong absorption of ultraviolet (UV) light.

2. *Frequency shift of optical absorption band*

   A frequency shift of the optical absorption band of nanoparticles is manifested both on the blue shift and the red shift of absorption spectra, depending on intra- and interparticle environments.

   After chemical modification of the surface of semiconductor nanoparticles, their absorption spectra will have a red shift. Quantum size effects or surface effects may be attributed to the blue shift of the absorption band whereas an increased overlap of electron wave functions will lead to a red shift of the absorption band.

3. *Quantum confinement effect*

   When the size of semiconductor nanoparticles is smaller than the Bohr radius a_B of an exciton, overlap of the wave functions of the electron and the hole will occur, resulting in an exciton absorption band. The overlapped electron and hole wave functions will increase with the reduction of the size of nanoparticles by $(a_B/r)^3$. Exciton absorption is enhanced and shows a blue shift.

4. *Luminescence of nanoparticles*

   Silicon with a particle size of less than 6 nm can emit visible light at room temperature. As the particle size decreases, the intensity of the emission band will increase and exhibit a blue shift. Tabagi holds that the luminescence of silicon nanoparticles is a result of the quantum confinement effect of the carriers. Brus believes that because silicon nanoparticles are smaller than 6 nm, the translational symmetry that can only be found in conventional bulk structures will disappear, and therefore the luminous phenomenon occurs.

5. *Optical properties of dispersion system*

   The Tyndal effect of nanoparticles will appear in dispersed systems [3] and is associated with the size of the dispersed particles and the wavelength of light projection.

For characterization of the optical properties of nanomaterials, the commonly used methods include infrared spectroscopy (IR), UV–visible spectroscopy, and scanning near-field optical microscope (SNOM) analysis.

Strictly speaking, in terms of size distribution, IR maps the electronic wave function as a function of depth penetration. Traditional infrared spectrophotometry is classified as the dispersive type and Fourier-transformed infrared spectroscopy (FTIR) is the nondispersive type. FTIR features a larger energy output, very high accuracy and precision in frequency (or wave number) measurement, as well as the feature of multiplicity.

UV–visible spectroscopy is based on the principle of electronic transitions in molecules. Molecular energy includes the electronic state of energy, vibrational energy, and rotation. The unique nature of nanomaterials with different optical absorption properties differentiates them from their bulk counterparts.

SNOM analysis is a combination of SPM and a fiber-optic probe, with the optical measurements and observations available in high spatial resolution beyond the diffraction limit. SNOM combines advantages of optical microscopy and electron microscopy.

SNOM analysis can provide clear optical images of the surface of nanomaterials for spectral analysis and can probe molecular aggregation. Combined with time-resolved fluorescence microscopy, SNOM can also be used to analyze optical phenomena at the nanoscale. At present, near-field optical microscopy has been applied in the fields of biology and medical research, and in research using semiconductor optoelectronic high-density optical media. The resolution of such a technique is ~50 nm.

2.7 Scanning Probe Microscopy [4–6]

SPM is extremely advantageous in the study of nanomaterials because atomic-scale resolution of nanomaterial surfaces can be examined. Various surface properties can be explored, including mechanical, magnetic, electrical, thermal, and optical properties. Many researchers have widely applied SPM in their nanotechnology research. The major advantages of SPM include lack of sample preparation and the ability to be used in various environments. Disadvantages are obvious and include low scanning speed, poor data reproduction, and the lack of elemental analysis capabilities. In scanning tunneling microscopy (STM), a small voltage is applied between the metal probe and the conductive samples at a distance between a few and tens of angstroms. Atoms at the probe tip and the sample surface can thus be maintained at a fixed value of tunneling current to measure the shape of the surface structure with atomic resolution. The STM technique can achieve the highest resolution of all known microscopy methods but cannot be used with insulating materials.

2.7.1 Working Principles of Scanning Tunneling Microscopy

The STM uses metal tips to scan the sample surface, which produces a quantum tunneling current under standard operating conditions. Because the tunneling current occurs more strongly on the tip of the atom, STM has atomic-scale lateral resolution. Using tunneling current as measurement signals, the image of the sample surface can be obtained. Therefore, both the sample and probe must be conductive, and the sample surface must be atomically planar for accurate measurements. Such requirements limit the use of STM (Figure 2.1).

2.7.2 Operating Mode of STM

In STM, as the tip scans the surface of the sample, the movement trajectory is recorded. The calculated density distribution from the surface morphology of the sample yields the atomic-scale image of the surface. This method can be used to observe the surface morphology of the sample, with many fluctuations. Meanwhile, the voltage applied in the z-drive is applied to calculate the height value of the surface fluctuations. However, for sample surfaces with little fluctuation, tip height can be controlled to perform a conservation scanning, and the changes in tunneling current are also recorded to obtain the distribution of surface morphology. This scanning mode is highlighted by high speed and the ability to reduce noise and thermal drift on the measurement; however, it is generally not able to be used to observe samples with surface fluctuations larger than 1 nm (Figure 2.2).

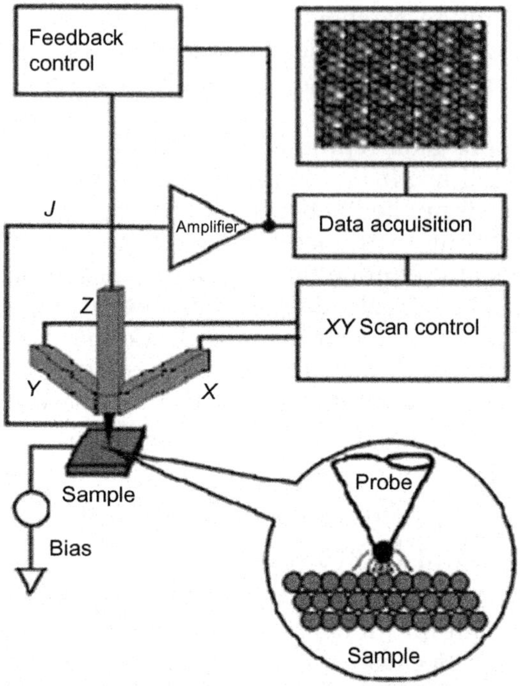

Figure 2.1 Basic structure of STM.

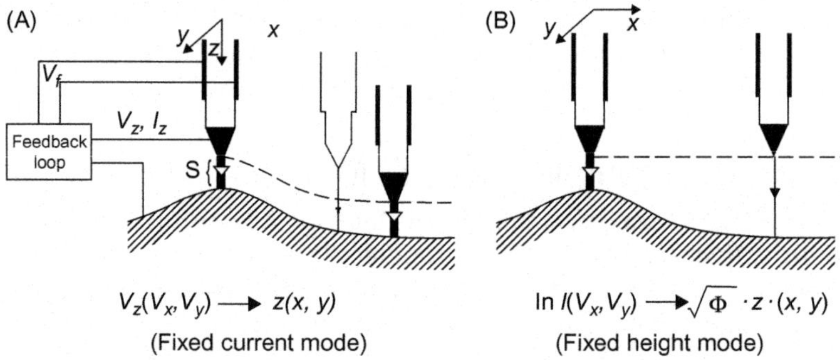

Figure 2.2 Schematic diagram of scanning modes: (A) fixed current mode and (B) fixed height mode. S is the spacing between the tip and sample, I and V_b are tunneling current and bias voltage, V_z is the feedback voltage to control the tip height in the z direction.

A. Fixed current imaging: The tunneling current (about 1 nA) is set as feedback signals. As the spacing between probe and sample surface is very sensitive to tunneling current, setting the value of the tunneling current means locking the spacing between probe and sample surface. While scanning over the surface of the sample, the probe must adjust its height (i.e., z values) according to the fluctuations of the surface. Therefore, the height variation of the probe can be presented so as to reflect the morphology of the sample surface.

Table 2.6 Advantages and Disadvantages of the STM Scanning Modes

Mode	Principle	Advantages and Disadvantages
Fixed current mode	Probe is maintained at a fixed current value height and adjusted with the morphology of the sample surface. The probe height variation is used to present the image of the sample surface.	Advantages: Larger depth of field. Disadvantages: Scan speed is slower and vulnerable to low-frequency noise.
Fixed height mode	With the probe being set at a fixed height, tunneling current values are directly used to present changes of surface morphology.	Advantages: Quick scanning rate, which can potentially observe surface dynamics. Disadvantages: Probe damage is possible if large fluctuations in the sample height are present.
Current density mode	Combination of the two above-mentioned materials, wherein bias modulation is used to gather images.	Disadvantages: Larger data sets; time consuming.

B. Fixed height imaging: Tunneling current values are directly used to present the image. While scanning the sample surface at a fixed height, surface fluctuations will lead to changes in spacing between probe and sample surface, and tunneling current values will also change accordingly (Table 2.6).

2.7.3 STM Application: Atomic Manipulation

Thirty years ago, a dream of many scientists was the ability to see individual atoms directly and to spatially manipulate individual atoms as desired. Thirty years later, this is no longer a dream, as many laboratories around the world have the knowledge and instrumentation to perform these activities that were once thought to be impossible. In 1990, IBM scientists Eigler and associates performed the first successful manipulation of atoms using an STM scanning needle to move Xe atoms at will on nickel surfaces. To demonstrate their capabilities, they arranged Xe atoms to spell the name of their company, *IBM* (Figure 2.3). The next year, Hosoki and colleagues, scientists for Japan's Hitachi Ltd., successfully used small single-atom vacancies to write the word *PEACE'91HCRL* on the surface of molybdenum disulfide (MoS_2). Later, other scientists

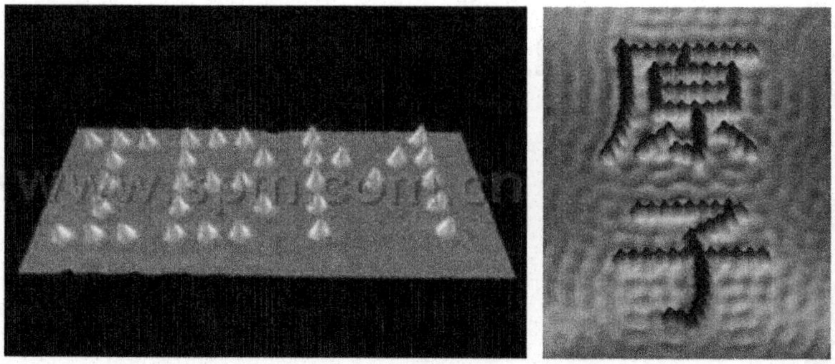

Figure 2.3 Examples of atomic manipulation.

manipulated nearly 100 iron atoms and formed two Chinese characters "原子" (Figure 2.3). These results captured the imagination of the scientific community and were commonly advertised throughout scientific media and conferences at that time.

In fact, there are three ways to manipulate individual atoms on surfaces. The first method is to use the interaction force between atoms on the STM tip and the surface atoms of the sample. Atoms on the surface can be moved from their original location to a specified new location, as demonstrated by Eigler. The second method is to use an electric field to evaporate the surface atom field onto the scan tip, or to manipulate the atoms on the tip and onto the sample surface. This method was used by Hosoki. The third method is to move the atoms with an electric field gradient. This method is currently undergoing thorough investigation and is viewed optimistically by many researchers in the field. The use of STM for atomic surface modification and single-atom manipulation shows wide application possibilities. Currently, it has significant applications in areas such as molecular and quantum devices, information storage, life sciences, and materials sciences. Single-atom manipulation mainly includes three basic steps, namely movement, extraction, and placement. These technologies are also essential for the future application of single-atom manipulation and the surface processing of structures for atomic-scale devices.

The main application of the atomic manipulation technique is in creating nano- or atomic-level structures. One such application is in high-density information storage. IBM scientists have demonstrated the ability in the aforementioned movement of xenon atoms, which can be directly used in the production of atomic-scale high-density memory. If each location without an atom on the surface represents 1 and the location with atoms represents 0, then this surface can be used as binary memory. Such storage density is unprecedented, far exceeding the memory density that is achievable in modern semiconductors and magnetic disks. Furthermore, STM can easily achieve the images of these atoms, which can be thought of as the equivalent to reading bits of data at an atomic level.

2.7.4 Advantages of STM

The scanning tunneling microscope apparatus itself has many advantages that make it a highly effective experimental tool for studying the surface structure of materials and biological samples of microelectronics-related surfaces. For example, STM instruments can be used by biologists in the study of individual protein molecules or DNA molecules, defects in crystals, and structure/morphology studies of microelectronic devices with nanoscale components. Before the invention of STM, the scale could only be potentially observed using destructive methods. STM is a nondestructive method, preserving the sample surface. In addition, STM is not limited to the diffraction limit of optical microscopy, which results in the incapability of imaging objects at half the size of the optical wavelength.

2.8 Atomic Force Microscopy [6,7]

Atomic force microscopy is mainly used in the measurement of atomic van der Waals forces to obtain images of the surface atoms. Because the AFM can observe any material regardless of electrical properties, it has advantages over STM. AFM technology was jointly developed in 1985 by Binnig from IBM and Quate from Stanford University [8] to scan nonconductor materials. The AFM probe, if coupled with a layer of magnetic material, can be used to detect the nanoscale magnetic force as well, known as *MFM*. If an AFM probe is coupled with a layer of metal, then it is possible to detect electric properties, which is known as *static electricity microscopy*. Similar approaches can be used to modify the probe for a wide variety of measurements such as near-field optical microscopy and scanning capacitance microscopy. The development of this series is referred to as *SPM*. Among the series, the atomic force microscope is the most widely used.

2.8.1 Working Principle of AFM

AFM works mainly on the principles of force between the probe and surface atoms, which are recorded by an optical offset to derive images and calculate various parameters. The key

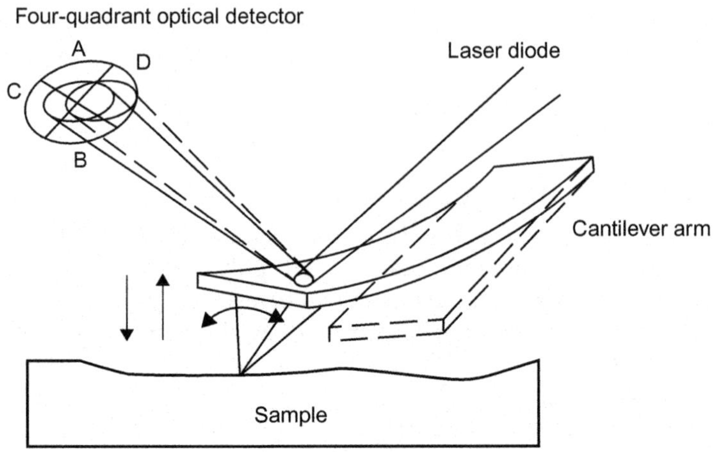

Figure 2.4 Diagram of the AFM working principle.

component of AFM is a microcantilever, the head of which is designed with a thin, pointed probe used to scan the sample surface. This cantilever is usually made from silicon or silicon nitride, with the curvature of its probe in the nanoscale. Because the probe is placed near the surface of the sample, the bending of the cantilever will comply with Hooke's law and have an offset as a result of the probe and the interaction with the surface. Typically, the offset can be observed by measuring the reflected laser beam that shoots onto the microcantilever with a light-sensitive diode array. When the probe scans over the sample surface, the laser reflection angle will change, along with the current of the light-sensitive diode. By measuring the current change, the bending or deflection of the cantilever can be calculated. The data will then be analyzed to produce three-dimensional images of the sample surface (Figures 2.4 and 2.5).

2.8.2 Comparison of the AFM Scanning Modes

There are three modes of AFM scanning, namely contact, noncontact, and tapping mode. They are compared in Table 2.7.

2.8.3 Application Examples of AFM

There are a wide range of AFM applications. One such example is in life sciences [9]. Because the AFM works in a very wide scope, direct imaging of biomedical samples can be made in the natural state (air or liquid) with high resolution. Thus, AFM

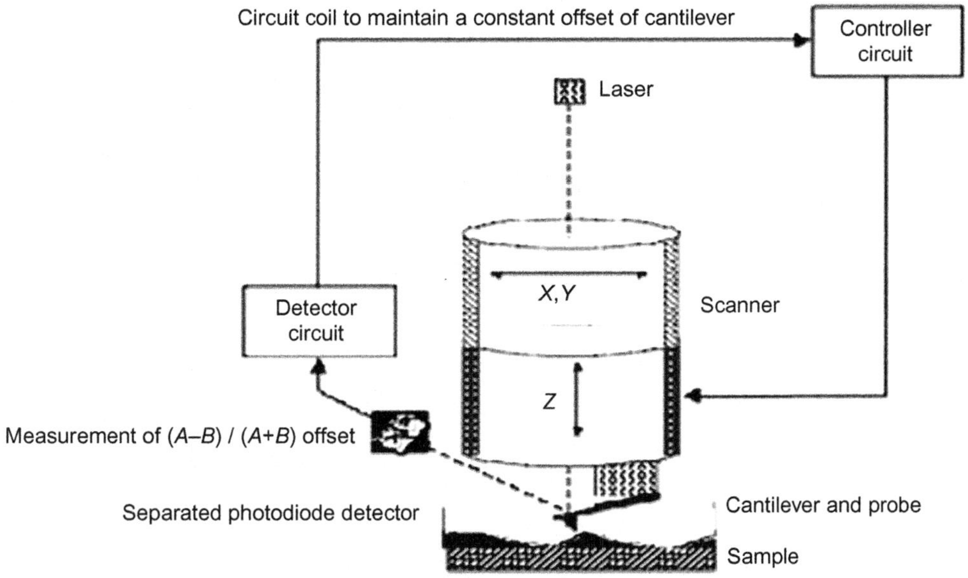

Figure 2.5 Basic structure of AFM.

Table 2.7 Comparison of the AFM Scanning Modes

Scanning Mode	Basic Principle	Advantages and Disadvantages
Contact (2–3 nm)	In the image scanning, the tip has a gentle contact with the sample surface. A very weak repulsion arises between atoms to produce the probe offset. Feedback is used to control a fixed value. The probe position change is obtained corresponding to all the scanning points, which renders the topographic images of the sample surface.	The contact area is small, but access force may damage the sample, especially for soft materials. However, the larger force can usually get better resolution. Therefore, an appropriate level of force is needed. As the repulsion is very sensitive to the distance, atomic resolution is easier to obtain.
Noncontact (<20 nm)	Detection is executed by using the long-range attraction between atoms—van der Waals forces. However, this attraction has a very small rate of change in the distance. Modulation techniques are required to increase the signal to noise ratio.	In the absence of contact between probe and sample, damage is avoided. But under the effects of the water film on the surface of the sample in air, the resolution in general is only 50 nm, while there is the availability of atomic-scale resolution in ultra-high vacuum.
Tapping (5 nm)	The probe repeatedly taps on the surface of the sample. As the probe is oscillated to a trough, there is a micro-contact with the sample. Due to the uneven surface of the samples, the atomic force may change the amplitude of the probe in beating. The use of feedback control will be able to obtain a high degree of image.	In the absence of contact between probe and sample, there is no need to worry that the sample may be damaged. But under the effects of the water film on the surface of the sample in air, the resolution in general is only 50 nm, while there is the availability of atomic-scale resolution in ultra-high vacuum.

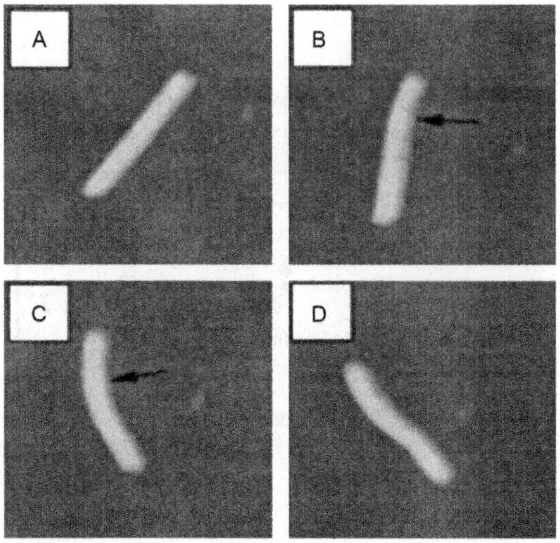

Figure 2.6 Experiment: Using the AFM tip in the bending and rotation movements of virus molecules. (A) The starting state. (B) The early stage of rotation. (C) The mid stage of rotation. (D) The end state after the rotation movement. From http://www.docin.com/p-474252844.html.

has become a vital tool for studying biomedical samples and biological macromolecules. Applications of AFM in biology include three aspects: morphological observations on biological cell surfaces; observation and study of the structure of biological macromolecules and other properties; and power spectrum curves observed between biological molecules.

Images here show experiments in which the AFM tip was used in the bending and rotational movement of virus particles (Figure 2.6).

References

[1] Y.F. Zhu, Characterization and Testing of Nano-Materials, Chemical Industry Press, Beijing, 2006.

[2] Y. Martin, H.K. Wickramasinghe, Magnetic image by "Force Microscopy" with 1000 Å resolution, Appl. Phys. Lett. 50 (1987) 1455–1457.

[3] M. Kerker, The Scattering of Light and Other Electromagnetic Radiation, Academic Press, New York, NY, 1969, pp. 54–59.

[4] J. Colchero, H. Bielefeldt, A. Ruf, M. Hipp, O. Marti, J. Mlynek, Scanning force and friction microscopy, Phys. Stat. Sol. (a) 131 (1992) 73–75.

[5] S.N. Magonov, M.-H. Whangbo, Surface Analysis with STM and AFM, VCH, Weinheim, 1996, 323p.

[6] Y.B. Cai, Y.C. Liang, An overview of nano-materials, and their preparation, Charact. Struct. Chem. 20 (6) (2001) 425–438.

[7] R.M. Overney, E. Meyer, J. Frommer, D. Brodbeck, R. Luthi, L. Howald, et al., Friction measurements on phase-separated thin films with a modified atomic force microscope, Nature 359 (1992) 133–134.
[8] G. Binnig, C.F. Quate, C. Gerber, Atomic force microscope, Phys. Rev. Lett. 56 (1986) 930–934.
[9] T.Q. Zhao, W.M. Zhou, Y.J. Liang, S. Dong, Progress in biological application of atomic force microscope, Lett. Biotechnol. 13 (4) (2002) 68–70.

3

CARBON NANOTUBES

CHAPTER OUTLINE

Nanomaterials and Devices. DOI: http://dx.doi.org/10.1016/B978-1-4557-7754-9.00003-2

3.1 Allotropes of Carbon and Structure

3.1.1 Allotropes of Carbon

As a nonmetallic element, carbon is a group 14 element located in the second period of the periodic table. Its chemical symbol is C, with the atomic number of 6 and the electronic configuration of $[He]2s^2 2p^2$. Carbon is widely distributed in nature, is found in all living organisms, and is in the atmosphere and the Earth's crust. It is well known that carbon exists in crystalline and amorphous forms in many different allotropes. Examples of amorphous carbon are anthracite, bituminous coal, peat, and lignite. Historically speaking, only two kinds of crystalline forms of carbon were previously known: diamond and graphite. However, things changed in 1985 with the discovery of C_{60} and C_{70} and with the discovery of carbon nanotubes (CNTs) in 1991. Recently, Andre Geim, a professor of astrophysics in Manchester, found the existence of a single atomic layer of graphite (graphene), which attracted interest of many research groups and won Andre Geim the Nobel Prize in 2010. Graphene is known to possess very interesting properties that can be applied to many fields, including electronic engineering, nanomechanics, and quantum computing. Diamond, graphite, fullerene (C_{60}), CNTs, and single-layer graphite (graphene) are in the carbon allotrope family.

3.1.2 Structures of Carbon Allotropes

There are differences between diamond, graphite, fullerene (C_{60}), and CNTs in their structure and bonding orbitals. Graphene is a single C atom layer in graphite. Here, we focus on the structural characteristics of diamond, graphite, fullerenes, and CNTs (Figure 3.1).

Graphite consists of sp^2 carbon bonding in a planar structure, where carbon is integrated in covalent bonds with the van der Waals force between layers. Because of this structural arrangement, graphite is a good conductor of electricity.

In diamond, carbon has a three-dimensional structure with sp^3 covalent bonding, and it is a poor conductor of electricity.

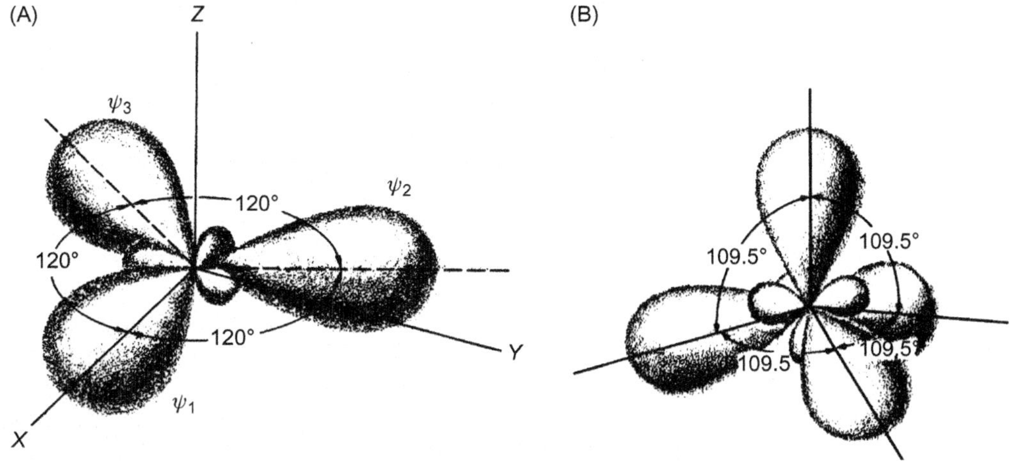

Figure 3.1 Bonding orbitals of carbon atoms: (A) graphite sp^2 and (B) diamond sp^3.

Diamond is a nonequilibrium form of C and will transform to graphite over time (although the timescale is approximately 5 million years). If heated or under impact of high-energy particles, then the change can be accelerated.

C_{60} was discovered in 1985 jointly by H. W. Kroto from the University of Sussex in the United Kingdom and R. E. Smalley and R. F. Carl from Rice University in the Unites States. For this discovery, they won the 1996 Nobel Prize in Chemistry. The structure of C_{60} has interesting features; the carbon atoms are in sp^2 bonding states, with 12 pentagons and 20 hexagons plus 30 π bonds and 90 σ bonds.

In January 1991, Sumio Iijima, a Japanese physicist from NEC Tsukuba Research Laboratory, found that multi-wall carbon nanotubes (MWNTs) can be made using high-resolution electron microscopy in the analysis of carbon fiber produced from arc discharge. An MWNT is a kind of tubular carbon molecule, with each carbon atom being sp^2-hybridized on the tube. Between them, the carbon–carbon σ bonds form a hexagonal honeycomb-like structure as the skeleton of the tubes. On each of the carbon atoms, there is a pair of π electrons not involved in the bonding that forms a conjugated π electron cloud across the entire CNT. Depending on the differences of layers, CNTs are categorized into single-wall carbon nanotubes (SWNTs) and MWNTs. Being ultrafine in the radial direction, nanotubes are strictly nanoscale in size. Tens of thousands of nanotubes can be equivalent to the width of a piece of hair. In the axial direction, the nanotubes can be up to tens or hundreds of microns. In 1993, both NEC and IBM discovered SWNTs.

The diameter of a CNT is generally between 0.4 and 50 nm, its length is 1 μm or more, and density is 1.3–1.4 g/cm^3 (these features are like that of wool or cotton). Its thermal conductivity is 23.2 W (cm K)$^{-1}$ (similar to diamond). Its electrical conductivity is dependent on its diameter and ranges from 10^{-3} to 10^{-4} Ω cm (similar to the semiconductor germanium) or 5.1×10^{-6} Ω cm (similar to copper).

Previous studies claimed that CNTs have limits in diameter. Generally speaking, SWNTs have a diameter from 1 to 6 nm, with some examples of smaller diameters of only 0.4 nm. Theoretically, 0.4 nm is the smallest possible diameter. If the size is reduced further, then the angle between the carbon atoms will become too small, and that may result in unstable structures. In addition, an SWNT with a large diameter is not a stable structure. Generally, if the diameter is larger than 6 nm, then the tube wall becomes unstable and is prone to collapse.

Regarding the length of the CNTs, the current laboratory preparation methods available can produce a length of SWNTs that is much shorter than that of MWNTs. SWNTs typically have a length of hundreds of nanometers to a few microns, whereas the length of MWNTs is generally in the tens of microns or more (the longest are up to a few mm). MWNTs have a layer of spacing of approximately 0.34 nm, whereas the diameter has a range of 0.4 nm to several thousand nm (depending on the number of layers of CNTs).

3.1.3 Graphene [1]

3.1.3.1 Single-Layer Graphite Material (Graphene) [1]

Graphene is a single layer of carbon atoms densely packed in a honeycomb crystal lattice. This material can be produced by the rapid separation of an atomic layer of graphite. It is a stable, flexible, and robust material with high conductivity. We know that graphite is a composition of carbon. Schematically, the structure of graphene is formed by layers of graphite atoms being vertically torn like paper. Graphene has become a model system in condensed matter physics, because its charged particles can move at relativistic speeds, as if they do not have any mass. This feature has led to some peculiar electronic transport properties. For example, in a graphene layer sandwiched between two superconducting electrodes, a superconducting current can flow at low temperatures, loaded by either electrons or the holes, depending on the gate voltage and also the charge density in the graphene layers.

Interestingly, a limited current can flow through graphene even if the charge density is zero. These observations provide clues to study the relativistic phenomenon, known as time-reversal symmetry, and the electron transport mechanism in graphene.

Regarding electronics in a graphene layer, there will be no scattering in the sub-microdistance. This provides the conditions for the manufacture of speed distribution transistors. At present, a graphene single-electron transistor (SET) has been realized in the laboratory.

3.2 Types and Nature of CNTs

3.2.1 Types of CNTs

As mentioned, CNTs are seamless tubular nanocrystals formed by the curved graphite layer. Graphite consists of sp^2 carbon bonding in a planar structure. In CNTs, carbon on the plane is combined with a covalent bond into a honeycomb-like structure, and carbon hexagons curl up into CNTs.

The hexagonal crimping of planar carbon atoms can start from different angles, resulting in different CNTs. There are three different forms, or chiralities, of SWNTs, namely the *Armchair, Zig-Zag,* and the *Chiral* types (see Figure 3.2 and 3.3).

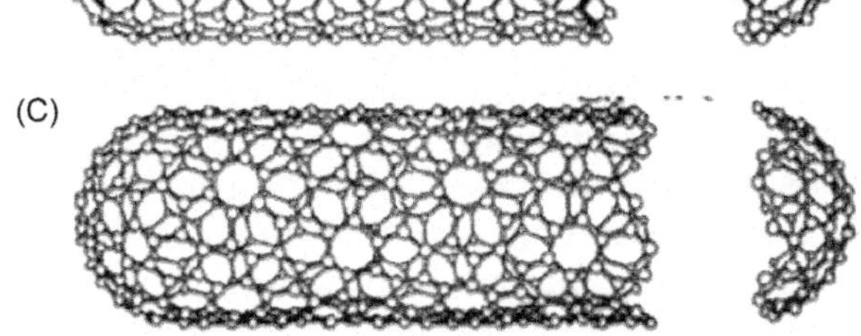

Figure 3.2 Three kinds of CNTs. (A) Armchair type. (B) Zig-zag type. (C) Chiral type.

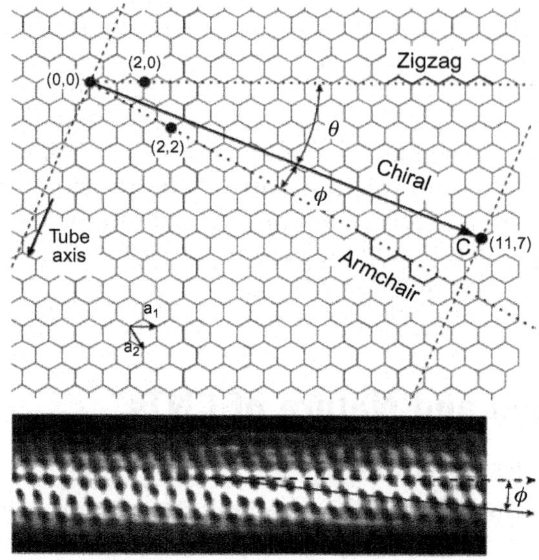

Figure 3.3 Curling of CNTs.

3.2.2 Characteristics of CNTs

3.2.2.1 Mechanical Properties

The sides of CNTs are composed of hexagonal carbon rings (graphite sheet); however, at the cap-shaped part where the tube body is bent and the tube is capped, carbon atoms are structured in a number of pentagonal and heptagonal rings. Because the carbon–carbon covalent bond that constitutes these different carbon ring structures is one of the most stable chemical bonds in nature, CNTs are expected to present very good mechanical properties, with strength close to that of the carbon–carbon bond. Studies have shown that the strength of CNTs is 100 times that steel, but with only one-sixth the density.

CNTs have outstanding flexible features (extension rate of up to 30%); they can be bent into a small angle or into a tiny ring structure. When the bending stress is removed, the CNTs can be completely restored to their original state from the large bending deformation. Under great external stress, a brittle fracture will not occur. Fibrous materials consisting of nanotubes with good mechanical properties will have many applications in the future in the industrial sector. One such example is for use as cables in a space elevator, a theoretical device that can transport goods and people from the Earth to space stations. CNT cables might be the only material that will not be broken because of its weight.

3.2.2.2 Electrical Characteristics

Different types of CNTs vary in their conductive properties: for single-wall nanotubes with metallic conductivity, the one-third saw tooth-shaped CNT comes with metal features, whereas the two-thirds palm-shaped CNT has a semiconductor character. The CNTs of semiconductor nature will have a narrower band gap as diameter increases. In the case of large diameters, the band gap is zero and shows metallic features.

Because the electronic flow of CNTs is restrained by the quantum confinement, electrons in CNTs usually have axial movement only along CNTs in the same direction as the layer of the graphite sheet, whereas radial movement will be greatly limited.

These electrical properties of CNTs can enable widespread applications in the future in nanoelectronics. For example, metallic CNTs can be used as connection lines in nanometer-integrated circuits (ICs), whereas the semiconductor nature of CNTs can be used to produce nanoelectronic switches and other nanometer quantum devices.

Tests of the field emission effect of MWNTs found that the measured resistance values of CNTs follow Ohm's law.

R. E. Smalley, the 1996 Nobel Prize winner, performed his study in 1995 by imposing −100- to −110-V bias voltages to open nanotubes, which produced 0.5- to 1.5-mA field emission currents. With the help of a CCD camera, the tips of nanotubes could be observed to emit very weak incandescent light. In the same year, De Heer, André Chatelain, and Daniel Urgate also first proposed the idea of using CNTs as field emission devices.

3.2.2.3 Thermal Properties

Studies revealed that CNTs have superior thermal conductivity.

David Tomanek at Michigan State University found in his experiments that the measurement of thermal conductivity of CNTs could be as high as 3,000 W/m/K, almost identical to the thermal conductivity of diamond, and he predicted that in the future it could reach 6,600 W/m/K. Thermal conductivity of this magnitude can be applied in future circuits by transporting heat away from dense circuits, a current issue in IC construction and design.

Similarly, researchers at the University of Pennsylvania found that CNTs are currently the world's best thermal conductivity material because of their transmission speed of up to 10,000 m/s by way of ultrasonic heat transfer. The study also revealed that even if they were bound together, the heat would not be transferred from one CNT to another. This shows that CNTs can only be one-dimensional in the direction of heat transfer.

In addition, research has shown that CNTs are ballistic wires with ballistic conductance. At room temperature, the electrons can pass through wires without any resistance and backward scattering to produce heat. In the tests performed at Georgia Institute of Technology, for example, the researchers applied 10 million amps/cm of current through the CNTs. They found that the resistance of MWNTs is a constant that is independent of their lengths and widths.

3.2.2.4 Superconducting Phenomenon of CNTs

Mathieu Kociak, who worked for CNRS laboratories in Paris-South University, claimed that the phenomenon of superconductivity was observed in nanotube ropes. This means that the phenomenon of superconductivity was first discovered in a system with minimal conduction channels, because super-current is required to pass CNTs of the substrate material in the two-dimensional network structure with lack of contents. The researchers expected to increase the transition temperature by means of proper ion implantation. The current transition temperature is only 300–400 mK.

3.2.2.5 Chemical Properties

The use of CNTs can improve the performance of an atomic force microscope. For example, Harvard's John Hafner placed a 0.9- to 2.8-nm diameter CNT at the tip of an atomic force microscope probe. Nanotubes used as microprobes will increase not only the resolution but also the applicability in the study of the nature of adhesion between the sample and the scanning tip. Hafner called such methodologies chemical force microscopy.

In addition, Masako Yudasaka at NEC Laboratories, Japan, placed C_{60} molecules into CNTs, resulting in very high pressure values. C_{60} can only generate a force of micro-newtons; however, divided by the area of CNTs, it will cause a pressure of 0.1 GPa. This discovery will lead to novel changes in fullerenes for chemical applications. Yudasaka also claimed that CNTs with a cone-shaped tip, which have a diameter of 2 nm, length of 50 nm, and an open angle of 20°, will be able to supersede activated carbon in filters used to absorb gases.

3.2.3 Electronic Structure of CNTs

3.2.3.1 π-Electron Orbital and the Energy of the Conjugated Molecule in Planar Structure

In 1931, the German physicist E. Hückel proposed the Hückel equation based on the Schrödinger equation. The solutions to eigenvalues and eigenvectors of the Hückel equation obtained the π electron orbital and the energy of conjugated molecules in planar structures. This showed that because of the way the carbon electron cloud of benzene is distributed in the ground state, it is surrounded on both sides, and this also explained some of the current-related phenomena. The Hückel equation is a useful tool for calculating the electron structure of graphite, graphene, C_{60}, and CNTs. The following is a brief description of this theory [2].

An electron in a molecule occupies a molecular orbital (MO). Assuming the wave function for an MO is expressed by Ψ, then Ψ can be expressed by a linear combination of atomic orbitals:

$$\Psi = \sum_{i=1}^{n} c_i \psi_i \tag{3.1}$$

Here, ψ_i is the atomic orbital wave function. The energy W of the MO can be obtained by multiplying by the left side of the Schrödinger equation:

$$\Psi \hat{H} \Psi = \Psi W \Psi \tag{3.2}$$

Using a volume integral

$$W = \frac{\int \Psi \hat{H} \Psi \mathrm{d}V}{\int \Psi^2 \mathrm{d}V} \tag{3.3}$$

we substitute Eq. (3.3) with Eq. (3.1) to get

$$W = \frac{\int \left(\sum_{i=1}^{n} c_i \psi_i \right) \hat{H} \left(\sum_{i=1}^{n} c_i \psi_i \right) \mathrm{d}V}{\int \left(\sum_{i=1}^{n} c_i \psi_i \right)^2 \mathrm{d}V} \tag{3.4}$$

Recombining the sum we get

$$W = \frac{\sum_j \sum_k c_j c_k \psi_j \hat{H} \psi_k \mathrm{d}V}{\sum_j \sum_k c_j c_k \int \psi_j \psi_k \mathrm{d}V} \tag{3.5}$$

Using the equations

$$H_{jk} = \int \psi_j \hat{H} \psi_k \, \mathrm{d}V \tag{3.6}$$

$$S_{jk} = \int \psi_j \psi_k \, \mathrm{d}V \tag{3.7}$$

Eq. (3.5) can be rewritten as

$$W \sum_j \sum_k c_j c_k S_{jk} = \sum_j \sum_k c_j c_k H_{jk} \tag{3.8}$$

Recombination of the sum terms gives

$$W \sum_j c_k S_{ik} + W \sum_k c_j S_{ji} = \sum_k c_k H_{ik} + \sum_j c_j H_{ji} \tag{3.9}$$

For $H_{jk} = H_{kj}, S_{jk} = S_{kj}$ Eq. (3.9) can be rewritten as

$$W \sum_k c_k S_{kj} = \sum_k c_k H_{jk} \tag{3.10}$$

or

$$\sum_k c_k (H_{kj} - W S_{kj}) = 0 \tag{3.11}$$

According to Eq. (3.11) and the conditions of a nonmediocre solution, the secular equation is obtained as

$$\begin{vmatrix} H_{11} - WS_{11} & H_{12} - WS_{12} & \cdots & H_{1n} - WS_{1n} \\ H_{21} - WS_{21} & H_{22} - WS_{22} & \cdots & H_{2n} - WS_{2n} \\ \vdots & \vdots & \vdots & \vdots \\ H_{n1} - WS_{n1} & H_{n2} - WS_{n2} & \cdots & H_{nn} - WS_{nn} \end{vmatrix} = 0 \tag{3.12}$$

The energy of the molecules is available from the solution of this secular equation.

Hückel introduced an approximate method that can be used to simplify the solution of the secular equation. It is called the

Hückel MO method (the HMO method) and has three assumptions, as follows:

1. $H_{ii} = \alpha$
2. For neighboring atoms, $H_{jk} = \beta$, for non-neighboring atoms, $H_{jk} = 0$
3. $S_{ii} = 1,\ S_{jk} = 0\ (j \neq k)$

Using these assumptions, for ethylene, secular Eq. (3.12) can be reduced as:

$$\begin{vmatrix} \alpha - W & \beta \\ \beta & \alpha - W \end{vmatrix} = 0 \tag{3.13}$$

As $x = (\alpha - W)/\beta$, then Eq. (3.13) is changed to

$$\begin{vmatrix} x & 1 \\ 1 & x \end{vmatrix} = 0 \tag{3.14}$$

That is,

$$x^2 - 1 = 0 \tag{3.15}$$

So, $x = \pm 1$. Thus,

$$W = \alpha \pm \beta \tag{3.16}$$

Here, $W = \alpha + \beta$ corresponds to the bonding state π-MO and $W = \alpha - \beta$ corresponds to the anti-bonding state π^*-MO.

To calculate the coefficient of the wave function, we substitute the energy value into the secular equation:

$$\begin{aligned} c_1(\alpha - W) + c_2\beta = 0 \\ c_1\beta + c_2(\alpha - W) = 0 \end{aligned} \tag{3.17}$$

As $W = \alpha + \beta$, $c_1 = c_2$ is given, and as $W = \alpha - \beta$, $c_1 = -c_2$ is given, by the use of normalization condition

$$\int \Psi^2 \mathrm{d}V = 1 \tag{3.18}$$

and as Eq. (3.1) is similar to the Hückel equation, Eq. (3.18) can also be expressed as:

$$\sum_j \sum_k c_j c_k S_{jk} = \sum c_i^2 = 1 \tag{3.19}$$

$|c_1| = |c_2| = 1/\sqrt{2}$ is given.

3.2.3.2 Electronic Structure of Graphite

Using the aforementioned method, we can calculate the energy dispersion relation for graphite electronics [3]:

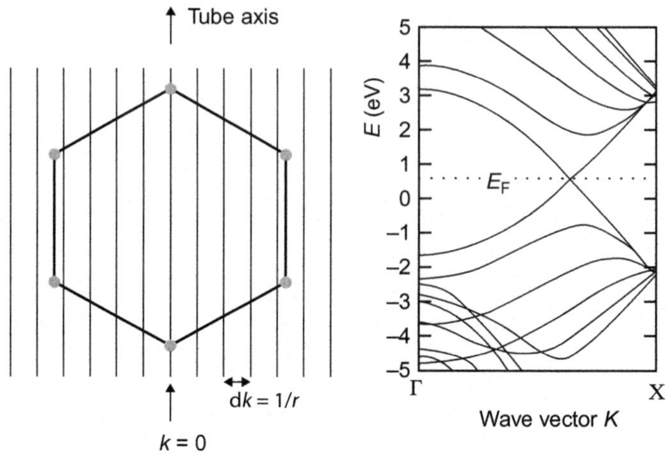

Figure 3.4 Electronic structure of armchair-shaped CNTs.

$$W_{2D}(k_x, k_y) = \pm\gamma_0\left[1+4\cos\left(\frac{\sqrt{3}k_x a}{2}\right)\cos\left(\frac{k_y a}{2}\right)+4\cos^2\left(\frac{k_y a}{2}\right)\right]^{1/2} \tag{3.20}$$

This relationship is essential for the calculation of the electronic structure of CNTs.

Energy dispersion relations for graphite are applicable in the calculation of the electronic structure of CNT electronics. Figure 3.4 is the electronic structure of armchair-shaped CNTs with periodic boundary conditions. Armchair-shaped CNTs are ideal materials for production of one-dimensional quantum wires.

3.3 Preparation of CNTs

The common methods used for the preparation of CNTs mainly include arc discharge, laser ablation, chemical vapor deposition, solid-phase pyrolysis, glow discharge, gas combustion, and polymerization synthesis.

The arc discharge method is the oldest major method for the production of CNTs (Figure 3.5). In 1991, Japanese physicist Sumio Iijima used this method to produce carbon fiber and discovered CNTs for the first time. The arc discharge method works as follows: a graphite electrode is placed in a reaction vessel filled with helium or argon gas to stimulate the arc between the two poles. At this point, the temperature can reach 4,000°C.

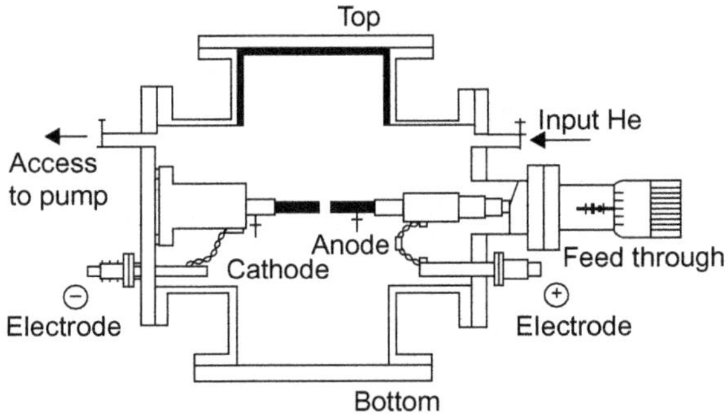

Figure 3.5 Diagram of arc discharge device.

Graphite will evaporate under this condition to generate products, including fullerenes (C_{60}), amorphous carbon, and SWNTs or MWNTs. By controlling the catalyst and hydrogen content in the container, the relative yields of several products can be adjusted. Technically, it is relatively simple to use this method to prepare CNTs. However, the CNTs produced in this way may be mixed with C_{60} and other products, and it may be difficult to obtain the products with high allotrope purity. Furthermore, the products are often MWNTs rather than SWNTs, limiting the scope of this method for practical research. In addition, the method tends to consume excessive energy in the reactions. In recent years, some researchers have found that if molten lithium chloride is used as the anode, it can effectively reduce the energy consumed in such reactions and make it easier to obtain higher purification.

In recent years, a chemical vapor deposition method has been developed to overcome the shortcomings using the arc discharge method. This approach enables gaseous hydrocarbons to pass through a template with catalyst particles attached. Gaseous hydrocarbons can then be decomposed to form CNTs at temperatures of 800–1,200°C. The highlighted advantage of this method is that the residual reactants of gas can leave the reaction system to obtain CNTs with high purity. At the same time, there is no need to have a very high temperature, so the energy required in this process is comparatively lower than that for arc discharge experiments. However, the CNTs made with this process have irregular diameters and shapes, and a catalyst must be used in the preparation process. Research is currently being performed to control the structure of CNTs generated by chemical vapor deposition, including controlling the arrangement of the catalyst template.

Other methods include solid-phase pyrolysis. It is a new method by which a conventional carbon-containing sub-solid body undergoes pyrolysis at high temperatures for the growth of CNTs. This approach allows a relatively stable process and does not require a catalyst. However, it is limited by the availability of raw materials, so scaled and continuous production cannot be achieved.

In addition, there are ions or laser sputtering methods. Although it is easy to achieve continuous production with these two methods, the scale is limited because of equipment limitations.

3.4 Applications of CNTs

3.4.1 CNT Electronics

3.4.1.1 The Limits of Microelectronics Technology and the Emergence of Nanoelectronics

The transistor is the most vital component in electronic circuits. Of all types of transistors, metal oxide semiconductor field-effect transistor (MOSFET) is particularly important because of ease of integration into ICs. As shown in Figure 3.6, MOSFET has three electrodes, namely source, drain, and gate; gate is responsible for controlling devices by manipulating the volume of current that flows from source to drain.

Many of these transistors, together with a number of other electronic devices such as resistors, capacitors, diodes, and wiring, are integrated in a single chip to constitute an IC. Further combination of ICs with different functions will become the basis of electronic devices in complex equipment such as computers, televisions, and communication equipment. This is analogous to organs in the body, which comprise numerous cells that constitute an organism.

At present, the size of the transistor continues to shrink, whereas the density of electronic devices on unit chips keeps

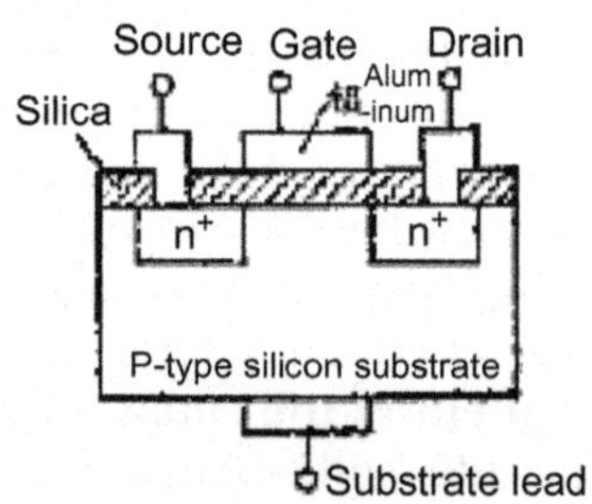

Figure 3.6 Basic structure of the traditional transistor.

increasing. This is demanded by ultra-miniaturization of devices as well as increases in computing capacity. Intel's co-founder Gordon Moore described the future development trend of computer chips in 1965. Moore's view is: "The number of transistors that can be placed inexpensively on an integrated circuit has increased exponentially, doubling approximately every two years." This is the famous Moore's law.

Facts have proved that Moore's law accurately forecasts the development trend of chip technology in the past 30 years. With the increasing integration of ICs, the size of transistors and the line width of ICs are getting smaller and smaller, resulting in a great challenge for Moore's law. Until 2010, the smallest size of microelectronic devices and the minimum width of IC both have been smaller than 100 nm. This approaches the physical limits of the lithographic processing technology in modern microelectronics, potentially halting the progress of Moore's law in future semiconductor technology.

Using advanced photolithographic techniques, transistors of narrow size and line width can be etched on silicon, successfully reducing the size of the transistor channel to 0.1 μm (100 nm). However, the further reduction of size scales will be difficult because of limitations in photolithography optics and chemistry. In the lithography process, the minimum line width of processing depends on the selected wavelength of light. At present, light used in current lithographic processes is in the *deep ultraviolet region*, so it is also known as deep ultraviolet (UV) lithography. Deep UV has a wavelength of 240 nm, and its theoretical limit in lithography is 100 nm. Therefore, current lithographic technology will be unable to process a minimum line width of less than 100 nm. To improve the accuracy of lithography, researchers have been constantly exploring and developing more stable light sources with shorter wavelengths. Many chip manufacturing companies in the United States and Japan and some research institutes in China are developing extreme UV lithography that applies the ultra-UV laser source. With shorter ultra-UV wavelengths, a minimum line width in lithography of 70 nm or less is possible. With the increased precision of lithography, surface uniformity between substrate and the photolithography mask are required to improve with the decrease in feature size. Another bottleneck may be required to improve the size scales and accuracy of lithography.

The beam spot diameter for electron-beam lithography can be made very small, with its etched precision achieving a minimum line width of 10 nm. However, such processes are too slow and expensive for large-scale production.

Scanning tunneling microscopy (STM) and atomic force microscopy (AFM) nanoprocessing technologies, unconstructed by the limitations of photolithography and e-beam lithography, are likely to be the preferred tool in the future in the processing of ultra-large-scale ICs.

At present, the key materials for making ICs have also been updated to further improve performance. For example, IBM has replaced aluminum with copper to connect transistors in the circuit. Copper has better conductivity than aluminum, which increases the transmission rate between devices and also makes chips smaller and less expensive.

Current microelectronics technology, in addition to the aforementioned restrictions in lithographic processing technology, is also subject to restrictions of quantum effects in devices and limitations from power dissipation of electronic devices.

At present, electronic devices are actually "current" devices that involve a large number of electronic phenomena that obey the classical theory of electron motion. In the next section, the microprocessor chip is used as an example for discussing the limitations of electronic behavior in microelectronics technology. The chip microprocessor works with the on/off state of logic gates, which is dependent on the availability of current flow. Smaller chip widths will lead to a substantial reduction in the number of electrons that flow through the logic gates in a certain amount of time. When the number of electrons is sustainably reduced, the logic gate will be in a state of uncertainty in determining the "on" or "off" and will not work properly. In addition, a device being reduced to nanoscale may lead to other quantum effects, resulting in some nonclassical behavior of electrons.

Another major limitation in the development of microelectronics technology is the large power consumption that will be demonstrated for such devices. Chip redesign and manufacturing, as well as improvements in the lithographic process, can improve the integration of ICs to a certain extent, but the large operating current power of current microelectronic devices comes with relatively high power consumption. With the substantial increase in chip integration and speed of communication, the speed of electronic flow in the circuit will also increase, which will eventually lead to an exponential increase in the power needed to run the chip properly, potentially resulting in chip failure.

To overcome the limits of microelectronic technology, molecular electronics and nanoelectronics are now of immense research importance. Nanoelectronics mainly involves the electronic, atomic, and molecular motion in the nanoscale patterns, focusing on nanoscale films and wires and various other assemblies, based on quantum properties with respect to the electronic properties, characteristics, and processing/assembly conditions. The performance involves some of the basic issues including amplification, oscillation, pulse technology, and computing processing. The new theory is mainly based on the quantum details of electronics, electronic quantum tunneling effect, noncontinuity of the electronic energy levels, quantum size effect, and statistical fluctuation characteristics.

Quantum effects in nanoelectronic devices exhibit operating currents of 1–10 electrons, and thus have very small energy consumption. Compared with the current microelectronic devices in ultra-large-scale ICs, their power consumption will be greatly reduced. Another significant advantage of nanoelectronic devices is that their working frequency can be increased significantly.

By virtue of today's microprocessing technology, solid-state electronics channels of very small size can be made to manipulate electron motion through voltage control. Conductance will show a step-like variation as G_0. This change is due to the single-electron access to the channel and is called the *single-electron effect*. The most promising application of the single-electron effect is its use in the electronic features of special conductance material, such as CNTs, C_{60}, and quantum dots (QDs). These materials can be used to create single-electron devices in molecular electronics to be used as SETs, single-electron switches, single-electron circuits, and single-electron logic devices. Therefore, a new discipline, molecular electronics, has emerged. The advent of the SET brings the hope of further reducing the size of electronic devices.

3.4.1.2 Single-Electron Transistor [4–6]

SETs are quite different from the traditional transistor regarding their principles of operation. SETs are novel nanoelectronic devices based on the basic physical principles of the Coulomb blockade effect and the single-electron tunnel effect.

As early as the 1950s, people discovered the Coulomb blockade effect and the single-electron tunneling effect. These effects were not taken advantage of until the 1980s, when they were successfully used to make electronic circuits. The main reason for the delay in practice is simply limitations in manufacturing.

Figure 3.7 is a diagram of an SET, wherein the biggest difference from the traditional transistor is that it allows only one electron from source to drain. This requires electronic access to

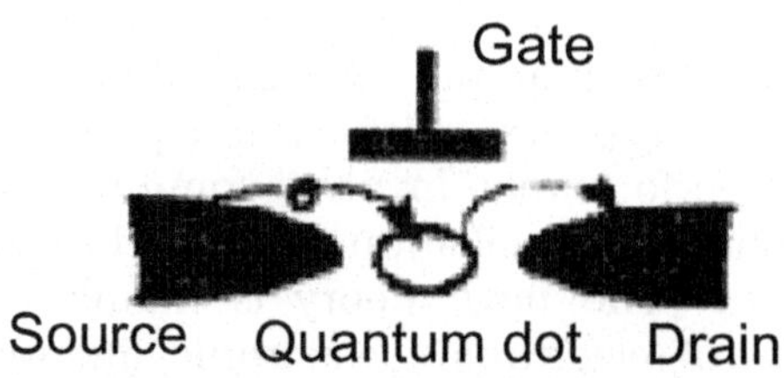

Figure 3.7 Diagram of SET.

the channel and the ability to turn off the source by adjusting applied voltage. This enables the election to temporarily stay on an island-like QD, followed by electron transport on an applied voltage to the drain. This is similar to the Panama Canal, where the water level is regulated through the gate opening and closing to allow a ship coming through the canal.

In actual devices, a thin layer of the tunnel junction is in between the QD island and the source and drain. Conductive devices are separated by an insulating material. As the thickness of the insulation between the devices decreases, the electrons will transmit the charge through the insulating layer, resulting in device failure. This is called the quantum tunneling effect in quantum mechanics. If two nanoparticles are connected by a tunnel junction (a very small gap or a very thin insulating layer), then a single electron will pass through the tunnel barrier (tunnel junction) from one nanometal particle to another. This process is called *quantum tunneling*. To make a single electron tunnel from one nanoparticle to another, the electron is required to have enough energy to overcome the Coulomb blockade of electrons. This process is the *single-electron tunneling effect*.

The island in the middle is known as the Coulomb island, which is usually a QD 10–100 nm in size. Because the size of the material is close to that of the electron wavelength, the electron confined in QDs comes with discrete energy levels. This is why the QD is called an "artificial atom." However, according to Heisenberg's uncertainty principle, the energy has no accuracy. Only when the accuracy of this energy, $\triangle E$, is far less than the energy gap, E_g, of the QD can the electrons be bound at a certain level. This condition is expressed as:

$$\Delta E \ll E_g = e^2/C \tag{3.21}$$

where e is the amount of the basic charge and C is the capacitance of QDs. $\triangle E$ can be estimated according to the uncertainty principle:

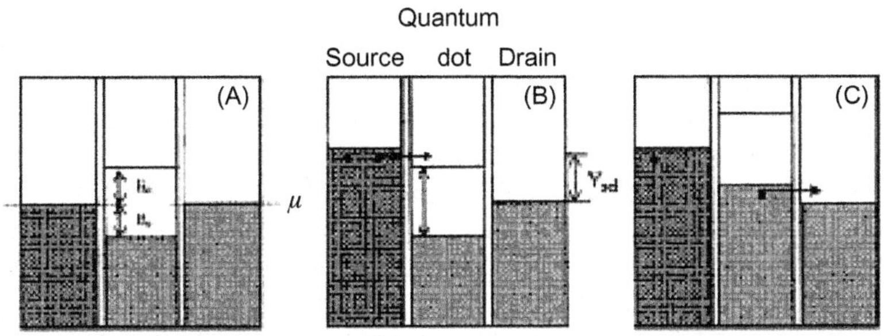

Figure 3.8 The quantum well formed by two tunnel junctions and a quantum dot.

$$\Delta E \Delta t \geq h \tag{3.22}$$

Here, h is Planck's constant. The uncertainty of time, Δt, can be approximated as the characteristic time of the equivalent RC circuit formed by the junction and QD:

$$\Delta t = RC \tag{3.23}$$

Here, R is the resistance of the junction. Substituting Eqs. (3.22) and (3.23) into (3.21), we get

$$R \gg h/e^2 \approx 26\ \text{k}\Omega \tag{3.24}$$

This means that the junction resistance must be greater than 26 kΩ to ensure the stability of the electron on the same level.

From the energy point of view, the QD and the tunneling junction with high energy on both sides can form a potential well, as shown in Figure 3.8A. Importantly, this energy barrier has limited height on both sides of the QD. Moreover, the voltage can be adjusted to individually increase or decrease the height value, as shown in Figure 3.8B. The source is applied with a negative bias so that the potential energy of the source can be increased to exceed the lowest unoccupied energy level of the QD. The electrons then have the opportunity to penetrate the barrier and flow from the source to the drain, as shown in Figure 3.8C.

In Figure 3.8, two tunneling junctions and the QD can form a quantum well of energy. By adjusting the applied voltage, the "depth" of the well will change accordingly so as to control the single-electron flow.

An SET is a transistor that has a separate source, drain, and gate. Like the classical FET, current flows between the source and the drain via gate-adjusted electric field control. A major difference in the SET is that only one electron at a time can come through inside the SET. This is strikingly different from the FET, which allows tens of thousands of electrons to flow through at the same time. This difference is caused by the Coulomb blockade effect.

The Coulomb blockade effect is one of the most important physical phenomena observed in the field of solid-state physics in the 1980s. When the size in the physical system reaches a nanoscale order of magnitude, the system will have a noncontinuous charge and discharge process; in other words, it is quantized. To charge an electron, energy, E_c, is required to be $e^2/2C$ (charging energy), where e is the electron charge and C is capacitance of the physical system. The smaller the system is, the smaller the capacitor C is, while the energy, E_c, will become greater. This energy can be referred to as Coulomb blockade. In the charging and discharging process in a nanoscale system, the electron cannot undergo continuous mass transfer, but rather single-electron transfer. Such single-electron transfer in a nanomaterial is known as the Coulomb blockade effect.

After an electron enters into a QD from a source, energy levels of the QD will increase accordingly (as compared in Figure 3.8B and C). The increased amount will be charging energy E_c:

$$E_c = e^2/2C \tag{3.25}$$

The entry of the first electron results in a level increase, which is equivalent to erecting a barrier to the second electron. If the charging energy is greater than the electron energy obtained by a thermal perturbation:

$$E_c > k_B T \tag{3.26}$$

where k_B is the Boltzman constant, then the second electron will not be able to obtain sufficient energy to cross the barrier into the QD until the first electron leaves and the energy levels decrease back to the original level. This is the principle of the Coulomb blockade effect.

The smaller the QD size, the smaller the capacitance and the greater the charging energy will become. For example, when a QD has a diameter of 100 nm, capacitance value of approximately 10–16 F, and relative charging energy of approximately 10^{-3} eV, the operating temperature is required to be below 4 K to satisfy Eq. (3.26) and to ensure the Coulomb blockade effect. Such a low temperature is clearly not conducive to practical applications. To raise the operating temperature, the size of QDs must be reduced to improve the charging energy. According to estimation, if you want to achieve the Coulomb blockade effect at room temperature (300 K), the QD is expected to have a capacitance value as small as 10^{-18} to 10^{-19} F. Accordingly, its size must be less than 10 nm. Based on the current level of technology, it is still rather difficult to prepare such an SET operation at room temperature. This is also a current challenge faced by those in this research field.

At present, SETs can be divided into four categories by their raw material: (1) semiconductor (silicon, GaAs) SETs; (2) metal SETs; (3) superconductor SETs; and (4) the SET from a CNT, in which the SET is made from semiconductor materials, is easy to integrate and can be used for very-large-scale ICs. Figure 3.9 shows the SET made from semiconductor materials. Compound semiconductors AlGaAs/GaAs with energy gaps of different widths are first stacked to form a quantum well. This is followed by electrode patterning made by lithography.

Unlike the aforementioned top-down production process, Figure 3.10 shows gold particles (~10 nm) deposited on the substrate of silicon dioxide, wherein a long chain of three gold particles links the metal source and drain at a distance of 30 nm and works as a Coulomb island required by an SET. This approach is not as good as that of semiconductor technology in terms of precise positioning of the QD. Nonetheless, this

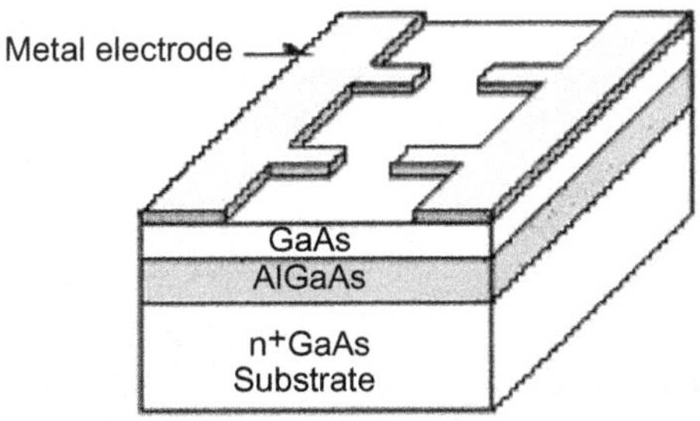

Figure 3.9 SET made from compound semiconductor heterostructure junction.

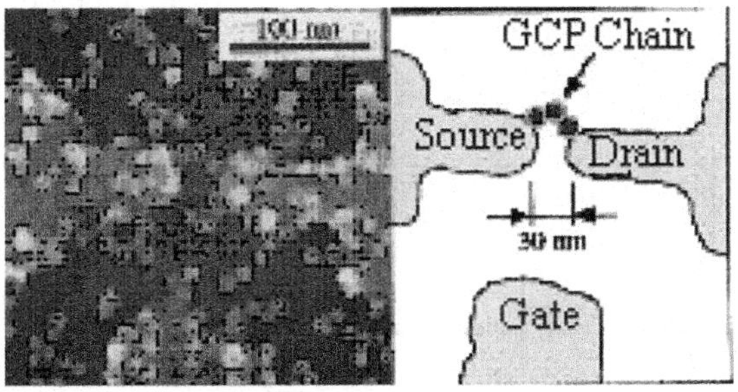

Figure 3.10 SET made from GCP chain on silica substrate.

technique uses very small nanoparticles to demonstrate the production of room-temperature SETs.

For SETs, the center island can also comprise a single molecule. For example, the center island in Figure 3.11 comprises C_{60}.

The smallest possible structure of the center island is simply one atom at the center of an SET. Such a configuration would have a minimal tunnel junction capacitance. Practically, it is difficult to place a single-electron atom between the two electrodes that is not connected with the outside world. When a single atom of the STM tip is very close to a sample, this system can be deemed as the basic structure of an SET on the single-atom scale (Figure 3.12).

Compared with conventional transistors, SETs have the advantages of small size, fast transmission, high sensitivity, and, most importantly, low power consumption. The microelectronics at the

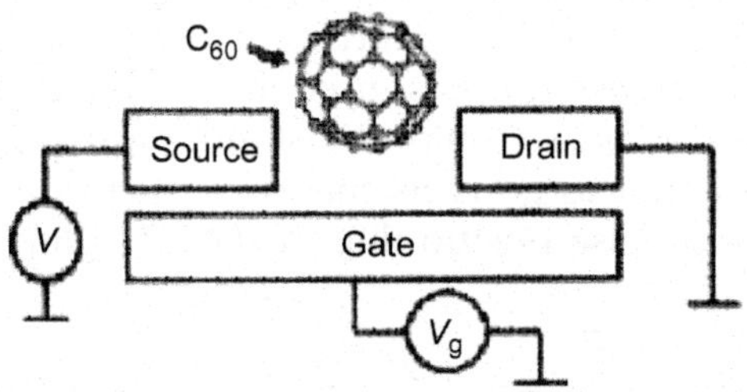

Figure 3.11 Basic pattern of SET at the molecular scale made from C_{60}.

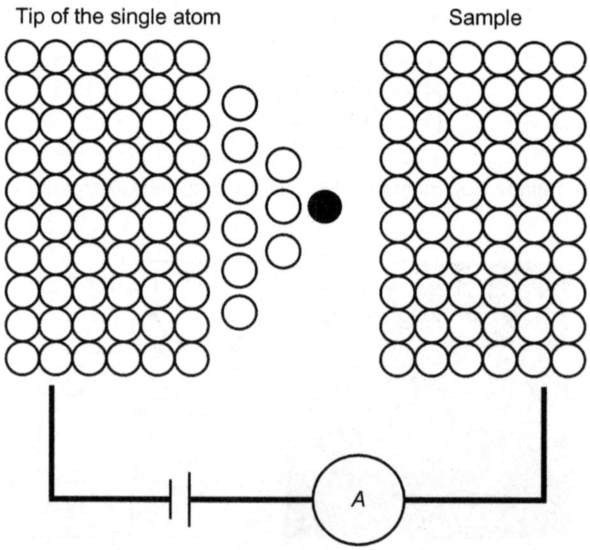

Figure 3.12 Basic structure of SET with the center island of a single atom.

present time have formally entered the domain of the nanoscale from the sub-micron scale, and the technical limits set by basic physical laws are already in sight. For instance, Moore's law is beginning to face serious challenges. The SET is likely to be a new generation of electronic devices in the mainstream of the twenty-first century.

3.4.1.3 CNT Electronics [7–9]

Diodes, FETs, and other elements of electronic devices form the basis of silicon-based microelectronics. Similarly, the development of nanoelectronics requires a new materials framework to be built, whereas the field of CNT-based electronics has been considered to be very promising.

The development of CNT electronics will mainly come through the following stages. First, the electrical properties of CNTs must be fully elucidated. Second, the base elements of CNT-based electronic devices need to be identified and developed (that is, those elements corresponding to silicon-based microelectronics, such as diodes and FETs). Third, these electronic devices based on CNTs are used to build a logic circuit with certain functions. Finally, the key issue to be resolved is how to achieve the integration of these circuits reliably on a mass-production scale. At present, CNT electronics research has made great progress in the first two stages. Meanwhile, the third stage of the research work is underway. There are reports that prototype CNT-based logic circuits have been built in laboratories.

3.4.1.3.1 Quantum Wire

The study of quantum wires mainly concentrates on a variety of CNTs. CNTs are the ideal material for quantum wires. Quantum wire made from CNTs is called *nanotube quantum wire*. Currently, the conductivity of such large molecules is being extensively studied.

Different CNTs can be candidates in the formation of a variety of nonlinear electronic devices. The chirality of CNTs can be used to determine whether these CNTs have the features of being a conductor or a semiconductor. Furthermore, different local defects, such as the existence of five-membered rings and/or six-membered rings, may affect their electronic properties. CNTs can be divided into SWNTs and MWNTs according to their structure. SWNTs are formed by a curling layer of graphite, whereas MWNTs consist of multiple rolled layers (concentric tubes) of graphite.

Conductivity of an SWNT At low temperatures, SWNTs have a resistance of approximately 1 MΩ between the two ends. Under different voltages, the typical current–voltage curves show a clear diode rectification effect: a one-way pass and the curve showing high nonlinearity at a higher bias.

As a voltage is applied between different electrodes, the same CNT can be found with different electronic behaviors at different locations.

Conductivity of a Single MWNT SWNTs have a conductance value of $2G_0$, which is not dependent on diameter or length. MWNT conductance should increase with the number of layers. However, experiments found that the conductance value of MWNTs is only $1G_0$, because only one layer of CNT contributes to the conductance measurement.

If MWNTs are conductors, then their first layer must be conductive. If the second layer is insulating, then the axial resistance is very large. If this is the case, then this and the following layers will not make any contribution to the conductivity.

It should be noted that CNTs are not the only option for preparing molecular wire; there are other methods available. For example, the method of electrochemical growth can be used to connect two metal nanoparticles with metallic copper wire.

3.4.1.3.2 CNT-Based Junction

CNTs can form a molecular junction, cross-junction, and/or p–n junction. A molecule junction is formed by introducing a pair of pentagon–heptagon (abbreviated as 5–7 pair) defects into an SWNT, and then two or more SWNT segments are connected to form the junction. Cross-junction refers to the junction formed by crossing two SWNTs. The p–n junction is an outcome of the specific modulation-doped action of SWNTs. They all show characteristics shared by the diodes in silicon-based microelectronics.

Studies found that the hexagonal network of SWNTs can be introduced with a 5–7 pair defect to acquire a seamless connection of two or more segments of CNTs with different atomic and electrical structures. Thus, a molecular junction of CNT is formed.

CNTs include two types: metal (abbreviated as M) and semiconductor (abbreviated as S). By introducing a 5–7 pair of defects, two sections of SWNTs are linked to form molecular junctions that may be classified into three categories: MM, MS, and SS. Reports of the electrical properties of MS and MM appeared as early as 1999. It is reported that the MS molecular junction has nonlinear current–voltage (I–V) characteristics like a rectifier diode, whereas the MM molecular junction exhibited conductance dependent on temperature. Figure 3.13A and B show the AFM image of 5–7 pair defects in the molecular junction with CNTs. The angle between the two CNTs is approximately 40°. Figure 3.13C shows the diagram of a molecular junction with a heptagonal recess and a pentagonal convex. As shown in Figure 3.13A, there are three electrodes that are mounted on a molecular junction and two straight CNTs. The CNT molecular junction has an important potential application in producing electronic switches.

The practice of doping to adjust the energy band plays a very important role in microelectronics. The semiconductor process is actually a "process of control doping." In today's microelectronics, the most basic components, such as diodes, bipolar transistors, and FETs, are all achieved through the intrinsic semiconductors being doped. Based on these traditional devices, doped CNTs show the same features and functionality. In 2000, the Dai group reported on the preparation and properties of the p–n junction based on the modulation-doped CNTs. In their experiment, a semiconductor-based SWNT received the axial modulation doping, so that half of it is kept as the p-type while the other half is doped with potassium to form the n-type, forming a p–n junction. I–V_g characteristics are shown in Figure 3.14. This I–V_g characteristic curve is divided into four zones. If the gate voltage V_g is between −20 and −12 V, then the conductance is close to zero (state I); if it is between −12 and −7 V, then the conductance clearly has a peak (state II); if it is between −7 and −1 V, then it decreases to zero again (state III); and if V_g increases to between −1 and 20 V, then

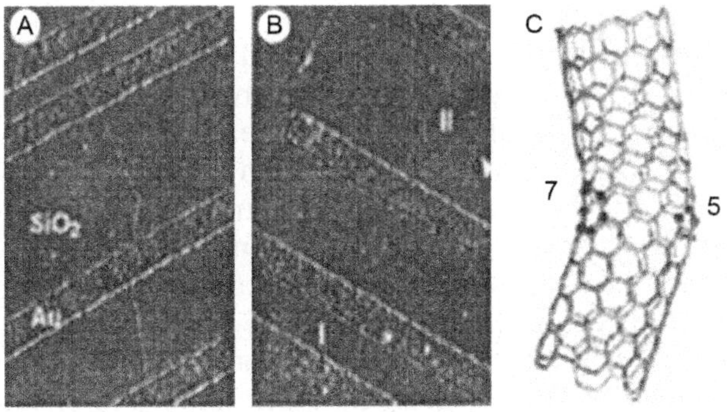

Figure 3.13 Molecular junction with CNTs. (A) An AFM image of 5–7 pair defects in a molecular junction with a carbon nanotube. (B) Another example of 5–7 pair defects in a molecular junction. (C) Diagram of a molecular junction, which has a heptagonal recess and a pentagonal convex. From Ref. [10].

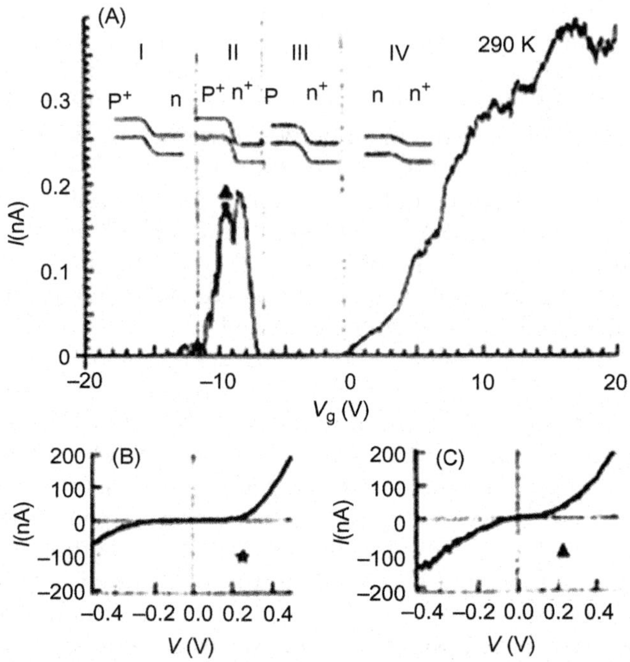

Figure 3.14 $I-V_g$ and $I-V$ characteristic curves of CNT p–n junction. (A) $I-V_g$ characteristic curve. (B) $I-V$ characteristic curve for gate voltage at the "star" point. (C) $I-V$ characteristic curve for gate voltage at the "triangle" point. From Ref. [11].

the conductance begins to increase almost monotonically as the gate voltage increases (state IV).

3.4.1.3.3 SET with CNTs

The nature of CNTs is closely related to their structures. In terms of electrical conductivity, CNTs can be either a metal or a semiconductor in nature. Even in different parts of a single CNT, changes in the structure can also bring about different conductivities. In addition, the radial motion of electrons in CNTs is limited, showing a typical quantum confinement effect, whereas the axial movement of electrons is not impeded. Therefore, we can say that CNTs are one-dimensional quantum wires. As a typical one-dimensional quantum transport material, metallic CNTs show typical Coulomb blockade effect at low temperatures. As the outer electron enters such a tiny capacitor as the CNT (voltage change: $\Delta V = Q/C$, where Q is the injected charge and C is capacitance of CNTs), capacitors that are small enough to be injected with an electron will produce sufficiently

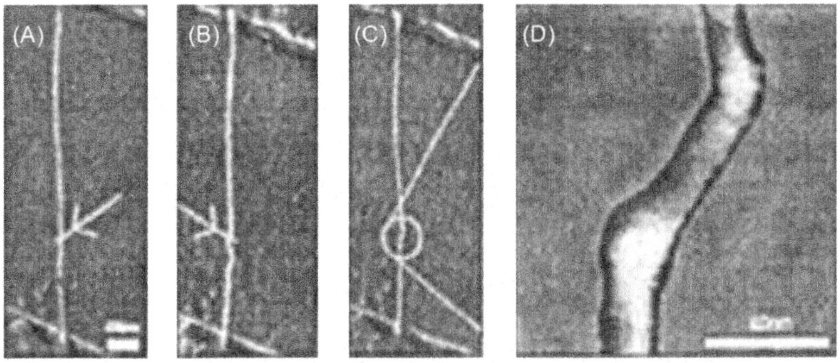

Figure 3.15 SETs with CNTs. (A) The AFM probe is used to drag a part of the carbon nanotube along the direction of the arrow to produce strong bending. (B) The AFM probe drags another part of the carbon nanotube to produce a second strong bend. (C) Image of the single-electron transistor. (D) A larger version of the "Coulomb island" formed in (C). From Ref. [6].

high reverse voltage for the circuit block. After the injected electron comes through the CNT, reverse-blocking voltage will disappear to continue to receive the injected electrons.

As mentioned, the future prospects of traditional silicon-based microelectronics could be encouraging with SETs. However, the majority of SETs need to work at very low temperatures, which severely limits their possible use in practice. Recently, scientists have found that metallic CNTs have strong bending ability, which is manifested as a nanometer-sized tunneling barrier for electron transport. Then, according to the structure of the SET, it can be conceived that the two strong bends of metallic CNTs are combined to form an SET.

In 2001, Postma and associates reported such SETs, which were based on a single-CNT metallic molecule and worked at room temperature. As shown in Figure 3.15A, a metallic-type CNT grows on an Au electrode located on the Si/SiO_2 substrate. Then, the AFM probe is used to drag a part of the CNT in the direction of the arrow to produce strong bending. As shown in Figure 3.15B, in the direction of the arrow, the AFM probe drags another part of the CNT to produce strong bending a second time. Figure 3.15C and D present the image of this SET. The "Coulomb island" between the two bending parts forms a segment of CNT of approximately 25 nm.

3.4.1.3.4 CNT-Based FET

In early 1998, Cees Dekker's group reported an FET that comprised an SWNT that operated at room temperature. The FET included a semiconductor SWNT, to which three metal

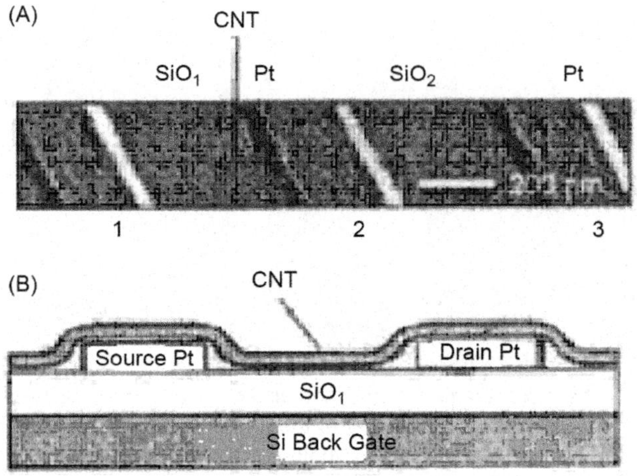

Figure 3.16 CNT-based FET. (A) AFM image. (B) Side sectional view of the FET. From Ref. [12].

electrodes were connected. By adjusting the voltage applied to the gate, the CNT could be converted between on-state and insulating state. There had been reports of similar characteristics of metallic SWNTs that worked at extremely low temperatures. Successful construction of the CNT-based FET makes it possible to take a major step forward in CNT-based electronics.

Figure 3.16A shows the AFM image of the connection of an SWNT and three Pt electrodes. On the Si substrate of the semiconductor, the thermal method is used to allow the growth of a layer of SiO_2 with a thickness of 300 nm while the Si substrate is used as a back gate, as shown in Figure 3.16B.

The CNT-based FET described is made from semiconductor-type CNTs, whereas the CNT-based SET is made by using metallic CNTs. A key issue is how to effectively separate these two types.

3.4.1.3.5 Complementary Nongate (Inverter) Circuit with CNTs

In August 2001, Avouris and team (from IBM) reported the successful development of two complementary CNT-based nongate circuits, with one being based on a structure of two CNTs and the other being based on a single CNT.

As mentioned, an ordinary FET composed of intrinsic CNTs is p type, but the potassium-doped CNTs can have n-type characteristics. The latest research by Avouris and team also found that the p-type CNT FET can be placed in a vacuum for annealing to become n type, and the oxygen environment can make CNTs

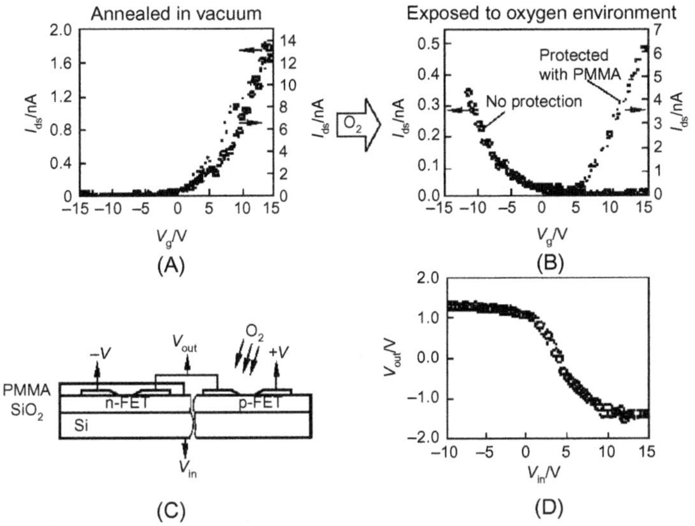

Figure 3.17 Complementary nongate circuit based on two CNTs. (A) $I_{ds}-V_g$ curve for the case annealed in a vacuum. (B) $I_{ds}-V_g$ curve for the case exposed to an oxygen environment. (C) The complementary nongate circuit. (D) The transmission characteristic curve. From Ref. [13].

turn into p type. Therefore, by applying these properties, the same substrate can be placed with p-type and n-type CNT FETs through appropriate procedures for the composition of a complementary inverter.

As shown in Figure 3.17C, an SWNT is first placed on Au electrodes deposited on an SiO_2 layer, with the Si substrate being the gate electrode. This constitutes a 2-p-type CNT FET (the experiment is performed in a vacuum environment of less than 10^{-4} Pa at room temperature). Then, a polymethyl methacrylate (PMMA) photoresist is used to cover the CNT FET on the left side, while the right one is exposed in the vacuum for annealing. After that, the two CNT FETs are transformed into n type. The two annealed CNT FETs have $I_{ds}-V_g$ curves as shown in Figure 3.17A. The circuit is exposed to the 10^{-1} Pa oxygen environment for a certain time, the right CNT FET that is not protected with the PMMA returns to p type, and the CNT FET not covered by PMMA remains as n type. Here, $I_{ds}-Vg$ curves of the two CNT FETs become as shown in Figure 3.17B. This constitutes a CNT inverter, which works in exactly the same way as the ordinary complementary metal oxide semiconductor (CMOS) inverter, and its transmission characteristics are shown in Figure 3.17D.

Figure 3.18A shows the AFM image of an inverter comprising a single nanotube on the three Au electrodes. Photoresist PMMA

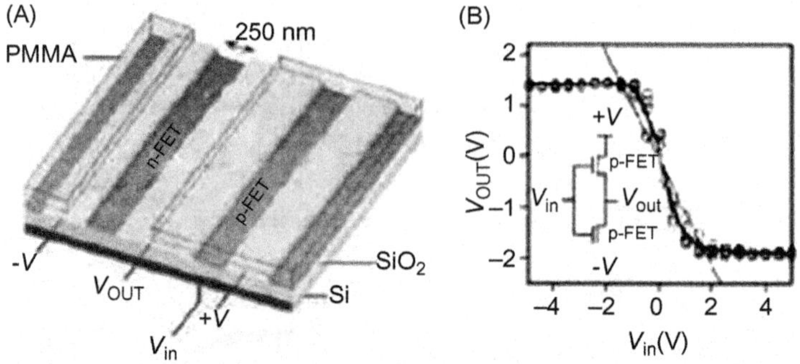

Figure 3.18 Complementary nongate circuit based on a single CNT. From Ref. [13].

covers part of the CNT, with the other part being exposed. The exposed part of the CNT is doped with potassium, so that it becomes n type, whereas the part not covered by the PMMA remains p type. This also constitutes a CNT inverter. Figure 3.18B shows the equivalent inverter circuit and transmission characteristics. Its equivalent circuit can also be expressed as a CMOS circuit. When the input (V_{in}) is high (positive voltage), n-FET is turned on, and p-FET is off, the output (V_{out}) is low (negative voltage); when the input is low (positive voltage), p-FET is on, and n-FET is off, the output is high (negative voltage).

To summarize, the concepts and structures of some of the CNT-based logic circuits have been explained, but there are still many problems that remain unsolved. First, we must implement and verify all the logical elements of CNTs. For example, an important logical storage unit used to build DRAM (dynamic RAM) has not been realized. Previously described CNTs are mostly of the resistor–transistor logic, which is not a good logic style for current ICs. In principle, we expect to have complementary logic elements such as CMOS. Second, two key issues in the future that will constrain the development of CNT electronics need to be addressed: although the semiconductor-type and metallic CNTs can be screened when currently building devices and single-logic circuits, it is difficult to ensure the consistency of electrical properties of CNTs in large-scale production; and there are issues regarding positioning and manipulating massive amounts of CNTs in large-scale circuits. When these issues are resolved, CNT electronics will develop rapidly.

3.4.2 Other Applications of CNTs

The unique nature of CNTs provides an almost unlimited amount of potential applications, as discussed.

3.4.2.1 Nano Test Tubes

CNTs provide the smallest capillary to study capillarity mechanisms, and the smallest test tubes for chemists to perform chemical reactions.

3.4.2.2 Nanobalance

The microparticles on CNTs can cause changes of beat frequency of associated current. By taking advantage of this, scientists invented a "nanobalance," with accuracy of 10^{-17} kg in 1999, which is able to weigh the quality of individual viruses. Further research on the nanobalance demonstrated that the weight of single atoms could be determined.

3.4.2.3 Nanomolds

Materials like metals and oxides can be filled inside CNTs, effectively using CNTs as molds. Specifically, the CNTs are first filled with metal and other materials, and then the carbon layer will be etched away. Thus, the smallest nanoscale wires or novel one-dimensional materials can be prepared and used in future molecular electronic devices or nanoelectronic devices.

3.4.2.4 CNTs: Field Emission Cathode Materials

Using CNTs as field emission cathode materials is expected to address problems limiting cathode-ray tubes.

The cathode field emitter in a display is made from silicon or metal and is usually processed by methods such as chemical etching and electron-beam deposition. Because of restrictions on microprocessing technology, the manufactured silicon or metal field emission device has a minimum size of approximately 20–50 nm with uniformity issues.

Field emission devices have a large maximum size, requiring their threshold voltage to reach 100 V or more to eject electrons. Such cathode field emission devices that have high operating voltages and inferior uniformity will not be applicable in the next generation of flat-panel displays.

CNTs provide an excellent choice for field emission cathode materials because the single-wall CNTs have an average diameter of 1–2 nm, only approximately one-tenth to one-twentieth of the maximum diameter of the silicon field emission cathode device that is widely used. The field emission cathode device made from SWNTs will be a very sharp and good electron emission source, with a threshold value of voltage for field emission that is able to be reduced to approximately 10 V, making it possible to produce a new generation of wall-mounted flat-panel displays with low energy consumption.

In 1999, Japan and South Korea successively developed CNT-based displays with thickness of a few centimeters. These products, compared with traditional displays, feature small size, light weight, good quality, low power consumption, fast dynamic response (only a few microseconds), and have a wide operating temperature range (able to work at a range of $-45°C$ to $+85°C$). It is clear that CNTs may have great potential for application in flat-panel displays.

3.4.2.5 Application of CNTs in Hydrogen Storage

Hydrogen has been largely regarded as the future of clean energy. However, hydrogen itself is low in density, and compressing it into liquid is not convenient or safe. Lightweight CNTs with hollow structures can serve as an excellent hydrogen storage container, with a stored hydrogen density even higher than that of liquid or solid hydrogen. Through thermal control, the hydrogen inside can be slowly released and used as an energy source. Researchers are trying to use CNTs to produce lightweight and portable hydrogen storage containers.

As an absorption material with the largest hydrogen storage capacity, CNTs will also contribute to the development of hydrogen-fueled vehicles. The world's four major automobile companies (General Motors, Ford, Toyota, and Daimler-Benz AG) are accelerating the development of hydrogen-fueled vehicles. For example, Toyota successfully developed the second-generation hydrogen fuel cell in March 2000, which is a novel solid polymer fuel cell using hydrogen as fuel. This car has an output power of up to 90 kW and maximum speed of 150 km/h, with the ability to be driven more than 300 km with one tank of hydrogen.

If we want to use hydrogen as a power source for vehicles, then some key conditions need to be met. A vehicle using hydrogen as fuel is expected to drive 500 km on 3.2 kg of hydrogen. This requires such vehicles to have containers with very large hydrogen storage density. Regarding two indicators of hydrogen storage, weight and volume, the container must, respectively, reach 6.5 wt% (weight percentage of hydrogen accounting for sum of hydrogen storage container and the hydrogen weight) and 62 kg H_2/m^3 (stored hydrogen weight per unit volume).

What is the future of CNT storage of hydrogen? The National Renewable Energy Laboratory and IBM performed a test of the ability of CNTs in adsorption of hydrogen (hydrogen storage capacity) in 1997 and found that the capacity of CNTs in adsorption of hydrogen increased as diameter increased. Thus, it is inferred that high-purity SWNTs have a hydrogen absorption capacity in the range of 5–10 wt%, of which the high-purity SWNTs with a diameter of 1.63–2.0 nm can attain hydrogen storage technical specifications of 6.5 wt%.

3.4.2.6 High-Energy Microbattery

With good electrical conductivity properties, CNTs can also serve as a cathode or supersede the conductive polymer material as a conductive medium to create high-energy batteries. These batteries feature a small size, high energy, and long life, and are the best choice for computers and automobiles.

3.4.2.7 High-Energy Capacitor

CNTs, after being pressed into sheets, can be used to make high-energy capacitors, whereas a small amount of CNTs added to other materials also can significantly improve the conductivity of the material (for example, polymer materials, by adding a certain amount of CNT, can make the resistivity lower by three orders of magnitude or more).

3.4.2.8 Chip Thermal/Heat Protection

With excellent thermal conductivity, CNTs are expected to become the heat plate for computer chips in the next generation of high-speed computers and can also be used as protective materials for a variety of high-temperature devices, such as engines and rockets.

3.4.2.9 Nanoreactor

Some researchers have also found that by virtue of a unique pore-like structure, large specific surface area (surface area per gram of CNTs up to several hundred square meters), and other characteristics, CNTs can be made into nanoreactors, which can make a chemical reaction confined to a small range.

3.4.2.10 Nanocomposite Materials

Based on the nature of CNTs, composite materials with high performance can be produced. For example, CNT-reinforced plastics have excellent mechanical properties, good conductivity, corrosion resistance, and the capacity to shield radio waves. The composite materials of CNTs with a cement matrix have good impact resistance as well as anti-static, anti-wear, highly stability, and environmentally friendly features. Ceramic composite materials, once reinforced with CNTs, can have high strength and good resistance to impact. The existence of the five-membered ring defects enhanced the reactivity of CNTs. Under the conditions of heat and the existence of other substances, CNTs tend to be at the end sides to form a tube, which is prone to metal infiltration to form metal matrix composite materials. Such materials, in addition to having high strength, high modulus, and high temperature, have very small thermal expansion coefficients and strong resistance to thermal changes.

References

[1] A.K. Geim, K.S. Novoselov, The rise of graphene, Nat. Mater. 6 (2007) 183.
[2] R. Waser, Nanoelectronics and Information Technology-Advanced Electronic Materials and Novel Devices, second, corrected ed., Wiley-VCH Verlag, Weinheim, Germany, 2005.
[3] T. Chakraborty, F. Peeters, U. Sivan (Eds.), Nano-Physics & Bio-Electronics: A New Odyssey, Elsevier, Amsterdam–London–New York–Oxford–Paris–Shannon–Tokyo, 2002, pp. 4–19.
[4] Q.D. Chen, An introduction to single electron transistors, Physics (bimonthly) 26 (3) (2004) 483–490.
[5] Jin-Yan Zhang, Ya-Zhi Cai, An introduction to single-electron transistor (SET), Nanosci. Nanotechnol. 19 (2006) World, Dec.
[6] H.W.C. Postma, T. Teepen, Z. Yao, et al., Carbon nanotube single-electron transistors at room temperature, Science 293 (2001) 76–79.
[7] J.F. Jiang, Q.Y. Cai, Nano-electronics, Nature 23 (1) (2001) 10–14.
[8] Y. Zheng, S. Yi, G.U. Shu-Lin, S.N Bo, Z. Rong, Z. You-Ke, Carbon nanotube-based electronics, Physics 31 (10) (2002) 624–628.
[9] Z. Yang, Y. Shi, S.L. Gu, B. Shen, R. Zhang, Y.D. Zheng, Carbon nanotube-based electronic devices, Res. Progress Solid State Electron. 22 (2) (2002) 131–136.
[10] Z. Yao, H.W.C. Postma, B. Leon, C. Dekker, Carbon nanotube intramolecular junctions, Nature 402 (1999) 273–276.
[11] C. Zhou, J. Kong, E. Yenilmez, H. Dai, Modulated chemical doping of individual carbon nanotube, Science 290 (1999) 1552–1555.
[12] S.J. Tans, A.R.M. Verschueren, Dekker, Room temperature transistor based on a single carbon nanotubes, Nature 393 (1997) 49–52.
[13] V. Derycke, R. Martel, J. Appenzeller, P. Avouris, Carbon nanotube inter- and intramolecular logic gates, Nano Lett 9 (2001) 453–456.

4

SEMICONDUCTOR QUANTUM DOTS

CHAPTER OUTLINE

In the middle of the past century, the advent of silicon-based semiconductors and transistors combined with the successful development of silicon-based integrated circuits triggered the electronics industry revolution. In the early 1970s, quartz fiber-optic materials and GaAs lasers were invented to promote the rapid development of fiber-optic communications technology, effectively catalyzing the start of the information age. In 1970, L. Esaki and R. Tsu (both working for IBM) proposed the concept of superlattices [1] and contributed to the quick development of low-dimensional semiconductor materials. This completely changed the design concepts of optoelectronic devices, such that the thought process for design and manufacture of semiconductor devices went from "impurity management" to the area of "band engineering." The development and application of nanoscience and technology will enable the capability to control, manipulate, and manufacture powerful novel devices and circuits at the atom, molecule, and/or nanoscale, potentially changing our current standards of life as we know it.

Nanomaterials and Devices. DOI: http://dx.doi.org/10.1016/B978-1-4557-7754-9.00004-4

Beginning in the 1970s, research on low-dimensional semiconductor quantum structures has had a rich 30-year history. Currently, the electronic or optoelectronic devices in a III–V quantum well structure have been commercialized. These materials helped achieve high-speed electronic devices and high-performance luminescence detectors. Quantum dots (QDs) are a class of zero-dimensional material, sometimes referred to as *artificial atoms*. Research on QDs has led to applications in light-emitting devices and detectors.

In addition to its application in electronic or optoelectronic devices, QDs can be used to manufacture qubits. For example, the charge of a QD can be used in the definition of quantum bits: one extra electron stands for "1" and none stands for "0." Some have suggested that the electron spin states (spin up or spin down) can be considered as "0" and "1." Recent research is more focused on the use of the exciton state in QDs; when the QDs absorbed the energy that is precisely from the photon excitation, a self-excited electron will enter the conduction band while leaving a hole in the valence band. Electrons and holes are paired to reduce energy by Coulomb attraction, resulting in the formation of excitons. Excitons have fairly long lifetimes, with relatively sharp radiation lines, and are easy to excite and detect with optical characterization methods.

There is a great deal of optimism surrounding the potential of QDs in quantum computing. However, despite all the excitement surrounding QD-based computing along with the various theoretical work put forward by researchers, no one has been able to test the quantum-bit logic operations. Experimental techniques are still lacking regarding the control and measurement of the QD process.

4.1 The Physical Basis of Semiconductor QDs [2]

4.1.1 Quantum Confinement Effect

Because electrons are limited in movement along certain directions, the electron energy in that direction would have been quantized. This is because in this direction, the bound electrons would lead to the formation of standing waves. According to the number of binding dimensions, we can classify materials into bulk material (0 binding dimension, 0D), quantum well (one binding dimension, 1D), quantum wire (two binding dimensions, 2D), and QD (3 binding dimensions, 3D). Energy quantization may have a great impact on materials in terms of the electronic density of states (DoS), which is defined as

$$\mathrm{DoS} = \frac{\mathrm{d}N}{\mathrm{d}E} = \frac{\mathrm{d}N}{\mathrm{d}k}\frac{\mathrm{d}k}{\mathrm{d}E}$$

For example, for the 3D case, $N(k) = k$ space volume/volume of each state, then

$$N(k) = \frac{4/3\pi k^3}{(2\pi)^3/V}$$

The carrier density difference in the structure with different dimensions is as shown in Table 4.1 and Figure 4.1.

Discrete states are caused by quantum confinement and are obtained with the energy-level solution of the Schrödinger equation:

$$-\frac{\hbar^2}{2m}\nabla^2\Psi + V(r)\Psi = E\Psi$$

Table 4.1 Expressions of Carrier DoS in the Structure with Different Dimensions

Structure	Limited Number of Dimensions	$\mathrm{d}N/\mathrm{d}E$
Bulk material	0D	$\sqrt{E}$
Quantum well	1D	1
Quantum wire	2D	$1/\sqrt{E}$
Quantum dots	3D	$\delta(E)$

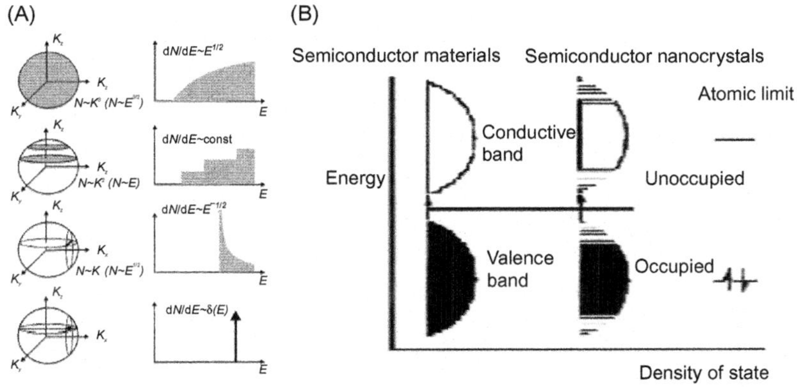

Figure 4.1 (A) Carrier DoS in the structure with different dimensions and (B) comparison of bulk materials and nanocrystals in electronic DoS.

For example, in the case of the one-dimensional infinite square potential well, its solution is

$$\Psi(x) \sim \sin\left(\frac{n\pi x}{L}\right)$$

where n is an integer. Here, the ground-state wave function is as shown in Figure 4.2. If the restrictions apply only in one direction (e.g., x), then the other two directions of energy are still continuous. The total energy is

$$\frac{n^2 h^2}{8mL^2} + \frac{p_y^2}{2m} + \frac{p_z^2}{2m}$$

For the 3D infinitely deep square potential well (quantum box),

$$\Psi(x, y, z) \sim \sin\left(\frac{n\pi x}{L_x}\right) \sin\left(\frac{m\pi y}{L_y}\right) \sin\left(\frac{q\pi z}{L_z}\right)$$

Here, n, m, and q are integers. Energy level is

$$E_{n,m,q} = \frac{n^2 h^2}{8mL_x{}^2} + \frac{m^2 h^2}{8mL_y{}^2} + \frac{q^2 h^2}{8mL_z{}^2}$$

This is the simplest case. Rather than an infinite depth of the box, the actual potential barrier can also be spherical, restricted, or exhibit harmonic oscillator potential limitations. Note that we considered only one electron in this case. In the actual situation, it is often required to deal with multiple particles and electron–hole pairs. Particle mass also requires consideration, along with potential mismatch at boundaries of particles.

The size of QDs is in the range of approximately 10–100 nm, which is equivalent to the de Broglie wavelength of electrons in semiconductors. Thus, the electron or hole would be subject to three-dimensional quantum confinement effects and have

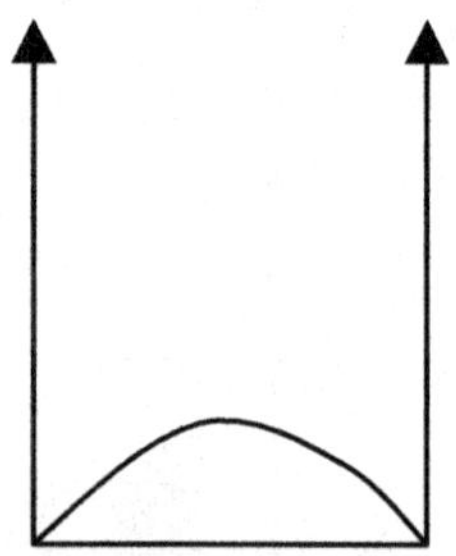

Figure 4.2 Electronic ground-state wave function of one-dimensional infinitely deep square potential well.

quantized energy levels, forming the so-called zero-dimensional electronic system. As mentioned, QDs are often referred to as artificial atoms. The main reason is that QDs have an electronic configuration that is very similar to that of atoms, so energy levels of QDs are often introduced with symbols such as s, p, and d, which are referred to as the ground states of QDs and the excited-state levels.

Take the InAs QDs, for example. Their geometric shape is similar to a flat convex lens; therefore, the potential energy in the *r* direction can be approximated with a two-dimensional paraboloid. This approximation tells us that the ground state and the excited state have a number of degeneracy, respectively, *2, 4, 6*, and so on (including spin degeneracy). In other words, QDs can be filled with two electrons, *p* energy levels can be filled with four electrons, and so on. Clearly, the QDs are not the same in electronic configurations with the atoms. In the orbit as *s, p, d*, the atomic energy levels can be filled with the numbers *2, 6, 10*, and so on. This difference is mainly rooted in the QDs with essentially different forms of potential energy from the atoms. The atomic potential energy is mainly formed in the Coulomb interaction, with three-dimensional symmetry. However, the potential energy form of QDs is directly related to the geometric shape. Because its shape is similar to a convex lens, its height is much smaller than the diameter. Therefore, electrons in QDs have only two-dimensional (plane) symmetry, with the degeneracy form of energy levels unlike that of atoms.

4.1.2 Excitons and Luminescence [3]

4.1.2.1 The Concept of Excitons

Due to Coulomb forces, excited electrons and holes can be tied together to form bound electron–hole pairs, for example the excitons. In a semiconductor, the electron and hole are of attracting potential, where the hole has an effective mass greater than the electron effective mass, leading to the formation of a hydrogen-like atomic system. Exciton binding energy is determined by the Bohr theory:

$$E_n = -\frac{e^2}{2\varepsilon a_0 n^2}; \quad a_0 = \frac{\varepsilon h^2}{4\pi^2 \mu e^2} \quad \text{here, } \mu \text{ is the reduced mass.}$$

In the QDs, excitons may arise from within the QDs and are subject to them, with the restriction extent decided by their size. An exciton also has discrete energy levels, so an absorption peak similar to the δ function is presented in the exciton absorption spectrum.

4.1.2.2 Energy Band Structure of Excitons

At zero temperature, the semiconductor band is structured with a filled valence band and an empty conduction band. Between the top of the valence band and the bottom of the conduction band, there is a gap with no electrons, called the *band gap*. In terms of the symmetry of the cubic lattice, the top of the valence band is the state equivalent to angular momentum $l = 1$. Due to the interaction of spin and orbital, a fourfold degeneracy of state is formed as

$$(j = l \oplus s = 3/2, \quad m_j = \pm 1/2, \quad \pm 3/2)$$

commonly known as heavy hole,

$$\left|\frac{3}{2}, \pm\frac{3}{2}\right\rangle$$

and light hole,

$$\left|\frac{3}{2}, \pm\frac{1}{2}\right\rangle$$

The light and heavy holes are classified according to the size of the effective mass in the vertical interface. Taking into account the role of limitations in the vertical direction, usually the side width of QDs is much larger than that in the vertical direction, exhibiting strong limitations in the vertical direction. The light-hole band is held down because of a small effective mass. So, the absorption of light close to the energy gap comes from a major contribution of the heavy holes.

An electron, when excited from the valence band into the conduction band, will leave a similar positively charged hole. Driven by Coulomb attraction between the hole and electronics, the electrons are combined into an exciton state. Binding energy released in the formation of the exciton is only 6 meV in bulk materials. Regarding two-dimensional quantum wells, the effects of the limitations can increase the binding energy of the exciton by 15 meV (simply interpreted as the positive and negative electron distance being compressed, thus resulting in the increase of the Coulomb force). In the QD structure, exciton binding energy will be further released by approximately 20 meV. Exciton band energy is equal to semiconductor band gap E_g, coupled with the binding energy of electron and hole from limited effects E_e, E_h and minus the binding energy to form excitons E_b:

$$E_{ex} = E_g + E_e + E_h - E_b \tag{4.1}$$

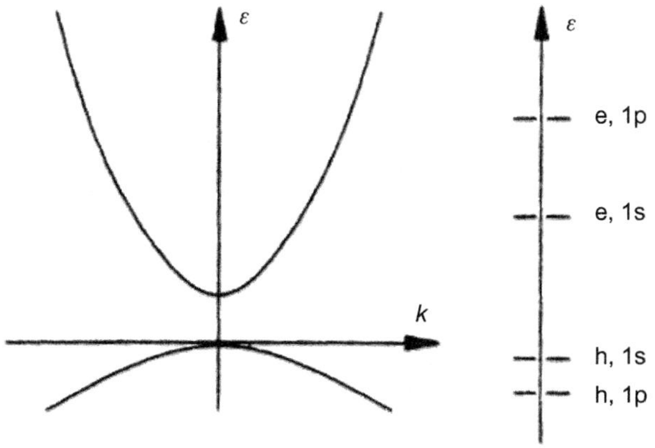

Figure 4.3 Diagram of single-particle energy spectrum in bulk materials (left) and small QDs (right).

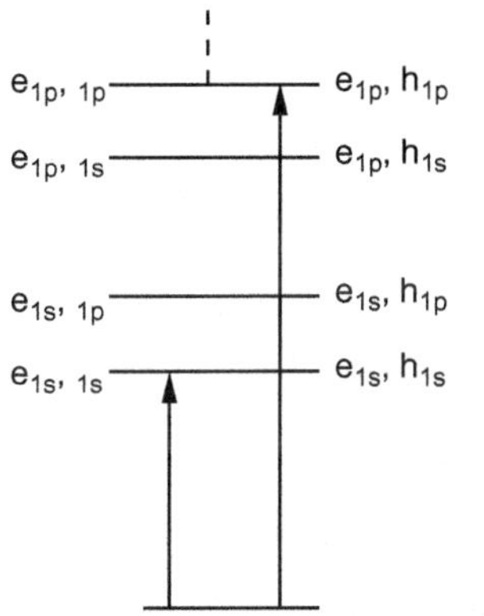

Figure 4.4 Diagram of transition of a single electron–hole pair in semiconductor QDs.

Specifically, Figure 4.3 shows exciton e 1s and h 1s

$$E_{ex}^{1s1s} = E_g + E_e^{1s} + E_h^{1s} - E_b^{1s1s}$$

Figure 4.4 shows exciton e 1p and h 1p

$$E_{ex}^{1p1p} = E_g + E_e^{1p} + E_h^{1p} - E_b^{1p1p}$$

When the electron and hole in the excitons recombine, a photon is emitted that is equal to the energy difference between valence and conduction states. The exciton lifetime can be as long as a nanosecond (10^{-9} s) and, as such, is readily observable with various experimental techniques, allowing the possibility to study the quantum evolution of the excitons.

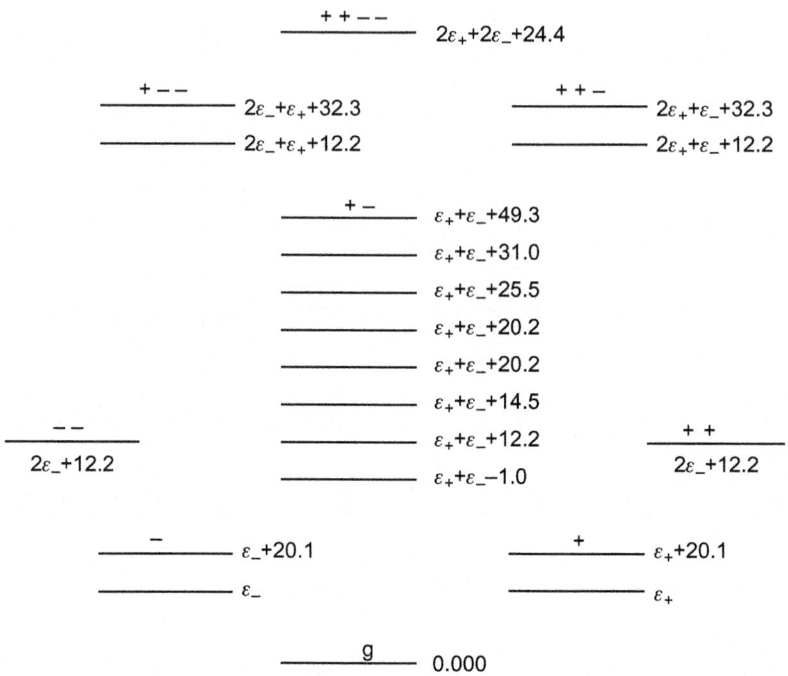

Figure 4.5 Energy spectra of GaAs QDs in multiple-exciton states of regular hexahedron $40 \times 35 \times 5$ nm $\varepsilon_+ = \varepsilon_- = 1764$ meV. From Ref. [4].

When the excitation light to generate the electron–hole increases, we need to consider the multi-exciton states. Because there are a number of electrons and holes, two separate excitons may be close to each other to form coupled pairs of exciton states, which have energy of 1 meV, which is lower than that of the independent pairs of exciton states. Now laser can have a frequency stabilized to 4 neV (4×10^{-9} eV). Thus, it is not difficult to distinguish between these two states in experiments. Figure 4.5 shows an energy band diagram of multiple excitons in typical GaAs QDs: g is all electronic valence bands in the vacuum state and ε_+ is an exciton state excited by the D-light:

$$\text{hole}\left|\frac{3}{2}, -\frac{3}{2}\right\rangle \text{and electron}\left|\frac{1}{2}, -\frac{1}{2}\right\rangle$$

ε_- is an exciton state excited by the L-light:

$$\text{hole}\left|\frac{3}{2}, +\frac{3}{2}\right\rangle \text{and electron}\left|\frac{1}{2}, +\frac{1}{2}\right\rangle$$

Here, $\varepsilon_+ + \varepsilon_- - 1.0$ for the combination state of a D-exciton and an L-exciton, with energy less by 1.0 meV compared with that of the two separate excitons.

4.1.3 Calculations of the Exciton Binding Energy [5]

For spherical QDs, without considering the case of Coulomb interaction, the Schrödinger equation of an electron or hole can be written as

$$-\frac{\hbar^2}{2m_i}\nabla^2\varsigma_i(r) = \varepsilon_i\varsigma_i(r) \tag{4.2}$$

Here, i is electron or hole. Under the ideal quantum constraints, the boundary condition is

$$\varsigma_i(r) = 0 \quad \text{for } r = R \tag{4.3}$$

With the boundary conditions for Eq. (4.3), the Schrödinger equation (4.2) is solved as

$$\varsigma_i(r) = \sqrt{\frac{1}{4\pi R^3}}\frac{j_l(\alpha_{nl}r/R)}{j_{l+1}(\alpha_{nl})}Y_l^m(\theta,\phi) \tag{4.4}$$

Here, j_l is l-order spherical Bessel function. α_{nl} is the n th root of l-order spherical Bessel function. With the spherical Bessel function, boundary conditions for Eq. (4.3) can be written as

$$j_l\left(\alpha_{nl}\frac{r}{R}\right)\bigg|_{r=R} = 0 \tag{4.5}$$

that is

$$j_l(\alpha_{nl}) = 0 \tag{4.6}$$

We substitute Eq. (4.4) into Eq. (4.2) to get a discrete eigenvalue

$$\varepsilon_i = \frac{\hbar^2}{2m_i}\left(\frac{\alpha_{nl}}{R}\right)^2 \tag{4.7}$$

Using the band zero point at the valence band maximum, with Eq. (4.7) type, the energy levels of electrons and holes are

$$\varepsilon^{\mathrm{e}} = E_{\mathrm{g}} + \frac{\hbar^2}{2m_{\mathrm{e}}}\left(\frac{\alpha_{n_{\mathrm{e}}l_{\mathrm{e}}}}{R}\right)^2 \tag{4.8}$$

$$\varepsilon^{\mathrm{h}} = -\frac{\hbar^2}{2m_{\mathrm{h}}}\left(\frac{\alpha_{n_{\mathrm{h}}l_{\mathrm{h}}}}{R}\right)^2 \tag{4.9}$$

As mentioned, the single-particle spectrum is not in line with the optical absorption spectrum, because it does not take into account the Coulomb interaction between electrons and holes. The Schrödinger equation to describe electron–hole pairs is

$$\left(-\frac{\hbar^2}{2m_{\mathrm{e}}}\nabla_{\mathrm{e}}^2 - \frac{\hbar^2}{2m_{\mathrm{h}}}\nabla_{\mathrm{h}}^2 + V_{\mathrm{c}}\right)\phi(r) = \varepsilon\phi(r) \tag{4.10}$$

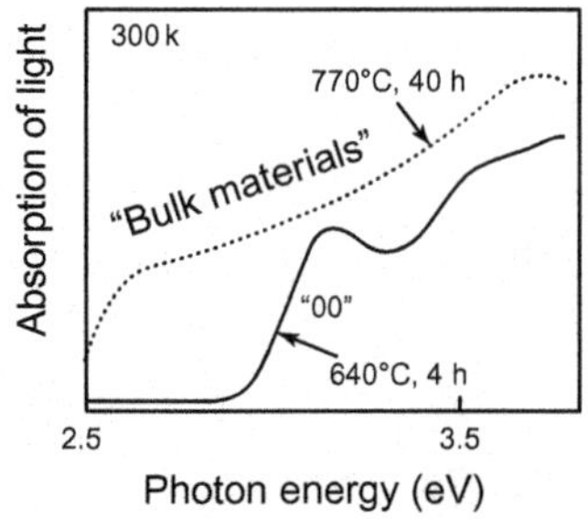

Figure 4.6 Linear absorption spectra of CdS crystals and QD in glass at room temperature.

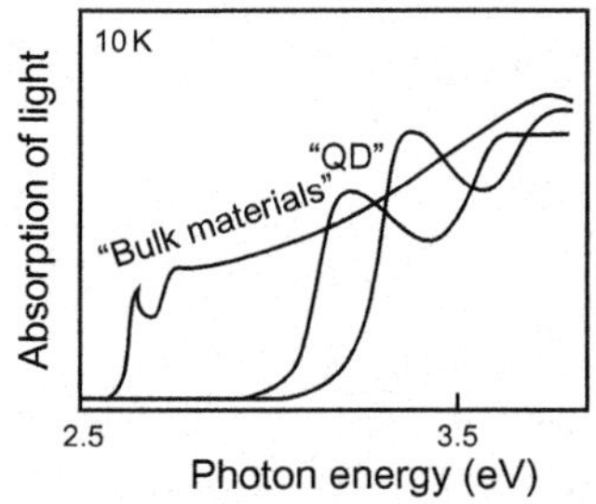

Figure 4.7 Linear absorption spectra of CdS crystals and QD in glass at 10 K.

Applied here are the usual spherical coordinates with the boundary conditions $\phi(r = R) = 0$, and V_c is Coulomb potential. If the impact of Coulomb potential is not included, then Eq. (4.10) can be resolved to get the energy of electron–hole pairs using

$$\varepsilon = \varepsilon^{e} + \varepsilon^{h} = E_g + \frac{\hbar^2}{2m_e}\left(\frac{\alpha_{n_e l_e}}{R}\right)^2 + \frac{\hbar^2}{2m_h}\left(\frac{\alpha_{n_h l_h}}{R}\right)^2 \tag{4.11}$$

and the wave function

$$\phi(r_e, r_h) = \varsigma(r_e)\varsigma(r_h) \tag{4.12}$$

where

$$\varsigma(r) = \sqrt{\frac{1}{4\pi R^3}}\frac{j_l(\alpha_{nl}r/R)}{j_{l+1}(\alpha_{nl})}Y_l^m(\theta, \phi) \tag{4.13}$$

When considering the Coulomb potential, Eq. (4.10) cannot yield an exact numerical solution. The variation method is commonly used [6].

Figures 4.6 and 4.7 show the linear absorption spectra of CdS crystals and QD in glass at room temperature and 10 K. Clearly, for QD, the spectrum line is more acute and easy to distinguish.

4.2 Preparation of Semiconductor QDs [7,8]

Semiconductor QDs have a heterogeneous structure in the nanoscale, which is coated by a lower energy gap of the semiconductor nanostructures in another material with higher energy gap. At present, methods for QD synthesis are mainly divided into the following.

1. *Chemical colloidal method*: By way of chemical synthesis of sol, this method can be used to produce multilayered QDs. This is a simple process that can be used in mass production.
2. *Self-assembly method*: Molecular beam epitaxy (MBE) or a chemical vapor deposition process is used. The principle of lattice mismatch is used so that the growth of QDs can undergo self-polymerization in a particular substrate. This method is applicable for mass production of QDs that are regularly arranged.
3. *Lithography and etching*: A beam or electron beam is applied directly on the substrate, which can be etched to produce the pattern. This process is time consuming and not applicable for mass production.
4. *Split-gate approach*: An external voltage is applied to generate a two-dimensional confinement to the two-dimensional plane of the quantum well. The gate is controllable regarding changing the shape and size of QDs and is suitable for academic research rather than for mass production.

Preparation of QD technology is established on the basis of a thin film growth process. Current literature often cites three kinds of classic models for thin film growth:

1. **Layered growth (Frank–van der Merwe (F-M)) mode**

   The F-M model is based on single start, and then there is growth of a second layer; that is, layer-by-layer growth. The growth of crystalline thin films is basically decided by the crystalline direction of the first layer. In the case of heteroepitaxy, more or less lattice mismatch may exist between the growth film and substrate, resulting in thin film stress.
2. **Island growth (Volmer–Weber (V-W)) mode**

   In V-W mode, the atoms are first deposited on the exposed surface of the substrate and will gradually form a small island, which may also increase at the same time, or decompose into single atoms. Meanwhile, prior to the formation of thin film, rearrangement over a wide range is possible, causing the so-called coarsening. The crystalline direction of the growth film cannot be quickly determined in the growth process.

3. **Mixed growth (Stranski–Krastanov (S-K)) mode**
   The S-K model is a mixture of layered growth and island growth. This model generally starts from layered growth. After that, island growth takes over on one or a number of growth layers.

Because self-organized growth can produce QDs with superior optical properties and is compatible with conventional device technology, it has been highlighted in recent years. Self-organized growth can be obtained via MBE or the metal organic chemical vapor deposition (MOCVD) method. Three-dimensional island substances can be naturally formed between the two kinds of materials with mismatched lattices in a specific growth pattern.

The following is an example of the sample growth method for indium arsenide QDs. It illustrates the preparation technology of semiconductor QDs.

By use of the MBE technology and S-K growth mode, the growth of indium arsenide QDs is as follows. Normally, the n-doped 100 surface of GaAs is used as substrate. Hydrogen-ion plasma is first used to remove the surface oxide at approximately 610°C, and then growth of a buffer layer with thickness of approximately several hundred nanometers occurs. The temperature is adjusted down to 500°C (usually between 450°C and 550°C) for the growth of the epitaxial layer of indium arsenide, and its thickness increases along with the generation rate and time. Before obtaining 1.8 monolayers (ML), obvious QDs will not appear. However, when the thickness exceeds 1.8 ML, that is, when it exceeds the threshold thickness, an indium arsenide layer will form the pattern of self-assembled QDs (SAQDs). The quality of QDs, such as the dot size, density, uniformity, morphology (shape), and so on, is greatly relevant to the growth methods, conditions, and parameter settings. High-quality growth technology of QDs can be state of the art. The QD covered with a layer of gallium arsenide may lead to the formation of a QD device with a sandwich structure. This part is the active layer in laser diodes. Also, to effectively confine the carriers in small spaces (carriers confinement) and for optical confinement where resonant cavities can be generated to induce luminescence, both sides of the active layer can, respectively, experience growth of a barrier layer of the part with the different n (or p) concentration of arsenic aluminum gallium alloy composition, enabling the generation of effective carriers that can reach the active layer to allow the occurrence of compound-emitting radiation. The outermost p+ (or n+) highly doped layer is mainly for the formation of ohmic contacts with electrodes.

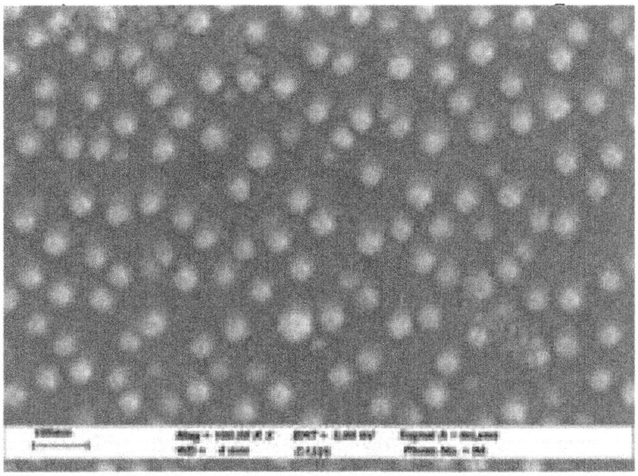

Figure 4.8 FE-SEM images of InAs/GaAs QDs. From Ref. [7].

After the growth of a QD sample, a field-emission scanning electron microscope (FE-SEM) can be used to observe their geometric morphology and measure the surface density (QD/cm^2) of QDs. The growth quality of QDs is essential for the performance of QD lasers and is quite sensitive to the growth methods and conditions. Take gas-phase MBE growth as an example. The crystalline dimension of the gallium arsenide substrate (100 or 111), doped composition, growth temperature, the vapor pressure of arsenic, element ratio of III/V, growth rate, the interruption mode and time, composition and thickness of overgrowth layer or cap layer, and the growth pattern of the sub-monolayer of indium and gallium may affect the quality of the growth of QDs. Figure 4.8 shows a two-dimensional image captured by FE-SEM on InAs/GaAs QDs.

Apart from using FE-SEM to measure the density of QDs, high-resolution transmission electron microscopy can accurately measure the size of the QD (height $\times$ base area). Atomic force microscopy (AFM), scanning tunneling microscopy, or scanning probe microscopy can also be used for observing geometry and measuring QDs. Figure 4.9 shows a two-dimensional image obtained by atomic force microscopy on InAs/GaAs QDs. QD density is generally between 1×10^9 and 1×10^{11} cm^{-2}. The bottom diameter is approximately tens of nanometers, and the height is approximately a few nanometers. The QD size, density, and the growth conditions and the parameter correlation are described in detail here. If you are interested in this subject, please refer to the recent literature of nanotechnology.

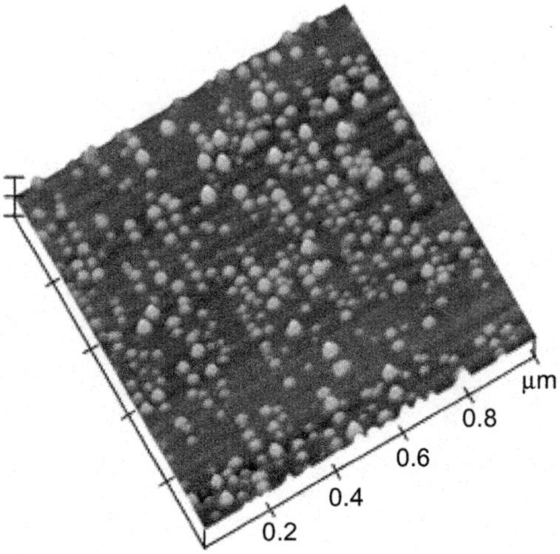

Figure 4.9 Two-dimensional AFM images of InAs/GaAs QDs. From Ref. [7].

4.3 Laser Devices Based on QDs [7]

Semiconductor QDs have many important applications. Here, we use an example to introduce the current status of their application in the field of laser devices.

Since the inception of quantum well structures [1], researchers around the world have carried out much theoretical work, expecting to apply quantum mechanism in semiconductor laser technology, and then their interests shifted to quantum wires and QDs of lower dimensions. In 1982, Arakawa and Sakaki at the University of Tokyo performed theoretical calculations and indicated that QDs have three-dimensional electronic limitations and the energy DoS in the form of a δ function. Meanwhile, they theoretically predicted the performance qualities of the QD laser, which include low threshold current, high temperature characteristics, and high glow efficiency. Compared with conventional semiconductor lasers, QD lasers show great improvements in thermal stability. In 1986, Asada predicted through calculation of theoretical means that the threshold current density with the QD structure will be significantly lower compared with that with a one-dimensional quantum well structure, which indicates a new direction for solving the problem of having a threshold current density that is too high, which is a problem encountered with semiconductor lasers. At that time, however, the production of QDs was based on lithographic technology, which made it difficult to get high-quality, nanoscale QDs. Although the quantum size effect with experimental results has confirmed the theoretical predictions one-by-one, the production of high-performance QD lasers has always failed.

In the mid 1980s, with the development of semiconductor epitaxy techniques (e.g., MOCVD and MBE), it was possible to control the growth of semiconductor thin film materials precisely and to access high-quality two-dimensional restricted quantum well and superlattice materials. Made from these two-dimensional materials with confinement heterostructure, the performance of optoelectronic devices, such as lasers and detectors, has seen rapid improvement. Their products quickly entered the market and acquired a wide use. Tremendous success in two-dimensional quantum well materials encouraged researchers to continue to try to restrict electron movement in more dimensions and launched experimental studies of quantum wires and QDs.

In 1994, the research team of Marzin used stress (geometric deformation) resulting from the lattice mismatch between the arsenide and indium gallium arsenide heterostructure layers to form SAQDs [9]. In the same year, N. Kirstaedter and N. N. Ledentsov reported the world's first self-organized QD lasers on edge emission [10]. They inserted a gradient refractive rate into self-organized QDs with single-layer index InGaAS/GaAs, which were used for a separate confinement of heterostructure and the quantum well laser structure. The original quantum well served as an active medium, thus achieving a low current density threshold (120 A/cm^2) at low temperature (77 K). After that, SAQDs have been receiving great attention. Extensive studies have covered topics from the basic physical properties of quantum devices to fabrication studies that helped obtain rewarding results. From 1994 onward, the field of QD lasers underwent rapid development. Low threshold current density, excellent thermal stability, and a variety of other fascinating properties have attracted much attention in a growing number of laboratories.

The years 1996 and 1997 witnessed a rapid development of QD lasers. A large number of international research teams have joined the ranks to fuel intense research into QD lasers. To achieve the ground state of QD lasers, optimizing conditions for the growth process have to be provided as the foundation for significant improvement in QD size and shape uniformity.

With in-depth studies performed in recent years, the actual produced QD lasers could have a current density threshold much lower than those of conventional and quantum well lasers. In 1996, N. N. Ledentsov used 10 layers of $In_{0.5}Ga_{0.5}As/Al_{0.15}Ga_{0.85}As$ in a QD superlattice structure as the active region for QD lasers so that at room temperature, the threshold current density decreased to 90 A/cm^2. In 1999, G. T. Liu and colleagues successfully developed the InAS/$In_{0.15}Ga_{0.85}As$ QD laser with threshold current density of 26 A/cm^2 at room temperature. So far, QD lasers with HR coating on their sides have obtained a threshold current density of 10–20 A/cm^2, which is two to four times lower than that of the best quantum well lasers. Of the lasers with multilayer QDs as the active region, QDs on each layer may have a low threshold current density, even as low as 7–10 A/cm^2.

The temperature stability of QD lasers has also been increasing. In 1994, Kirstaedter reported the first electric-pumped QD lasers, which showed good temperature stability at low temperatures (150–180 K). At room temperature, however, the thermal stability of current density threshold was inferior to that of commercial GaAs quantum well devices. In 1997, Maximov and associates placed QDs into the GaAs/AlGaAs quantum wells. This approach led to an increase in the barrier height of escape of carriers in QDs, greatly reducing the escape probability of the carrier. Meanwhile, leakage current was reduced to yield a laser characteristic temperature T_0 of up to 385 K at an operating temperature between 80 K and 330 K, well above the characteristic temperature of the quantum well laser. Nonetheless, an increase in T_0 has led to a significant increase of the current density threshold. In 1999, Shernyakov reported the world's first GaAs-based QD laser with operating wavelength of 1.3 μm, which is the first such device that works at room temperature but at the same time has both a high characteristic temperature T_0 (160 K) and a low current density threshold J_{th} (65 A/cm^2, three-layer QD arrays). Although the highest characteristic temperature T_0 of InP-based quantum well lasers that work in the same band is 60–70 K, the lowest current density J_{th} threshold is 300–400 A/cm^2.

For ideal QD lasers, QDs should have the same size and shape; that is, the QDs should be of one single electron and hole energy level, and should be easy to achieve in a single-mode operation. In 1996, Kirstaedter and associates at the temperature of 77 K with the density value slightly higher than the current density threshold ($<1.1 \times J_{th}$) observed single-mode operation. By contrast, quantum well lasers to achieve single-mode operations are required to be well above the threshold current density.

In 2004, the University of Tokyo and Fujitsu succeeded in the trial development of QD lasers that worked in the 1.3-μm wavelength; the optical power fluctuation caused by the temperature can be adjusted to approximately one-sixth the original value. Within the range of 20–70°C, it can send out optical signals steadily at 10 Gbit/s without the compensation of temperature-caused optical power changes. Eliminating the need for external circuitry for temperature compensation helps reduce the size of the optical transmitter and the overall cost of the product. The research team is currently trying to extend the free-to-adjust operating temperature range of the laser to 0–85°C. This significant progress contributes to the development of small optical signal transmitters with low cost and power consumption. The Metropolitan Area Network (MAN) and high-speed optical Local Area Network (LAN) are expected to benefit from it.

At present, QD laser materials have been discovered, but the best material thus far is still the arsenide indium (or gallium indium arsenide alloy) QDs with the growth of the III–V family on GaAs substrates. Indium arsenide has a narrow energy gap, which can be adjusted through the x value after adding a small amount of gallium component ($In_xGa_{1-x}As$, $0 \leqq x \leqq 1$) for preparation of nanometer-sized QD structures. Furthermore, it can be used to produce laser light source or light-detecting devices with wavelengths in the range of 1.3–1.55 μm, which is suitable for use in optical communication. Here, we must point out that the laser light with wavelengths between 1.3 and 1.55 μm is vital because optical fiber has a very low energy loss in this band, quite suitable for long-distance optical communications.

QDs in the III–V family are valued mainly in the application of their optical and electronic properties, such as high-frequency (high-speed) electronic devices, high frequency light-emitting devices, and light detectors with high-efficiency. Measurement tools for study of QD optical properties include photoluminescence, time-resolved photoluminescence, and the temperature change experiments that come with the power changes in the optical excitation or pumping or the cryogenic systems, and these may obtain the radiation photon spectra of the e-hole recombination from spectral data and the relaxation mechanism and time of its carriers, life information, and other device-related parameters.

The development of QD lasers in recent years has made considerable progress and has launched a strong challenge to traditional semiconductor lasers, but there is still a large gap between its performance and theoretical prediction. The following problems must be addressed to further enhance the performance of QD lasers. First is the growth of size-uniform arrays of QDs. Although the QD material provides many potentially advantageous properties, its uneven size distribution makes the QD light-emitting peak have nonuniform broadening, whereas the luminescence peak is wide and much larger than that of the quantum well material (meV). In fact, only a small part of QDs contribute to the light-emitting process. This limits the optical gain and makes further reductions in the lasing threshold of the lasers difficult to achieve. Second is the ability to increase the surface density and volume density of QDs to maximize the material gain of QDs. Third is the ability to optimize the structural design of QD lasers so that it is conducive to QDs in the carrier capture and bondage. And last is the ability to control QD size or to select a novel material system for broadening the scope of work of the lasing wavelength of QD lasers to the band of 1.4–1.6 μm in Wavelength Division Multiplexing (WDM) networks.

For the indium arsenide QD materials, the current indium arsenide QDs have achieved considerable progress in laser diode devices. At a wavelength of 1.3 μm and a low critical current density (~19 A/cm^2), it can launch a continuous single-mode wave in the operation mode at room temperature, with power of up to 210 mW. At present, the performance of QD lasers has been better than that of the InP quantum well laser, but its performance is not entirely satisfactory. To improve the effectiveness of indium arsenide QD lasers, we must also perform studies focused on reducing the critical current, improving the temperature characteristics in device operation (increase of the temperature of the device), and preventing the switch between the ground-state module and the excited-state module, as well as to improve laser luminous efficiency. Indium arsenide QDs still have a number of issues that need further study, such as high density, size, uniform epitaxial layer growth of QDs, as well as precise observation of QD shape and size measurements, the component analysis of inside and outside QDs and stress distribution of measurements or estimates, theoretical calculations of QDs with different morphologies on the quantum band and energy levels, energy value of electrons in the optical transition between energy levels, the relaxation mechanism of carrier energy, and the measurements of carrier lifetime and other main physical quantities. With the efforts of scientific researchers, more efficient high-power and single-mode high-quality laser devices in high-temperature operation with a wavelength of 1.3–1.55 μm will be developed.

4.4 Single-Photon Source

Quantum optics is focused on the study of the nature of light and the basic interactions between light and matter. Semiconductors are mainly related to the physical principles and applications of electronic materials and devices. Over the past few decades, quantum optics and semiconductors have been developing along separate paths. Specifically, quantum optics research is focused on the field of atomic and molecular optics. Over the past decade, rapid progress in semiconductors has been made in nano-related science and technology. People began to notice that some of the semiconductor mesoscopic systems and quantum systems can also be found with some quantum optical phenomena. In this way, a new area of research was discovered, known as "semiconductor quantum optics." Even more specifically, there is a field in mesoscopic systems where quantum optical phenomena called "mesoscopic quantum optics" are discussed. In various types of semiconductor mesoscopic and quantum systems, the most notable is the QD system. QDs are called "artificial atoms" and have an energy level structure that is very similar to that of atoms. Therefore, it is not difficult to imagine that the quantum optical properties existing in a number of atomic and molecular systems can also be found in the semiconductor QD system.

Semiconductor light-emitting materials or devices (such as lasers and light-emitting diodes) serve as a medium to convert the carriers (electron and hole) into photons. If this conversion process is fast and efficient, then the statistical behavior of the carriers will be converted to that of photons. It is well known that electrons and holes are fermions, while photons are a kind of boson. Through proper device design, or in specific mesoscopic semiconductors and quantum systems, the photon statistics of radiation will be different from classical electromagnetic waves, and such light is also often referred to as "nonclassical light."

Quantum optics has extremely important or potential applications in many areas, and the most notable area is quantum information, including quantum computing and quantum communication. In particular, the use of a single-photon source is expected to bring about the very promising quantum cryptography. Semiconductor quantum optics is particularly vital for the development of practical applications of quantum information. For example, by virtue of current technology, semiconductor quantum optics has been successfully applied in the production of highly efficient single-photon radiation, whereas the wavelength of operations can also be extended to that in the optical fiber communication band (1.3 μm). Applications of quantum cryptography in optical communication have already made significant progress.

The theoretical basis of quantum optics was first established by Roy J. Glauber in 1963; he also won the 2005 Nobel Prize in

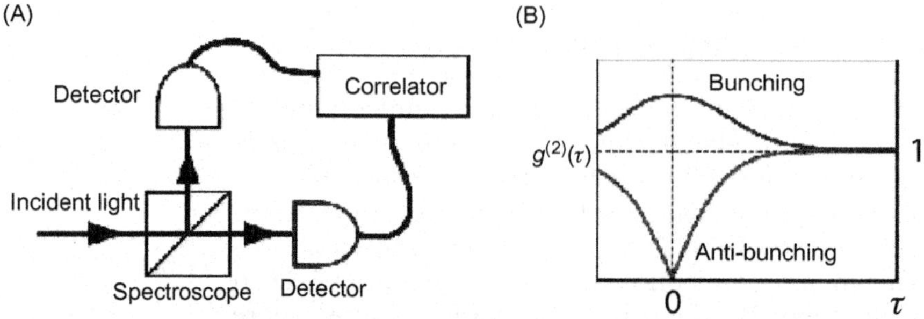

Figure 4.10 (A) Schematic diagram of an HB-T interferometer device. A beam splitter divides the incident light into two beams. (B) Diagram of second-order coherence function on the photon bunching and anti-bunching phenomenon. From Ref. [11].

Physics for his outstanding contributions to quantum optics. The concept of optical quantization can be traced back to the blackbody radiation theory created by Max Planck in 1900 and the interpretation by Einstein of the photoelectric effects in 1905. However, the real birth of quantum optics theory is symbolized by an optical interferometer (often referred to as an HB-T interferometer), which was jointly designed by R. Hanbury-Brown and R. Q. Twiss between the years 1952 and 1956. This interference device was originally used to observe Sirius, and later it was used to observe the coherent characteristics of mercury, which then led to the surprising discovery of some positive coherence between the detected photons.

Figure 4.10A shows a schematic diagram of an HB-T interferometer device. A beam splitter divides the incident light into two beams. Meanwhile, two-photon detectors were used to observe the intensity coherence properties of incident light, which is usually expressed by second-order coherence function $g^{(2)}(\tau)$. Here, τ indicates the time difference for two detectors in detecting photons, as shown in Figure 4.10B.

Coherent features can be distinguished by the use of the value $g^{(2)}(\tau = 0)$. If considering the incident light in particles, then $g^{(2)}(\tau = 0)$ can be regarded as the "possibility" of two detectors in simultaneously detecting photons. As $g^{(2)}(\tau = 0) > 1$, an increased likelihood that two detectors detect photons at the same time is observed. It can be said that incident light waves have positive coherence in time between them. E. M. Purcell thought at the time that such positive coherence could be explained using quantum statistics. Because photons are a kind of boson, when they have the same quantum state they tend to "gather" together to reach the two-photon detector at the same time. So, the phenomenon of light is also often referred to as

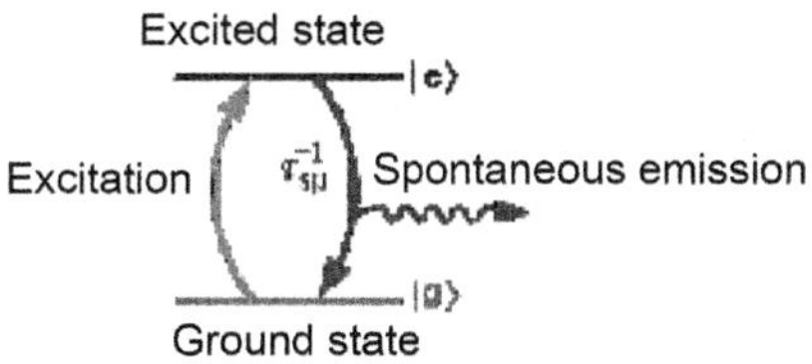

Figure 4.11 Schematic diagram of a separate two-level quantum system. From Ref. [11].

the coherent cluster (bunching) effect of photons. In fact, the coherent features of light can also be interpreted using classical electromagnetic theory, so arguments between the two theories were ongoing at that time. In 1963, Glauber proposed the quantum theory of optical coherence, giving a reasonable explanation for the photon bunching effect observed in HB-T experiments. Now physicists know that the photon bunching effect is actually associated with characteristics of photons from thermal radiation, and coherent light (e.g., laser) does not have any positive correlation (i.e., $g^{(2)}(0) = 1$).

The quantum optics theory gives an interpretation of the photon bunching effect, and it also expected photons to show an anti-bunching effect in some cases. The anti-bunching phenomenon of photons was first observed by Kimble, Dagenais, and Mandel with the fluorescence of a single sodium atom in 1977. Because the classical electromagnetic wave theory failed to explain the "anti-bunching" phenomenon of light, the phenomenon used to be considered direct evidence of the quantum nature of electromagnetic waves. Light with anti-bunching characteristics has also been classified as "nonclassical light." If a physical system has anti-bunching characteristics, then the system does not radiate two or more photons at the same time. In other words, the system only sends one photon at a time, so it is called a single-photon source.

The anti-bunching phenomenon is a characteristic of fermions. A single-photon light source radiates photons with anti-bunching features, indicating that a single-photon light source has the characteristics of fermions. In fact, any single independent quantum system can produce single-photon radiation. As shown in Figure 4.11 with a two-level system as an example, the system includes a ground state and an excited state. The electron in the ground state, once excited by light or electricity, may jump to the excited state, and then back to the ground state through spontaneous emission. This process sounds simple but implies a mechanism of nonclassical light. As the electron falls in the scope of fermions, when it occupies the excited state but has not yet produced spontaneous radiation, we cannot excite the next electron to the same excited state. The time for an electron to occupy the excited state is related to the lifetime of spontaneous radiation. Therefore, in this period of time, even if constant excitation is given to this system, it will still not produce photons. Thus, an independent two-level system will be unable to simultaneously produce two or more photons and only constitutes a single-photon source.

As already known, there are many systems that can produce a single-photon source, including single-atom or single-molecule systems. Stability control of a single atom or molecule, however, requires complex technologies that may cause difficulties in practical applications. In addition to atomic and molecular systems, many of the solid-state systems also can produce single-photon radiation, such as the nitrogen vacancy center in diamond chemical combination and semiconductor QDs. They share a feature of the system with similar energy levels of atoms or molecules.

References

[1] L. Esaki, R. Tsu, Superlattice and negative differential conductivity in semiconductors, IBM J. Res. Dev. 14 (1970) 61.

[2] A.P. Alivisatos, Semiconductor clusters, nanocrystals, and quantum dots, Science 271 (1996) 933–937.

[3] Y. Fu, W. Lu, Semiconductor Quantum Devices, first ed., vol. 1, China Science Press, Beijing, 2005, 102–109.

[4] Y.T. Lu, Semiconductor quantum dots and quantum information, Physics (Bimonthly) 25 (4) (2003) 497.

[5] A.S. Edelesten, R.C. Cammarata (Eds.), Nanomaterials—Synthesis, Properties and Applications, Institute of Physics Publishing, Bristol and Philadelphia, 1996, pp. 396–401.

[6] Z.Z. Guo, X.X. Liang, S.L. Ban, Pressure-induced increase of exciton-LO-phonon coupling a ZnCdSe/ZnSe quantum well, Phys. Stat. Sol. (b) 238 (1) (2003) 173–179.

[7] G.Z. Zhan, Optical properties of InAs quantum dots, Physics (Bimonthly) 25 (3) (2003) 1–7.

[8] M. Sugawara, Self-Assembled InGaAs/GaAs Quantum Dots, Academic Press, 1999.

[9] J.-Y. Marzin, J.-M. Gérard, A. Izraël, D. Barrier, G. Bastard, Photoluminescence of single InAs quantum dots obtained by self-organized growth on GaAs, Phys. Rev. Lett. 73 (1994) 716–719.

[10] N. Kirstaedter, N.N. Ledentsov, M. Grundmann, D. Bimberg, V.M. Ustinov, S.S. Ruvimov, et al., Low threshold, large T_0 injection laser emission from (InGa)As quantum dots, Electron. Lett. 30 (17) (1994) 1416–1417.

[11] W.H. Zhang, Z.M. Xu, Semiconductor quantum optics, Physics (Bimonthly) 28 (5) (2006) 851–856.

5

NANOMAGNETIC MATERIALS

CHAPTER OUTLINE

Nanomaterials and Devices. DOI: http://dx.doi.org/10.1016/B978-1-4557-7754-9.00005-6

5.1 Types of Nanomagnetic Materials [1]

5.1.1 Artificial and Natural Nanomagnetic Materials

Magnetic nanomaterials research and application can be traced back to the 1950s. Quasi-zero-dimensional magnetic nanoparticles were among the first group to be applied [2] and, principally, used in the following three areas. The first is trichoplusia technology for magnetic domain observation. The sample surface is coated with colloidal suspensions comprising iron magnetic powder that is thin. The bulk magnetic field at domain walls has the concentration of the magnetic powder here, which depicted the surface structure of the magnetic domain or the surface trajectory of domain wall. The second is the preparation of single-domain magnetic powder materials. Because the magnetization reversal process of single-domain particles is essentially the magnetic domain rotation without a motion process of the domain wall, the coercivity can be improved significantly. And the third is magnetic fluid for magnetic sealing, for example magnetic seal for spacesuit helmets in the 1960s. The superparamagnetic feature of nanoparticles is applied here. Astronaut helmets are sealed with nanomagnetic materials (magnetic fluid), and this was one of the earliest significant applications.

In 1988, amorphous FeSiB was annealed and doped with Cu and Nb control crystals to develop a novel nanocrystalline soft magnetic material. In the same year, magnetic multilayers were discovered with a giant magnetoresistance (GMR) effect, thus producing a novel discipline: *spin electronics*.

Theoretical research in 1993 demonstrated that the composition of soft and hard nanoscale magnetic particles could combine the advantages of higher M_s value of soft magnets with the higher H_c value of hard magnets, which could produce a novel nanohard magnetic material with magnetic energy levels twice that of the current top NdFeB material. Research in the twenty-first century has been very active regarding the use of template-based growth of one-dimensional magnetic nanowires, including materials such as single metal, alloy, compounds, multilayer materials, and composite materials, with potential applications ranging from storage medium to cell separation.

Nanomagnetic materials are mainly artificial. However, research has also noted that natural nanomagnetic particles might be found contained within many organisms, such as magnetic bacteria, pigeons, dolphins, stone turtles, bees, the human brain, and so on. In 1975, a string of magnetic nanoparticles was found within the body of magnetic bacteria. As mentioned in Chapter 1, magnetic nanoparticles in the abdomen of the bee have the function of navigation. Magnetic nanoparticles are still a research topic of great interest regarding their physical principles and biological processes. An average of 20 μg (approximately 5 million particles) of magnetic nanoparticles is contained in the human brain. Research is now exploring the relationship between magnetic nanoparticles and evolution, growth, and some brain functions.

A chiton contains many one-dimensional magnetic wires in its teeth and tongue. They are composed of many magnetic columns, inside which is a collection of single-domain particles, and their biological function may not be limited to enhancing the wear resistance of its teeth and tongue. Interestingly, it has a natural template for nanowire growth. In the mineralized teeth of organisms, there is a network of organic fibers that serve as the nanowire growth template. Figure 5.1 shows the template cross-section and the formation process of magnetic material within.

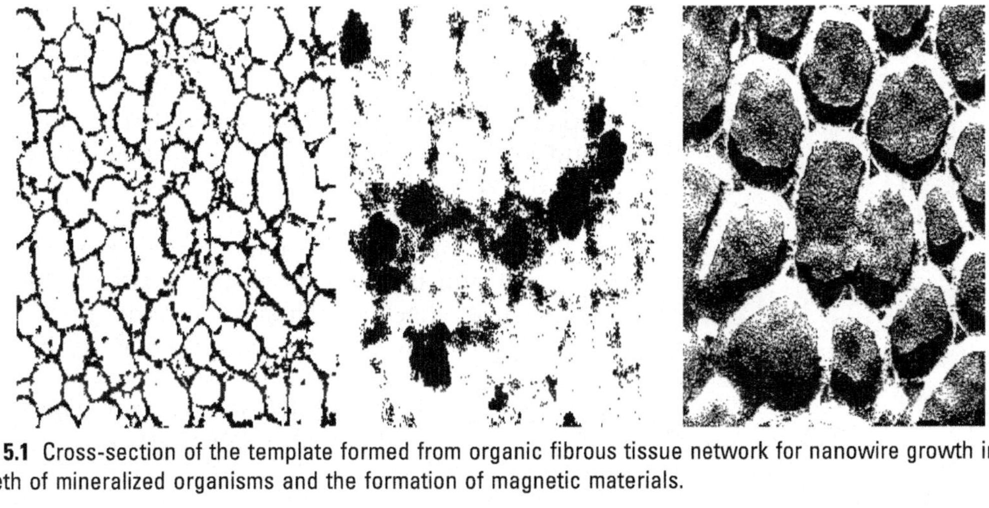

Figure 5.1 Cross-section of the template formed from organic fibrous tissue network for nanowire growth inside the teeth of mineralized organisms and the formation of magnetic materials.

At present, our understanding is insufficient regarding the synthesis of natural nanomagnetic particles *in vivo*; this is a topic that requires research efforts in the fields of biology, physics, and magnetics.

5.1.2 Classification of Magnetic Nanomaterials [3]

According to its applications, magnetic nanomaterials are divided into the following categories.

1. *Nanocrystalline soft magnetic materials*

   Nanocrystalline soft magnetic materials are produced in large quantities and varieties and are applied in electricity, telecommunications, and home appliances. Representative of the products are $Fe_{73.5}CuNbSi_{13.5}B_8$ (finement) with high saturated magnetization, as well as nanocrystalline Fe–M–B, Fe–M–C, Fe–M–N, Fe–M–O and other nanocrystalline soft magnetic series, where M can be Zr, Hf, Nb, Ta, V, and other elements. Their applications mainly include high-density magnetic heads, various types of transformers, magnetic switches, sensors, high-frequency micropower switches, and noise filter products.
2. *Nanocrystalline permanent magnetic materials*

   Nanocrystalline permanent magnetic materials are represented by $Nd_2Fe_{14}B$,$ThMn_{12}$, $Sm_2Fe_{17}N_x$, $Sm_2Fe_{17}C$, and Sm–Co–B (which has high thermal stability and good corrosion resistance). These materials are mainly used in electro-acoustic machines and mineral processing.
3. *Nanomagnetic recording materials*

   Depending on their form, magnetic recording materials include granular and continuous thin film materials. They can be divided into two categories by nature as metallic and nonmetallic materials. Typical products include CrO_2 and ferroalloy in the 1980s, Co-doped $BaFe_{12}O_{19}$ in the 1990s, and FeN, FeC, g-Fe_2O_3, CrO_2, Co-g-Fe_2O_3, and Fe, $BaFeO_x$, in the 2000s.
4. *Ferro fluid*

   Ferro fluid is composed of magnetic particles (<10 nm) plus surfactant and base solution. The main products are magnetic particles that have already matured in the 1960s, namely Fe_3O_4 (10 nm) Co, Ni (<6 nm), and its alloys or nitrides. Common base liquids are water, kerosene, alkyl, ester, polyphenyl, silicone, and Dover carbon. The special feature of magnetic fluid is that magnetic particles in it can be magnetized in magnetic fields, and they are movable. It is mainly used in high-speed rotating shaft seals, lubricants, loudspeakers, damping devices, and density separation.

5. *Granular perpendicular medium*

Granular perpendicular medium is currently in the developmental stage and may be used in high-density magnetic recording and magnetic sensors. The recording layers of hard disks that are commercially used today are mainly made from CoCrPt–SiO_2 materials. CoCrPt can be used as perpendicular recording media. When used as a perpendicular recording media, the recording layer is formed by epitaxial growth with Ru as the substrate, and the magnetic film formed is the granular film. It is now generally agreed that the role of exchange coupling between the internal film grains is a major source of noise, so nonmagnetic SiO_2 is used to isolate the magnetic particles, so that no interaction force from coupling exchange will exist between magnetic particles and, hence, reduce the noise.

6. *Nanomagnetic refrigeration working fluid*

Compared with the usual compressed gas refrigeration, magnetic refrigerators have advantages such as high efficiency, low power consumption and noise, small size, and no pollution. This opens a new avenue for food-freezing and refrigeration equipment. Magnetic refrigeration tends to develop from a low temperature to a high temperature. In the present studies, typical magnetic refrigeration fluids include Gd, La–Ca–Mn–O, and $Gd_5Si_2Ge_2$ materials.

7. *Giant magnetoresistance*

In a certain magnetic field, the resistance rapidly decreases at a ratio more than 10 times that of normal magnetic substances; this is a phenomenon is called GMR (see the following text). GMR materials have essential applications in magnetic sensor parts, high-density read heads, magnetic storage devices, numerical control machine tools, noncontact switches (magnetic switches: speed and position measurement for the automatic control anti-theft alarm system, car navigation, ignition devices), rotary encoders, MRAM (no power source to retain information), disks, and weak magnetic field detectors (10^{-2} to 10^{-6} T).

8. *Nanoceramic material*

Nanoceramic material mainly refers to nanometer-sized materials, such as Al_2O_3, ZrO_2, SiC, and Si_3N_4. Their features are hardness, good wear resistance, excellent chemical stability (anti-oxidation, anti-fatigue), low density, and resistance to high temperatures. Nanoceramic materials are mainly used in microelectronics and quantum devices (computer boards). This application requires materials to have a small size, spherical shape, narrow size distribution, and nonagglomeration with high purity.

Synthesis of nanoceramic materials is mainly performed by means of gas-phase synthesis and sol gel synthesis, which allow the materials to have sintering densification and can be performed under high pressure at low temperatures, resulting in *in situ* self-reinforcement. NanoSiC is one of the typical representatives.

9. *Raw materials for nanoprecision polishing*

   Raw materials for nanoprecision polishing can be used for some special window materials or multilayer interference film to improve optical performance, or can be used to improve electrical properties of piezoelectric materials, sensors, and so on. In addition, they can also be used in cell separation technology or used as a catalyst, as well as in specialty functional coatings (resistant to pollution, dust, wear, and fire).

10. *Ceramic-based nanocomposites*

    Nanoceramic particles dispersed in an inorganic matrix can enhance their performance. Table 5.1 shows the comparative data regarding the performance of Al_2O_3 before and after composition.

Nanoceramic particles dispersed in polymer matrix can be effective in modifying the performance. Table 5.2 shows a comparison between the data of the performance of Al before and after composition.

Nanoceramic particles are a candidate for carriers for the growth of nanowhiskers and carbon nanotubes. For example, Si–CN powder may be a base for growing Si–C whiskers and carbon nanotubes.

For ceramic-based nanocomposites, zeolite is a typical representative as the molecular sieve. Zeolite is an Si–Al-oxide

Table 5.1 Nanoceramic Particles Dispersed in Inorganic Matrix to Strengthen its Performance

Material	Strength	Toughness	High-Temperature Strength	Maximum Operating Temperature	Creep Life
Al_2O_3	350	3.5	–	800°C	–
SiC/Al_2O_3	1,520	4.8	800	1,200°C	Ten-fold increase

Table 5.2 Nanoceramic Particles Dispersed in Polymer Matrix Can Be Effective in Modifying Performance

Material	Tensile Strength (MPa)	Bending Strength (MPa)
Al + 1.0 wt% Si_3N_4	180	147
Al + 15 wt% Si_3N_4	176	94
Al	102	67.8

that has a certain pore size and abilities in reversible absorption–desorption and distribution of the same size molecular sieve, ion-exchange capacity, and surface acidity. In 1756, Cronstedt (a Swedish chemist) discovered zeolite from minerals and gave it the current name. In 1930, McBain (a US scientist) found its selective adsorption characteristics, calling it *molecular sieve*. Artificial zeolite synthesis began in the 1950s (86 structures are already known). By 1992, the MCM series (Mobil) had been synthesized, in which MCM-41 has a hexagonal array of phase, MCM-48 is in the cubic phase, and MCM-50 is in the lamellar phase.

MCM-41 has the following characteristics: hierarchically organized structure, nanoadjustable porosity (1.5–30 nm, typically 4 nm), large pore aspect ratio, high surface area (up to 1,200 m^2/g), as well as many changes in morphology and the composition of dense amorphous matrix (wall thickness of 1 nm), and higher thermal and water stability.

Figure 5.2 is a comparison of various zeolite pore sizes.

5.2 Basic Characteristics of Nanomagnetic Materials [1]

Among the three dimensions of a material, if any one dimension is within the range of 1–100 nm, then it is considered to be a nanomaterial. Nanomaterials are likely to exhibit one of the following common features: the quantum size effect, small size effect, surface effect, and macroscopic quantum tunneling effect (see Chapter 1).

In the magnetic domain structure and many other magnetic aspects, magnetic nanoparticles also show some other features.

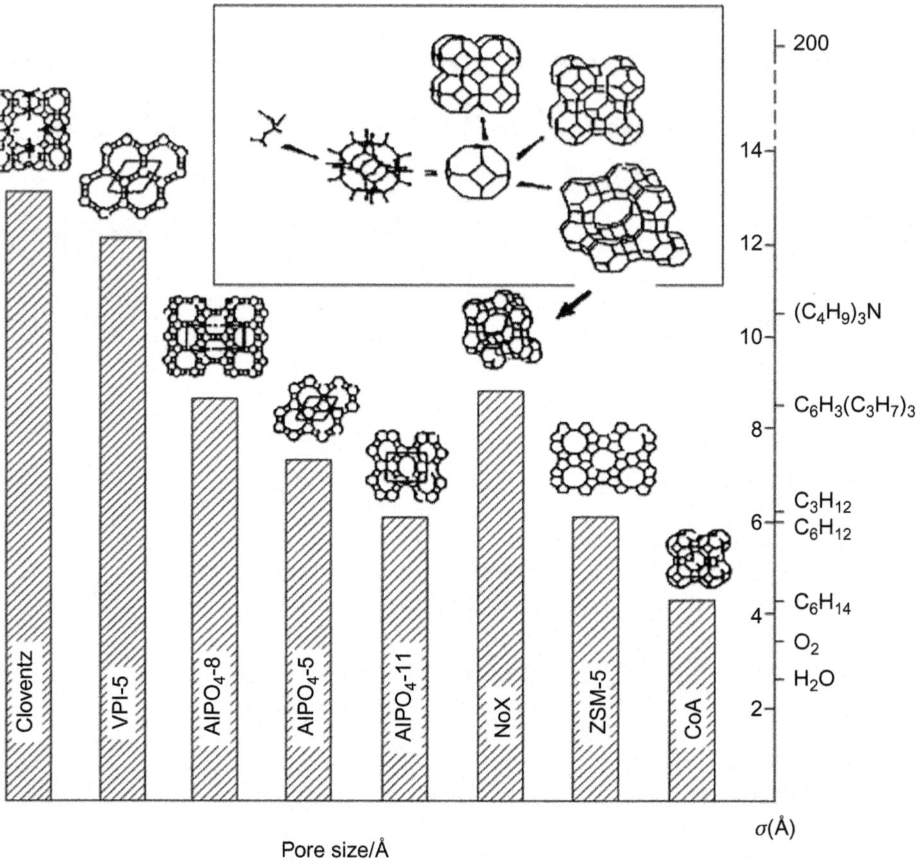

Figure 5.2 A comparison of various zeolite pore sizes.

5.2.1 Magnetic Domain

For bulk magnetic materials, the exchange interaction energy and magnetic anisotropy energy may cause parallel magnetic moments in their easy axis, but this will lead to a strong demagnetization energy.

For the radius R of a ball shape, demagnetization energy is

$$E_{\rm d} = \frac{4\pi}{3}\frac{\mu_0 R^3 M_{\rm s}^2}{6}$$

($M_{\rm s}$ is the saturation magnetization).

Obviously, the larger the size of R, the higher the demagnetization energy will be. To reduce energy, materials are bound to split into magnetic domains. But in the domain wall, as a transition zone between the two domains, magnetic moments are bound to deviate from the easy axis and adjacent magnetic moments are no longer parallel. The resulting domain walls can intervene in the total energy balance. For instance, a 180° domain wall may have an energy density given by

$$\gamma_{180} = 2\sqrt{A_1 K_1}$$

(K_1 is the anisotropy constant and A_1 is the reduced exchange integral part, and the same is true for the next equation.)

In the nanoscale, nanoparticles will become single-domain particles. This is because when the particle size R is very small, the energy of domain walls can be stronger relative to the demagnetization energy. There is no need to make the sub-magnetic domains, and thus a single-domain particle forms. The critical size for the formation of single-domain particles can be estimated through the following methods [4]. The demagnetization energy of a single domain is set equal to the sum of domain wall energy and demagnetization energy by which two magnetic domains are divided, namely:

$$\frac{4\pi}{3}\frac{\mu_0 R^3 M_s^2}{6} = \frac{4\pi}{3}\frac{\mu_0 R^3 M_s^2}{12} + \pi R^2 2\sqrt{A_1 K_1}$$

assuming that after the single domain is split, the demagnetization energy is one-half of the single domain; the critical value of R_c ratio of the single domain is available as $\sqrt{A_1 K_1}/M_s^2$.

As far as nanothin films are concerned, if the magnetic film has a larger thickness, D, with magnetic moments rotating in the domain wall plane, the magnetic charge will not occur in the domain wall. The surface magnetic charge produces little effect of the demagnetization field. Such a phenomenon is called the *Bloch* domain wall. When there is a thin film, the surface magnetic charge with a demagnetization field cannot be ignored. Magnetic moments will be rotating in the film surface, that is, the film surface does not produce a magnetic charge; it is found in the domain walls and on both sides, known as the *Neel* domain wall.

For Fe–Ni film, $D>100$ nm indicates the presence of the Bloch domain wall; $D<30$ nm indicates the Neel domain wall, between which is the cross wall in a transitional state.

In theory, when $D<12$ nm, the film is a single domain. But a uniform demagnetization field in the film is difficult, so there is always a magnetic domain that will be generated.

5.2.2 Superparamagnetic Feature

With a volume of V, single-domain magnetic particles have anisotropy approximated as K_1V (simplified as KV). When the magnetic particle size continues to decrease, the anisotropy barrier KV also will decrease accordingly. In this way, thermal motion energy k_BT (simplified as kT) may exceed the anisotropy barrier KV and cause the orientation turnover of the magnetic moment, so that the magnetization direction of particles is presented as a magnetic performance in "Brownian motion," with the total magnetization of particles aggregate as zero. This phenomenon is called superparamagnetism. It is characterized by coercivity:

$$H_c \to 0, \quad \text{as } \frac{\mu H}{k_B T} \ll 1$$

Magnetization strength is given by

$$M_P \approx \frac{\mu^2 H}{3k_B T}$$

where μ is particle magnetic moment. A superparamagnetic feature is typically manifested by the magnetic ordering small-size effect of nanomaterials. Superparamagnetism can also be described by the Langevin function. The exception is that particles do not include a single atomic or molecular moment, but rather a collection of magnetic ordering. Magnetic orientation between the collections is arranged in confusion, showing the macroperformance of the "paramagnetism."

For superparamagnetic colloidal particles in suspension, particles have only a weak effect of static magnetic and the Van der Waals force between them. As a result of the thermal motion, the magnetization vector within the particles can be rotated not only by overcoming the barrier of magnetic anisotropy energy but also by having motion as a whole. This is the magnetic fluid.

With a thermal motion energy kT, particle magnetic moment M_s with volume V can cross the barrier KV of anisotropy energy of K at a probability of $p = \exp(-KV/kT)$. That is, the original collection of consistent magnetic particles, after a period that is long enough, can be attenuated to the remnants of zero, with the relaxation time $\tau = (1/f_0)\exp(KV/kT)$, and frequency factor $f_0 = 10^9\ \text{s}^{-1}$.

If you have to wait a year (10^7 s) before a decay process becomes "paramagnetic", then this material cannot be superparamagnetic. So, τ is the relative standard.

For instance, $\tau<10^{-1}$ s can be used for the superparamagnetic standard. Clearly, τ is relevant to the material anisotropy K, the temperature T, and the particle diameter $D=V^{-3}$. In the case that the particle size V is fixed, only when particles attain a certain critical temperature T_0 or higher can they be expressed as superparamagnetic; here, T_0 is called the *cut-off temperature.* At a fixed temperature (e.g., room temperature), for the paramagnetic particles to show the performance of surplus they must have a size less than the critical size of V_0.

The following are some superparamagnetic data in actual cases:

$T=100$ K, $K=10^7$ J/m^3; if the material size is 6.3 nm, then the particle relaxation time $\tau=10^{-1}$ s; if size is 6.8 nm, then $\tau=10^1$ s, and if particle size is 7.6 nm, then $\tau=10^{+5}$ s (or 1 day!). It is shown that the range for ultra-paramagnetic performance is rather narrow.

The following material sizes are required to show superparamagnetic features at room temperature: spherical iron, 12 nm; ellipsoid iron, 3 nm; HCP cobalt, 4 nm; and face-centered cubic cobalt, 14 nm.

Acquiring standardized data for different materials is essential, because different measurement methods may lead to different results. Under the condition that measurement time for data acquisition $t<\tau$, the hot-fluctuation effect is not observed, and the material only shows the property of the usual single domain. Only when $t>\tau$ can superparamagnetism be clearly observed. For example, the time for Mossbauer measurement is 10^{-8} s, whereas part of the static magnetic measurements may have a time $t=1-100$ s.

5.2.3 Exchange Interaction

Exchange interaction is an equivalent interaction between particles in a microscopic identical multiparticle system. It reflects the indistinguishability of identical particles. Being a pure quantum effect, there is no corresponding classical concept for it. Nanomagnetic materials can be found within the normal magnetic exchange interaction; however, between the magnetic nanoparticles, nanomultilayer films, and one-dimensional magnetic wires, the availability of the exchange interaction and the relevant results are our concern.

Here is an example to further expand on this theory. For magnetic multilayers, magnetic layers may have exchange coupling through a nonmagnetic metal layer, such as iron whiskers, which use a wedge-shaped gold laminate to complete the exchange coupling with another iron layer. Scanning electron microscopy with polarization analysis shows that the ferromagnetic exchange coupling is performed because of shock from changes in the ferromagnetic and antiferromagnetic layers in accordance with the thickness of the gold laminate. This shocking cycle has two cycles: short cycle and long cycle. A short cycle is approximately half of the Fermi wavelength, that is, $\lambda_F/2$, with the same expectations from the RKKY exchange model. Here, RKKY stands for the exchange interaction model of Ruderman–Kitter–Kasuya–Yosida. The basic characteristics include *4f* electrons being localized, *6s* electrons swimming, and *f* and *s* electrons in exchange interaction, so that the *s* electron is polarized. The spin of the polarized *s* electron may affect the orientation of the *f* electron spin. The result is the formation of a swimming *s* electron as medium, so that the *4f* localized electron spin in magnetic atoms (or ions) produces exchange interaction with the *4f* electron spin of its neighboring magnetic atoms. This is an indirect exchange interaction. The RKKY model is applicable for rare earth. The reason for the formation of the long-period shocks may lie in the experimental methods, which were observed in a discontinuous nature, taking a single atomic layer as a single observation unit, rather than the continuous change of the thickness; the results are from intermittent sampling, as shown in Figure 5.3.

This research on exchange coupling led to the discovery of the GMR effect and its further development as the base for spintronics research. Between antiferromagnetic and ferromagnetic layers, exchange coupling can be firmly attached to

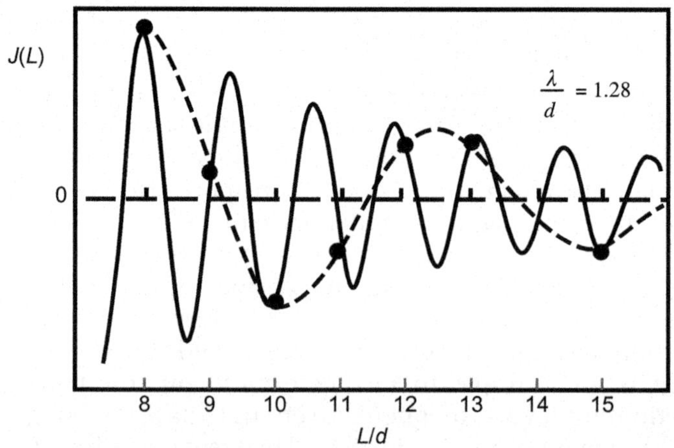

Figure 5.3 Diagram of long-period oscillations. Taken from Ref. [1].

the ferromagnetic layer. This method has been widely used in GMR device designs.

Nanocrystalline exchange may lead to random anisotropy. Because the magnetic nanograin size D is less than the exchange length L_{ex}, the exchange coupling between grains will effectively offset the local random anisotropy K, and the average density of anisotropy energy $<K>$ tends to be zero as the scale D gets smaller. As $<K> = K_1/N^{1/2}$, the number of grains contained in the context of the exchange length $N = (L_{ex}/D)^3$. So, $<K> = K_1(D/L_{ex})^{3/2}$, and the ferromagnetic exchange length is subject to the relation $L_{ex} = (A/\langle K\rangle)^{1/2}$. Therefore, the random anisotropy, $<K> = K_1{}^4D^6/A^3$, that is, $<K>$, declines in the six orders of magnitude as D decreases, and the corresponding magnetic permeability, μ, also increases by six orders of magnitude as the nanoscale of particles decreases.

The flexible exchange coupling between soft particles and hard magnetic particles can be used to maintain a high H_c and B_r, thereby enhancing the hard magnetic properties.

5.2.4 Coercivity H_c

Already introduced in the previous sections, the understanding of coercivity tends to be very complex with the changes in particle size. When the particles become as small as single-domain scale, the antimagnetization process is in either a homogeneous or inhomogeneous rotation process of the magnetic moment. H_c is proportional to (K/M_s) of the materials (K for the anisotropy constant). This is one of the ways to get a higher H_c, namely the hard magnetic properties of the material. However, nanomagnetic materials may have a smaller H_c simply by means of anisotropic exchange interaction that makes a very small average anisotropy $<K>$ with a scale less than the exchange length. As the particle size is reduced to the superparamagnetic critical size, the coercivity approaches zero.

Figure 5.4 shows several specific materials and the relationship between the coercivity and particle size.

5.2.5 Curie Temperature

Curie temperature T_c is an important parameter of magnetic materials, usually directly proportional to the exchange integral A, but also related to the material configuration and spacing of the atoms. In nanomaterials research, the Curie temperature is found to decline with the decrease of nanoscale particles or

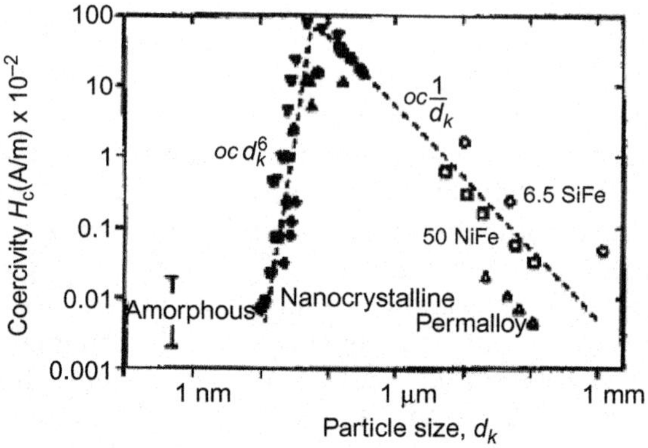

Figure 5.4 Nanocrystalline materials in the amorphous and crystalline state as well as the relationship between coercivity H_c and particle size d_k. From Ref. [1].

thin films because of the small size effect and the surface effect. The lack of interaction between surface atoms and the small scale may also lead to smaller atomic spacing. All this may cause a decline of the exchange integral A, resulting in a decline of Curie temperature. For example, for 5 nm Ni, lattice parameters shrink by 2.4%.

5.2.6 Susceptibility

Magnetic properties of nanoparticles are closely related with the parity of the total number of electrons contained. Electrons with an odd number follow the Curie–Weiss susceptibility law:

$$\chi = \frac{C}{T - T_c} \tag{5.1}$$

Because of the quantum size effects, susceptibility complies with the law of d^{-3} (d for an average particle diameter), whereas the magnetic susceptibility of electrons with an even number is subject to:

$$\chi \propto k_B T \tag{5.2}$$

Susceptibility complies with the d^2 law (see Chapter 1).

Obviously, the basic common characteristics of nanomaterials and several special aspects of magnetic materials are different from those of bulk materials. By harnessing these features, we can create many novel materials with special properties beyond that of the bulk materials.

5.3 Some Specific Nanomagnetic Materials

This section describes a number of specific nanomagnetic materials, including zero-dimensional magnetic fluids and magnetic microspheres, one-dimensional magnetic nanowires and their arrays, two-dimensional magnetic nanothin film and its applications, nanocrystalline soft magnetic materials and their industrialization, double-phase nanocomposite hard magnetic principles, and the idea of high-frequency and microwave nanomaterials.

5.3.1 Magnetic Fluids [5]

Magnetic fluids are a stable colloidal system formed by the nanoscale (10 nm or less) strong magnetic particles highly dispersed in a liquid. In the 1960s, the United States first used them in the aerospace industry, which then found their way into commercial applications. Now, a huge industry has been developed. Magnetic fluid businesses can be found in the United States, Japan, Germany, and other developed countries, with the world's annual production being millions of tons of magnetic fluid devices.

Magnetic fluid is composed of strong magnetic particles, base liquid, and surface-active agent. In it, the magnetic particles must be very small to render the chaotic Brownian motion in the base fluid. Such a thermal motion may offset the gravity sedimentation and produce a loss of electricity and magnetism in the role of the mutual cohesion between particles. Meanwhile, under the actions of gravity, electricity, and magnetic fields, it can be viable without producing precipitation and condensation. As with iron, the particle diameter should be less than 3 nm; for Fe_3O_4, the diameter is not allowed to exceed 10 nm. Magnetic particles and the base fluid are mixed into one, presenting combined features that are shared by general magnetic materials and the liquid with mobility. It boasts several unique properties.

There are different ways for the preparation of nanoparticles, of which chemical coprecipitation technology features the following advantages: easy operation, low cost, less demanding on the equipment, and so on. Selection of suitable surface-active agents is the key to preparation of magnetic fluids. Surfactant-coated surfaces of the particles can bring about the following effects: preventing the oxidation of magnetic particles;

Table 5.3 Composition and Properties of Magnetic Fluid

Magnetic Particles	Surfactant	Containing Liquid	Density (gcm^{-3})	Degree of Particles within a Unit of $/cm^{-3}$ Volume	Saturation Magnetization $M_g/T(G)$	Chemical Stability of the Atmospheric Environment
Fe_3O_4	Oleic acid	Synthetic oil	1.23		0.0438 (438)	Better
$Mn_{0.2}Fe_{2.8}O_4$	Oleic acid	Kerosene	1.31		0.052 (520)	Better
Co	Polybutadiene	Toluene		2.77×10^{17}	0.0384 (384)	Poor
	amber imide			11.3×10^{17}	0.107 (1,070)	
Fe	Polybutadiene	Lubricants	1.07		0.0448 (448)	Poor
	amber imide		1.24		0.0767 (767)	
ε-Fe_3N	Polyamines	Kerosene	1.19	8.8×10^{16}	0.117 (1,170)	Better
			2.04	3.4×10^{17}	0.233 (2,330)	Better

overcoming the cohesion of the particles caused by van der Waals forces; weakening the static magnetic attraction; and changing the nature of the surface of magnetic particles, so that particles and the base can be mixed into a single fluid. The general requirement for the surface-active agent is that one side of the active agent can be adsorbed onto the surface of particles, forming strong chemical bonds, whereas the other side can become soluble with the base liquid. Each of the base fluids requires different surface-active agents, and sometimes they even require two or more surfactants. Table 5.3 shows a variety of magnetic fluid properties and compositions.

Magnetic fluids are magnetic and have mobility, and have unique magnetism, fluid mechanics, and optical and acoustic properties. Magnetic fluid is presented with superparamagnetic zero intrinsic coercivity, and there is no residual magnetism. Magnetic fluid magnetized in an outer magnetic field satisfies the modified Bernoulli equation. Compared with the conventional Bernoulli equation, magnetic energy is added so that magnetic fluid accompanies novel magnetic-associated properties that make it unique. For example, the apparent density of magnetic fluid increases as the external magnetic field strength increases; as light passes through the diluted magnetic fluid, it will produce the light birefringence effect and a two-way color phenomenon. When the magnetic fluid is magnetized, the optical anisotropy will occur relative to the magnetic fields. Polarized light with the electric vector parallel to the direction of the external magnetic field is absorbed more perpendicular to the direction, with a higher refractive index. When ultrasound is transmitted in magnetic fluid, its speed and attenuation are associated with the external magnetic field, showing anisotropy; magnetic liquids in an alternating field have a frequency dispersion of magnetic permeability, magnetic viscosity, and so on.

The special nature of magnetic fluid has opened up many new fields of application. Some of the technical difficulties in the past have been resolved because of its emergence. The following is a brief introduction to the application of several of its principles.

1. *Dynamic sealing technology*

   The dynamic sealing technology in the rotation axis is one of the comparatively mature and the most important applications of magnetic fluid, now widely used in shaft seals for precision instruments, such as X-ray rotating anode diffractometer, crystal furnace, high-power lasers, computers, and so on. In the nonuniform magnetic field, magnetic fluid will be gathered in the largest magnetic field gradient. Therefore, the outer magnetic field can be used to constrain the magnetic fluid in a sealed part, forming an "O" ring of magnetic liquid, showing properties of no leakage, no wear and tear, self-lubrication, and long life.

   Currently, precision instruments all include a magnetic fluid sealing component. Such an instrument generally costs $2,000 to $3,000, and the magnetic fluids used in them are not sold separately. Magnetic fluid sealing technology is important for sealing of vacuums and gas dynamic instruments and sealing out dust. However, it is seldom used in sealing water because of some practical difficulties. If breakthroughs could be made regarding the closure of water and oil sealing, then its applications would be extremely broad and bring about huge economic and social benefits.

2. *Audio speakers*

   Injecting magnetic fluid into a speaker's voice coil gap may have a certain damping effect on the movement of voice coil and enable it to automatically locate. At the same time, the heat generated by it can be dissipated through the magnetic fluid. Therefore, adding magnetic fluid can help improve the speaker's power load. With the same structure design, the input power can be increased twofold, while the frequency response and fidelity can be enhanced. Using magnetic fluids in metal brane speakers can enhance the performance more significantly. At present, the production lines and magnetic fluids that many manufacturers of magnetic fluid speakers in China use are imported from abroad. If we can produce magnetic fluid ourselves, then the cost of such products can be substantially reduced.

3. *Damping devices*

   Magnetic fluid can be used as a rotary and linear damper to damp the additional oscillation modes of the system. Compared with general damping media, magnetic fluids can be localized with the help of the external magnetic field. For example, in a stepper motor, magnetic fluid damping can be used to eliminate the vibration and resonance in the system to ensure precise positioning of the motor. In addition, the use of magnetic fluid damping in the vibration isolation tables is able to eliminate noise interference from the external vibration to ensure the proper work of precision instruments (scales, optical equipment, and so on).

4. *Mineral separation*

   The apparent proportion of magnetic fluid may change with the changes of an external magnetic field. This characteristic can be used to filter nonmagnetic minerals in different proportions. Minerals with the proportion difference of 10% can be well separated using this technique. In this process, water-based magnetic fluids are normally used, and they can be reused.

5. *Switching function*

   Mercury and magnetic fluid can be placed in a nonconductive container, where the magnetic field is applied to change the location of mercury to achieve the purpose of switching on and off the electric current. Magnetic fluid can be sealed in a nonmagnetic container on the shaft. While the rotor is stationary, the magnetic fluid stays in the lower part of the container, so the sensor cannot detect it; when the shaft rotates, centrifugal force distributes the magnetic fluid on the container wall. Then, the sensor detects the magnetic fluid and causes a switching action.

6. *Precision grinding and polishing*

   Magnetic fluid grinding harnesses magnetic fluid's buoyancy: micron abrasions are suspended on a liquid surface, allowing close contact with the work piece to be polished. No matter how special the surface shape of a work piece is, this technology is applicable for precision polishing. It may also be used to grind high-quality Si_3N_4 ceramic balls with efficiency 40 times higher than that of traditional methods.

7. *Magnetic fluid sensors*

   There are two kinds of commercially available magnetic fluid sensors: one is used in the oil exploration industry to measure acceleration and tilting of drill bits, and the other is used in the construction industry to detect the tilting of underground pipes.

8. *Other applications*

   In addition, magnetic fluid has a bright future in many areas, such as magnetic fluid printing, magnetic fluid film bearings, sonar systems, magnetic substances, magnetic cell separation, artificial heaters with magnetic fluid, turbine power generation with magnetic fluid, optical switches, magnetic fluid brakes, and so on. Properties and applications of magnetic fluids are shown in Table 5.4.

Table 5.4 Properties and Applications of Magnetic Fluid

Performance Available	Function	Applications
Magnetic properties	Sealing	Pump, shaft, pipe and valve seals
	Fluid driving	Hydraulic transmission devices
	Pressure changes	Pressure sensors, flow sensors
	Visualization	β detection of magnetic domain, disk, tape testing, testing
Viscous properties, magnetic properties	Lubrication	Bearing lubricants
	Damping	Rotating damping, damping measuring instruments, speakers
	Load maintaining	Accelerometers, angular velocity meters, grinding, hydrometer, dressing and bearings
Magnetic properties, temperature properties	Heat exchange	Energy converters, heat ducts and frequency converters
Viscous properties, rheological properties	Guidance	Oil–water separation, contrast media and cancer agent
	Confirming the location	Level meters, liquid level meters, thickness meters
	Position control	Displays, optical fiber couplers
	Film deformation	Interfacial layer control devices
Acoustic performance	Sound change	Ultrasound focusing devices, sound-absorbing body
Optical properties	Light change	Optical shutters, optical switches

5.3.2 Magnetic Microspheres

With the help of the surfactant, nanomagnetic particles can be combined with monoclonal antibodies, enzymes, drugs, and genes known as magnetic microspheres, which have exhibited important application prospects in bioengineering, biochips, biomolecular labels, and so on. In addition, they can create targeting mechanisms in drugs, turning them into so-called biological missiles. Such drugs are effective in the treatment of cancers.

5.3.3 One-Dimensional Nanowires

The overall trend of the development of computer disks is decreasing in size while increasing rapidly in storage density. Common disks have a storage density of 10^6 to 10^7 bits/in.^{2}. With the advent of CD-ROM, the storage density increased to 10^9 bits/in.^{2}. Because of limitations of the properties of materials, 10^{11} bits/in.^{2} was once considered to be the limit of disk storage density.

Research and development of a new generation of ultrahigh-density magnetic recording materials have attracted the attention of many researchers. One-dimensional magnetic nanowires have developed rapidly in recent years and are a hot research topic. These materials can be made of a single metal, alloy, compound, complex, or multilayer film. Figure 5.5 shows typical examples of one-dimensional magnetic nanowires. Nanoquantum storage medium prepared with these provides a direction in the development of high-density and low-noise hard disk medium. In recent years, quantum dot-matrix thin film has become popular in this area. Specifically, magnetic

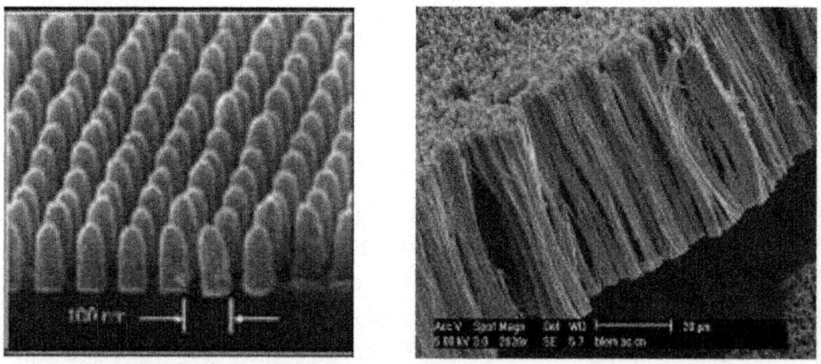

(A) (B)

Figure 5.5 (A) NiFe wire and (B) Co single-crystal wire.

metal or alloy is injected into the nonmagnetic media to obtain a quantum dot-matrix thin film isolated from each other. The basic idea of this approach is as follows. A template is first made to have a high pore density and controllable aspect ratio; then, magnetic metal or composite magnetic materials are injected into its microholes to form ordered nanowire arrays. As nanomagnetic units on the film are separated from each other, the film structure in this array is also known as a *quantum magnetic disk* (QMD). In comparison with conventional disks, QMDs possess the following advantages: the spontaneously quantified magnetic moment of each bit; precise location is not required on the write head in the quantization process; small, isolated, smooth transfer area that ensures high recording density and transmission noise approximating 0; and precise positioning of the read–write head to overcome the superparamagnetic limit. Chou et al. [6] injected magnetic material of 50 nm or less into the ordered nonmagnetic porous medium and obtained the QMD structured nanowire arrays with corresponding recording density up to 400 Gbit/in.^{2} or higher. In 1997, the first nanostructured disk was successfully developed by means of nanolithography in the Laboratory of NanoStructures in the Department of Electronic Engineering, University of Minnesota. This kind of disk has a size of 100×100 nm^2, which is the quantum wire array of cobalt rod based on a cycle of 40 nm by diameter of 100 nm and length of 40 nm. It has a storage density of 4×10^{11} bits/in.^{2}. After that, other scientists, such as those at MIT, also claimed that they made quantum disks of nanostructures [7].

Traditionally, magnetic data storage uses the technique of horizontal recording. As its name implies, the data bits are arranged horizontally (parallel to the disk surface). The use and evolution of this recording mode have lasted 50 years. However, the increased rate of storage density of this method slowed since the beginning of the twenty-first century because it has physical limits, for example the superparamagnetic effect. This has resulted in the recording density decreasing to an annual growth of 50–60% or even slower. In contrast with this mode, the data bits in perpendicular magnetic recording are aligned vertically (data bits perpendicular to the disk). Perpendicular recording mode (PRM) (Figure 5.6) can easily overcome the superparamagnetic effect, making it possible to achieve larger disk space for more data and, hence, to reach a higher areal density. In recent years, perpendicular magnetic recording mode has attracted widespread concern. Nanowire array is an important candidate for perpendicular magnetic recording material.

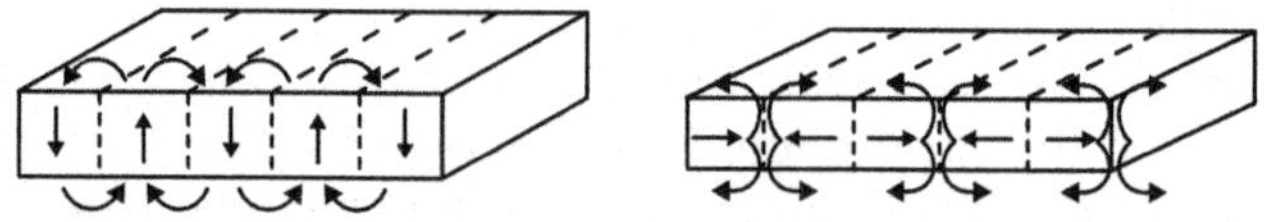

Figure 5.6 Comparison of PRM and the traditional longitudinal recording mode.

Quantum wire arrays can also be used for cell separation, and have shown purity of 80% and yield up to 85%.

5.3.4 Two-Dimensional Films [1]

Research on two-dimensional magnetic nanothin film has attracted much attention in recent decades. In the 1960s, there was a boom of studies on NiFe film, which was mainly used in magnetic memory and later replaced by semiconductors; in the 1970s, the research was focused on the magnetic bubble film, which had trial applications in magnetic storage and was later eliminated for various reasons. The 1980s witnessed the climax of a research campaign for magneto-optical film, which serves as the basis for the rewritable CD-ROM in the current market. By the 1990s, research on magnetic multilayers led to the discovery of the GMR effect and the emergence of a new discipline, *spintronics*.

In magnetic multilayers, the GMR effect works via a mechanism that spin-polarizes electrons; within the scope of its coherence length, it may have different scattering probabilities in different spin orientations of the film, that is, different resistances. The GMR spin-valve structure is one practical design. The spin-valve structure with nano-oxide layers is a finalized product (readout heads).

After the GMR discovery, tunneling magnetoresistance (TMR) was discovered in 1995, with magnetoresistive random access memory (MRAM) as one of its important application prospects. MRAM is a nonvolatile computer memory (NVRAM) technology that began its development in the 1990s, and amazing progress has been made ever since. Supporters of the MRAM technology believe that MRAM will eventually become dominant and supersede all other types of memory, becoming a true "universal memory" (Table 5.5).

5.3.5 Magnetic Nanocomposite Materials [1]

Magnetic nanocomposite material is one research target with very practical values. Because of the flexibility and availability

Table 5.5 MRAM Compared with the Current Memories (*F* is Feature Size)

Technology	DRAM	FLASH	SRAM	MRAM
Density	256 GB	256 GB	180 MB/cm^2	>256 GB
Rate	150 MHz	150 MHz	913 MHz	>500 MHz
Unit size	25 F^2/bit	2 F^2/bit		2 F^2/bit
Connecting time	10 ns	10 ns	1.1 ns	<2 ns
Write time	10 ns	10 μs		<10 ns
Erase time	<1 ns	10 μs		<10 ns
Hold time	2.4 s	10 years		Permanent
Cycling frequency of use	Infinite	100,000	Infinite	Infinite
Operating voltage (V)	0.5–0.6	5	0.5–0.6	<1
Switching voltage	0.2 V	5 V		<50 mV

of materials and performance tuning, suitable material recipes can be found in various areas. Stealthy materials in nanoresearch also take priority and include nanomagnetic composite particle films and nanomagnetic multilayers. A few examples of other magnetic materials are also described in this chapter.

After annealing the amorphous FeSiB, Cu and Nb are doped to control the grain size, and the resulting novel nanocrystalline soft magnetic materials may have a changing law that is totally different from that in conventional soft magnetic materials—small coercivity plus high magnetic permeability rate.

The mechanism in nanocrystalline soft magnetic material is represented in the exchange coupling interaction between nanograins, which can effectively offset the partial and random anisotropy. The average anisotropy energy density $<K> = K_1/N$, and the length scope of this exchange contains the number of grains $N = (L_{ex}/D)^3$. So, $<K> = K_1(D/L_{ex})^{3/2}$, and the ferromagnetic exchange length is subject to the relation $L_{ex} = (A/\langle K\rangle)^{1/2}$. Because the grain size D is less than L_{ex}, its random anisotropy can be expressed as $<K> = K_1{}^4D^6/A^3$, for example $<K>$ and D are subject to six orders of magnitude. Accordingly, magnetic permeability μ, with the decrease of the nanoparticles in the size D, also increased six orders of magnitude (Table 5.6).

Table 5.6 Comparison of the Novel Soft Magnetic Alloy with Traditional Materials on Magnetic Properties

Type	Composition	B_s (T)	μ_c ($f = 1$ kHz)	H_c (A/m)	p (μΩ cm)	λ_0	Loss (W/kg) 1 T/50 Hz	0.2 T/100 kHz
Amorphous alloy	$Co_{70.5}Fe_{4.5}Si_{10}B_{15}$	0.85	56,000	1.4	147	0	0	124
	$Fe_{78}Si_9B_{13}$	1.58	14,000	4.5	132	30.0×10^{-6}	0.130	140
Nanocrystalline	$Fe_{23.5}Si_{13.5}B_9Nb_3Cu_1$	1.24	10,0000	0.53	80	2.3×10^{-6}		40
alloy	$Fe_{91}Si_9B_{13}Nb_3Cu_1$	1.55	9,000	12.8		6.0×10^{-6}		
	$Fe_{91}Zr_7B_2$	1.70	40,000	4.4		0	0.085	79
	$Fe_{89}Hf_7B_4$	1.60	45,000	4.3		0	0.066	61
	$Fe_{84}Nb_7B_9$	1.50	30,000	6.5		0	0.085	79
	$Fe_{85}Zr_7B_6Cu_1$	1.52	41,000	3.2	56	1.0×10^{-6}	0.066	120
Ferrite	Mn–Zn	0.50	2,500	16.0		1.0×10^{-6}	0.210	120
Silicon	3.5% Si–Fe	1.97	770	41.0	48	23.0×10^{-6}	1.300	
Permalloy	$Ni_{80}Mo_5$	0.70	20,000	0.8	56	0		

5.3.6 Double-Phase Nanocomposite Hard Magnets

Theory shows that the soft and hard nanocomposite magnetic particles will be integrated with the strengths of both higher soft magnetic M_s and hard magnetic H_c, thus forming a novel hard magnetic material with a nanomagnetic energy level twice as high as the best available NdFeB. That is, its magnetic energy level is expected to be almost twice as high as the current value. Its mechanism is based on the production of flexible exchange-coupled magnetic particles in the hard and soft particle interface.

For example, with diameter of D, soft magnetic balls surrounded by a hard magnetic medium are made as a model to SmCo/Fe. The relationship between the reversal magnetization nucleation field and the particle diameter is calculated. The result indicates that if particles are less than 3 nm in diameter, the nucleation field can be up to 19.5 T.

5.3.7 High-Frequency Microwave Nanomagnetic Materials

The particles or thin film in nanoscale, which is less than the thickness of the skin of electricity, is very conducive to high-frequency applications. Requiring more permeability with a higher real part μ', the higher imaginary part μ'' may help to serve as stealth absorbing materials. If μ'' is so low that consumption is small, then it can be made to have similarly high-frequency soft magnetic and inductance materials.

Studies have shown that there is a Snoek limit in bulk materials regarding the frequency and magnetic permeability [8]. We provide simple proof of this here. We assume that its resonance mechanism is the natural resonance determined by the material anisotropy magnetic field $\omega = \gamma H_k$.

From $H_k = 2K_1/I_s$, we get

$$\omega = \frac{2\lambda K_1}{I_s}(K_1 > 0) \tag{5.3a}$$

From $H_k = -4K_1/3I_s$, we get

$$\omega = -\frac{4\lambda K_1}{3I_s}(K_1 < 0) \tag{5.3b}$$

Meanwhile, the magnetic permeability caused by the magnetization rotation, with a random distribution of the easy axis, can be expressed as

$$\overline{\mu}_\alpha = \frac{I_s^2}{3K_1\mu_0}(K_1>0) \tag{5.4a}$$

$$\overline{\mu}_\alpha = -\frac{I_s^2}{2K_1\mu_0}(K_1<0) \tag{5.4b}$$

Equation (5.3) is multiplied by Eq. (5.4), and we get

$$\omega\overline{\mu}_\alpha = \frac{2\gamma I_s}{3\mu_0} \tag{5.5}$$

Obviously, for an identified material $\omega\overline{\mu}_\alpha \propto I_s$ is a constant.

With the presence of cubic anisotropy, no ferrite can have permeability higher than the limits of Snoek (Table 5.7).

This absolute limit can be overcome with the support of planar anisotropy of nanothin film.

The easy plane in planar anisotropy is set to be vertical to axis c and magnetization anisotropy field in planar rotation is $H_{\alpha 1}$; the field is the anisotropy for magnetization to transfer out of the plane $H_{\alpha 2}$. Then, bulk materials in this planar anisotropy may have natural resonance frequency as follows:

$$\omega = \gamma\sqrt{H_{\alpha 1}H_{\alpha 2}} \tag{5.6}$$

For polycrystalline materials with an in-plane random distribution of the easy axis, rotational magnetic susceptibility is

$$\overline{\mu}_\alpha = \frac{I_s^2}{3K_1\mu_0} = \frac{2I_s}{3\mu_0 H_{\alpha 1}} \tag{5.7}$$

Equation (5.6) is multiplied by Eq. (5.7), and we get

$$\omega\overline{\mu}_\alpha = \frac{2\gamma I_s}{3\mu_0}\sqrt{\frac{H_{\alpha 2}}{K_{\alpha 1}}} \tag{5.8}$$

Experiments confirmed that the plane ferrite had an occurrence of natural resonance frequency that is higher than the Snoek limit. Therefore, the use of nanometer thin film materials

Table 5.7 Restrictions of the Operating Frequency on the Magnetic Permeability

f (MHz)	1,120	560	280	140
μ	5	10	20	40

enables the Snoek limit $\omega\bar{\mu}_\alpha$ to be increased by the fold of $\sqrt{H_{\alpha 2}/K_{\alpha 1}}$. For the FeNi alloy film,

$$I_s = 1.16\ \mathrm{T},\ K_1 = -2\times 10^2\ \mathrm{J\,m^{-3}}$$

And the line ratio

$$\eta = 2000,\ \sqrt{\frac{H_{\alpha 2}}{K_{\alpha 1}}} = 63.345$$

If the film thickness is 10 nm, then the diameter of a circular thin film should be 20 μm. The use of material with high saturation magnetization can have $\omega\bar{\mu}_\alpha$ further enhanced.

Because both the particle size and spacing are less than the length of exchange coupling, the exchange interaction between particles will average out the anisotropy of particles. Particle magnetization will be coupled together so that the whole film H_c is greatly reduced, and μ is greatly increased.

For instance, the nanogranular film (FeCo)-Zr_2O_5, at 1 GHz, $\mu' \sim 260$, $\mu'' \sim 320$, is improved by two orders of magnitude compared with the conventional expected value (Figure 5.7).

Nanoparticles or thin films with high-frequency characteristics can be used to produce high-frequency thin film inductors. It is important to address the miniaturization and integration of inductors that nanoparticles or thin film have higher high-frequency magnetic permeability μ' and lower μ'' to produce a small loss. The specific requirements of the material are as follows: high electrical resistivity ρ; high magnetic permeability and low loss at 2,000 MHz or more; and good temperature stability up to 200°C. This requires the material resonance frequency f_r to be higher than 2 GHz: $f_r = \gamma/2\pi(4\pi M_s H_k)^{1/2}$, where γ is the gyromagnetic ratio. Therefore, we must choose the right M_s, H_k and ρ. The following example illustrates the curve of the

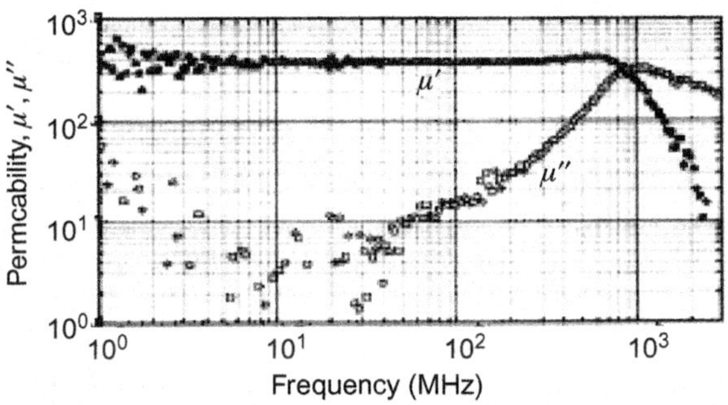

Figure 5.7 Nanoparticles (FeCo)-Zr_2O_5 of changes in film permeability with the frequency. From Ref. [1].

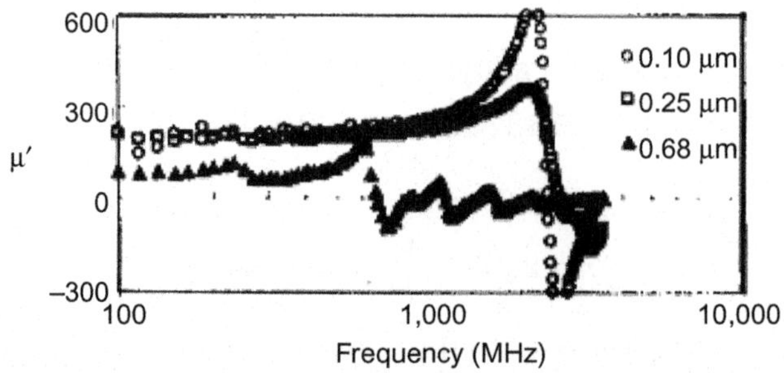

Figure 5.8 Real part μ' of permeability and frequency on the hard axis of CoZrNb film with 0.1/0.25/0.68 μm thickness. From Ref. [1].

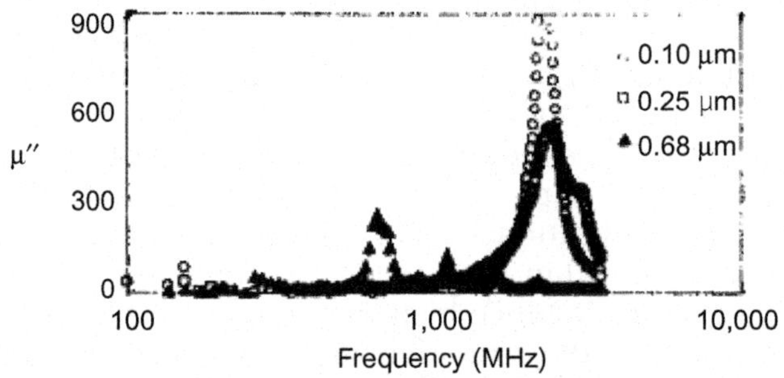

Figure 5.9 Imaginary part μ' of permeability and frequency on the hard axis of CoZrNb film with 0.1/0.25/0.68 μm thickness. From Ref. [1].

material CoZrNb on the real and imaginary parts of permeability versus frequency (Figures 5.8 and 5.9).

5.4 Preparation of Nanomagnetic Materials

5.4.1 Classification

Nanomagnetic materials can be expressed on a number of levels, for example zero-dimensional magnetic nanoparticles, one-dimensional magnetic nanowires, two-dimensional magnetic nanofilms, and the bulk of the magnetic nanoparticles complex. Preparation of nanomaterials can be divided into two categories: (1) top-down, that is, from big to small, whereby bulk materials are broken into nanoparticles or large areas are etched into the nanopatterns; and (2) bottom-up, namely small to large, whereby atoms and molecules are developed per the

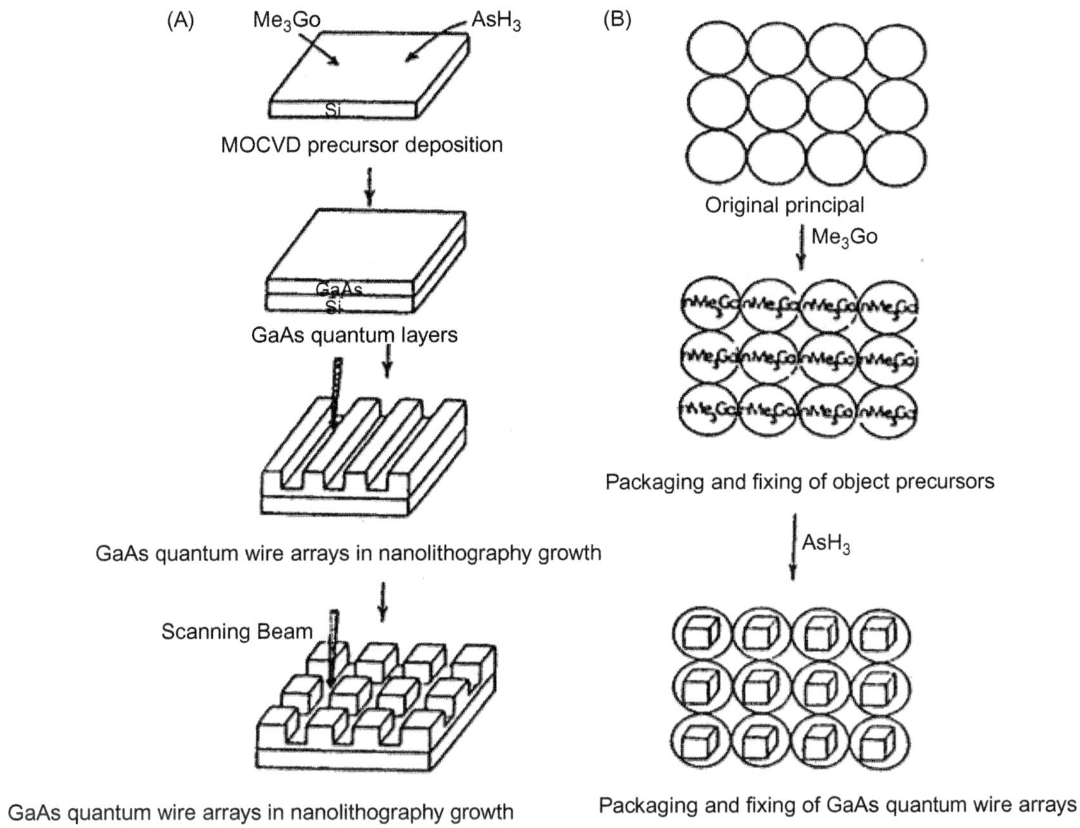

Figure 5.10 Two types of methods for preparation of nanomaterials (GaAs quantum dot arrays): (A) from top to bottom (etching) and (B) from bottom to top.

growth requirements into nanoparticles, nanowires, nanocomposite films, or nanoparticles (Figure 5.10).

Table 5.8 shows another classification. More changes can be created for practical use.

5.4.2 Specific Instances [1]

Examples for the specific preparation of nanomaterials are described here.

5.4.2.1 *Mechanical Crushing Method*

Mechanical means, such as high-energy ball milling, ultrasonic or jet milling, and others, can have powder prepared into nanoparticles. This is an example of a top-down approach,

Table 5.8 Classified Methods for Synthesis and Preparation of Some Commonly Used Nanomaterials

Method	Examples
Gas phase	Gas condensation, active hydrogen–molten metal reaction, sputtering, ohmic heating evaporation, hybrid plasma, laser-induced chemical vapor deposition
Liquid phase	Coprecipitation, spray, hydrothermal mode, microemulsion, sol–gel mode, electro-deposition, solvent evaporation decomposition, high-pressure quenching
Solid phase	High-energy ball milling, noncrystallization, combustion synthesis

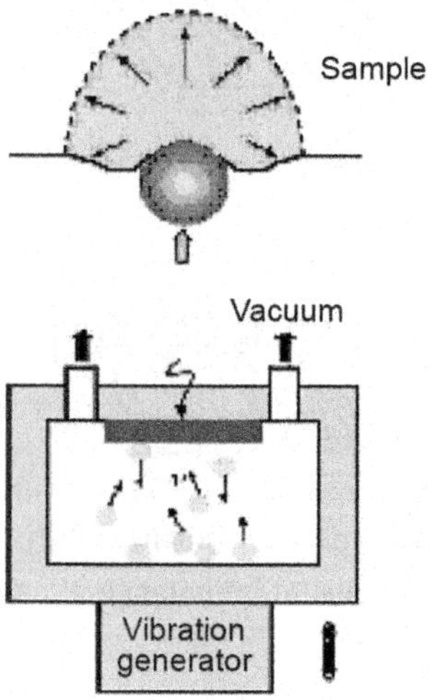

Figure 5.11 Nanosurface treated using a mechanical method, with each collision producing a group of dislocations.

which is suitable for refractory metals or materials beyond the use of chemical reactions. The disadvantages include the difficulties in classification according to the particle size and serious surface contamination.

Bombarding a metal surface with high-energy balls makes it possible to turn the surface structure into nanoscale; this can improve the abrasion and corrosion resistance of the processed material. Meanwhile, the surface is identical to the bulk material, and thus it does not peel off like nanocoating material. The main mechanism of this method is to produce a large number of defects and dislocations, which further develop into dislocation walls, and thus cut the large crystals into nanocrystalline grains (Figure 5.11).

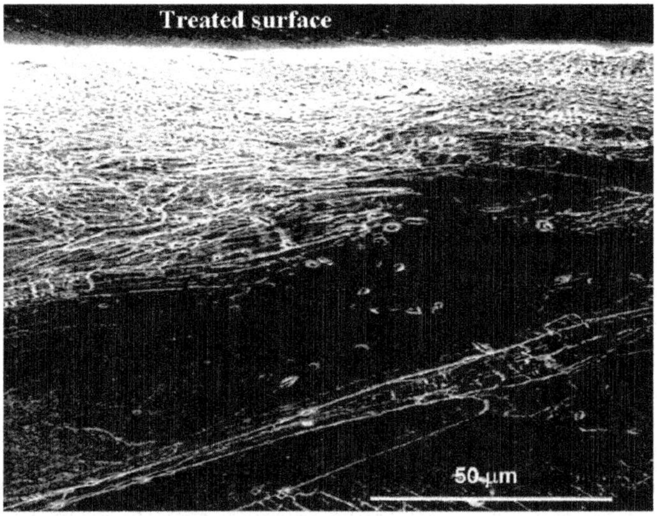

Figure 5.12 Cross-section under observation.

We can see from Figure 5.12 that dislocation density gradually decreases from the surface to the interior. It is the dislocation wall that cuts the large crystals into nanocrystalline grains.

5.4.2.2 Etching Method

Etching also follows the top-down approach, by which large areas of film, through chemical, electron beam, ion beam etching, or even atoms moving with the help of scanning tunneling microscopes and other equipment, are prepared into nanodots, nanolines, or other nanopatterns.

Etching treatment is commonly executed using lithography or an ion beam etching machine. SPM devices can be used for the atomic handling work, such as scanning tunneling electron microscopy and atomic force microscopy (see Chapter 2).

Corresponding to the top-down approach to manufacturing, nanomaterials are also able to begin growth from bottom to top, for example from atoms and molecules.

5.4.2.3 Physical Method

By using physical methods, there will be no chemical reactions in the preparation process. Commonly used methods are atomization, sputtering, evaporation, and noncrystallization. Gas coagulation is completed as follows: in a vacuum chamber filled with inert gas, the metal is heated, evaporating into the mist of atoms to collide with inert gas. This makes the metal lose momentum, and it is precipitated onto the liquid nitrogen-cooled rods. Then, this powder can be scraped off and collected. Atomization refers to a molten metal jet in a vacuum, under the impact of the surrounding ultrasound airflow, scattered into atomized tiny droplets and then solidified into the nanoparticles on the cooled substrate or the collector. This is an effective method of mass production of metal nanoparticles, where the design of an ultrasonic nozzle is an essential step. In the evaporation method, a metal is heated in low-pressure inert gas to form a metal vapor, which then is solidified in the frozen substrate to form nanoparticles or nanothin film forms onto other monocrystalline or polycrystalline substrates. With the heating method, evaporation treatment can be divided into e-beam heating (e.g., molecular beam epitaxy (MBE)), pulsed laser deposition, and resistance wire or resistor chip heating. The sputtering method is one of the most commonly used methods for the preparation of nanofilm. In an argon-filled vacuum chamber, the required target metal is set as the cathode, and the thin film substrate is set as the anode. Argon ions, formed by glow discharge between two poles, will impact on the cathode target under the electric field and are sputtered onto the substrate in the formation of thin film. The noncrystallization method is based on the presence of an amorphous thin strip or film and then by the control of the annealing conditions, so that crystals can be treated in the nanocrystalline state. For instance, amorphous soft magnetic alloy FeSiB doped with Nb and Cu may achieve control of the nucleation and grain growth in the crystallization process. It is an essential method that is easy for mass production of nanocrystalline soft magnetic materials.

Amorphous preparation involves the rapid cooling of molten metal at a rate of 1 million degrees per second to prevent its crystallization (Figure 5.13).

5.4.2.4 Chemical Method

If a preparation process involves chemical reactions, then it can be called a chemical method. Chemical methods mainly include metal organic chemical vapor deposition (MOCVD), sol–gel, hydrothermal method, codeposition method, and others.

The sol–gel method was developed in the 1960s for preparation of glass–ceramic products and is now commonly used in the preparation of nanoparticles. Its basic principle is that in

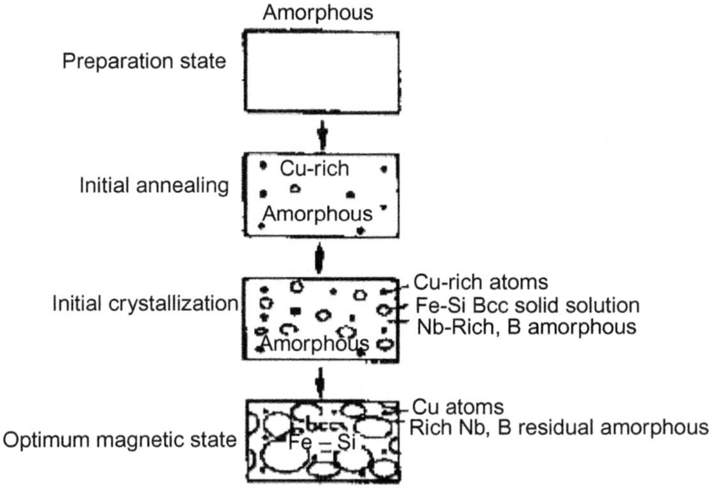

Figure 5.13 Crystallization process of amorphous alloy $Fe_{73.5}Cu_1Nb(Mo)_3Si_{13.5}B_9$.

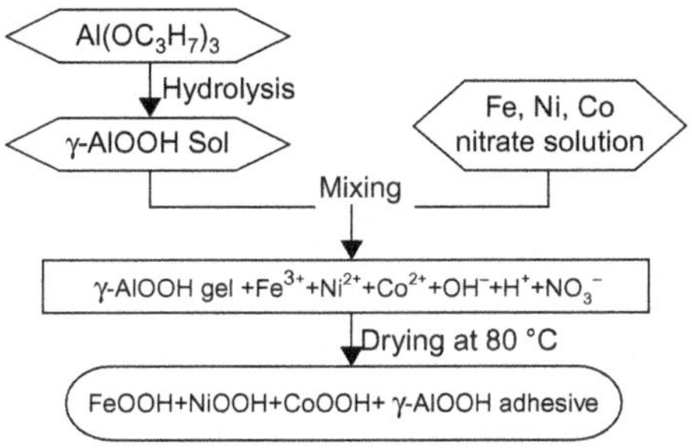

Figure 5.14 Example of an organic sol–gel process.

certain solvents and conditions, metal alkoxides or inorganic salts are controlled in hydrolysis to form a sol rather than to produce deposition. Then, the solute is condensed to gelation, forming inside a 3D network structure. The gel is then dried to remove organic components and, in the end, we obtain the required nanopowder material. Alternatively, if the sol is attached to the bottom, then the nanofilm is available.

Metal alkoxides are organometallic compounds M(OR)n with M–O–C bonds generated from the metal and ethanol reaction (M is a metal, R is alkyl or propenyl); with these, it is easy to obtain hydrolysis.

Figure 5.14 shows an example of a sol–gel method using an organic method (aluminum isopropoxide) to prepare the nanocomposite particles $Ni_{65}Fe_{31}Co_4/Al_2O_3$.

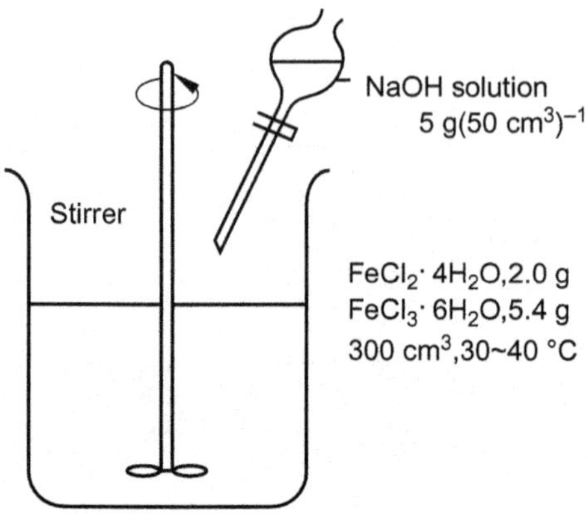

Figure 5.15 Preparation of Fe_3O_4 nanoparticles.

Chemical codeposition treatment depends on the codeposition of metal ions in the solution through a chemical reaction. First, metal salts are prepared with a good proportion and mixed evenly in the solution, followed by using alkali as the precipitating agent to achieve codeposition of a variety of metal ions.

Figure 5.15 shows Fe_3O_4 nanoparticles prepared by codeposition. Chloride solutions with divalent iron ions and trivalent iron ions are deposited under the action of the alkali sodium hydroxide.

In MOCVD, the evaporated metal organic matter is mixed and introduced into a vacuum chamber. Under the effect of the heat, the gas-phase reaction is induced to promote the decomposition of organic matter to form metal nanoparticles or thin films. In the presence of oxygen, metal oxides can be formed. Commonly used metal organic compounds are M-(tmhd) 2,3 M-(thd), and others, and the organic part of 3 M-(thd) is a 4-methyl-G flavanone.

5.4.2.5 Preparation of Magnetic Nanoparticles in the Magnetic Fluid

Preparation of magnetic fluids involves the full use of the nanoparticle surface effect, namely adsorption and variation of surface composition. Long chains, such as the hydrophilic carboxyl-COOH in fatty acids, are adsorbed on the surface of magnetic nanoparticles, whereas the lipophilic alkyl C_nH_{2n+1} is linked with the base magnetic fluid, such as polyphenylene ether, which acts as surfactant. Typical surfactants include oleic acid, polyimide, polyethylene amine, and others.

5.4.2.6 Two-Dimensional Nanowire Array: Template Method

Preparation of a high-quality template is a prerequisite for this approach. Commonly used templates are polymer, nanotube, molecular, porous alumina film, reverse micelle, block copolymer, and life body template.

1. *Polymer track etched template*

   Nuclear irradiation of the fission fragment can produce traumatic pits in a polymer polyester or polycarbonate film, and then it is chemically treated to form a random distribution of cylindrical nanotemplates with uniform diameter. The minimum diameter of these templates is 10 nm. However, they are divergent because of nuclear radiation and cannot guarantee that the hole is vertical to the template surface.

2. *Zeolite-type ordered template*

   As mentioned, the surfactant polymer is hydrophilic at one end and lipophilic at the other. When added to the solution with inorganic precursors, the lipophilic end will float to the surface because of the exclusion of water. With the increase of surfactant concentration, the lipophilic side floating to the surface will be saturated, followed only by assembly into micelles within solution. The lipophilic side is the inner side, with the hydrophilic outer side reducing the energy. Micelles can be formed in spherical, columnar, and laminar periodic arrangements. The space between micelles is surrounded by the solution with inorganic precursors. After this, the drying treatment is used to remove the solution water, which is heated to burn off the organic matter, forming the mesoporous molecular sieves composed of the inorganic wall. This process is also used as a template.

3. *Carbon nanotube template method*

   Carbon nanotube itself as the carbon source can take part in the reaction forming one-dimensional solid nanowires with a basic shape of the original carbon nanotube. This method is still not used for preparation of magnetic nanowires.

4. *The second anodization to prepare the Al_2O_3 template*

   High-purity aluminum chips with thickness of 0.5 mm or less are treated with 2 h of 500-degree vacuum annealing, followed by electrochemical polishing. In the H_2SO_4 or $H_2C_2O_4$ electrolyte, these chip electrodes undergo anodic oxidation at constant temperature and pressure for more than 10 hours. In a mixture of 6 wt% H_2PO_4 and 1.8 wt% H_2CrO_4, the formed oxide film can be completely dissolved.

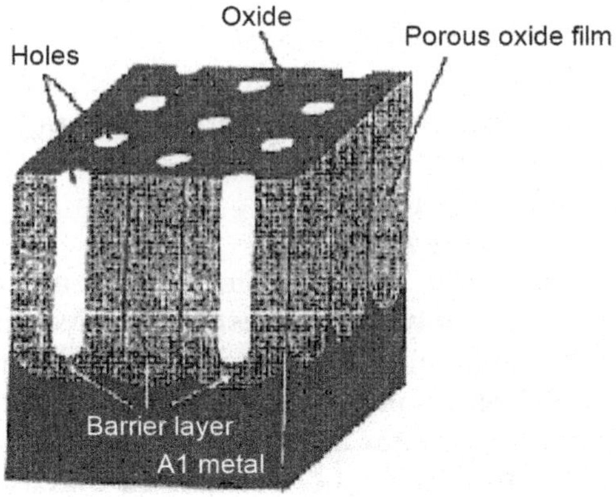

Figure 5.16 Schematic diagram of the template in an orderly Al_2O_3 array.

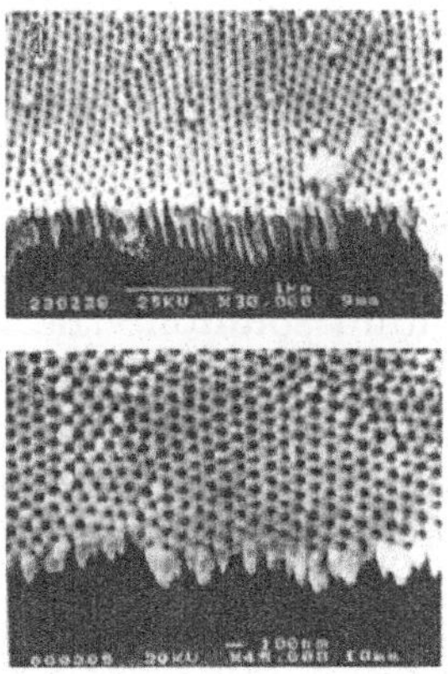

Figure 5.17 SEM oblique view of the mask template in an orderly array of ultra-thin Al_2O_3.

Under the same conditions, these aluminum chips are given a second anodic oxidation of 2–4 h to obtain the template as shown in Figures 5.16 and 5.17.

5. *Double-pass template*

   If the remaining aluminum is removed from the template in a mixture of 20% HCl and 0.1 M $CuCl_2$, followed by use of 5 wt% H_2PO_4 to remove the dense alumina barrier layer at the bottom, then a double-pass template can be obtained as shown in Figure 5.18.

Table 5.9 shows a condition for the actual control of pore size.

We can see from Figure 5.18 that the Al_2O_3 template generally self-assembles as a hexagonal close-packed ordered array,

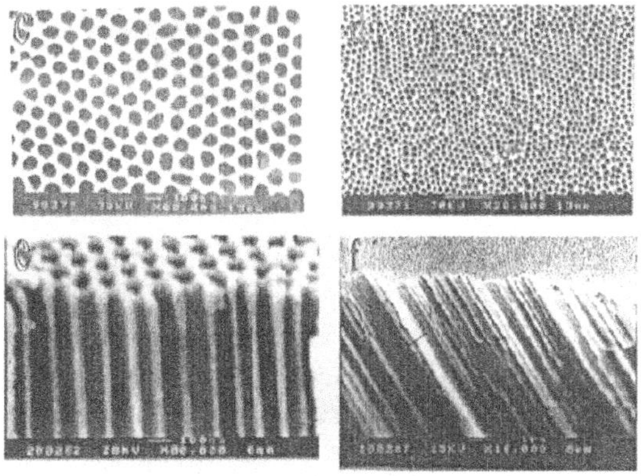

Figure 5.18 SEM images of the orderly array Al_2O_3 template with pore size in the 60–95 nm range.

Table 5.9 Conditions for Preparation and the Aperture Size of the Template of the Synthesized Al_2O_3 Ordered Array

Electrolyte Type	Electrolyte Temperature (°C)	Anode Voltage (V)	Aperture Size of Oxide Film (nm)
0.25 M $H_2C_2O_4$	3	70	80–95
0.25 M $H_2C_2O_4$	5	60	65–70
0.3 M $H_2C_2O_4$	17	40	40–45
0.3 M $H_2C_2O_4$	17	36	32–38
0.3 M H_2SO_4	3	27	22–25
1.2 M H_2SO_4	2	15	12–14
1.5 M H_2SO_4	1	12	10–12
2.0 M H_2SO_4	0–1	8	8–10
3.0 M H_2SO_4	0–1	6	6–8
3.5 M H_2SO_4	0–1	5	5–6

and its mechanism can be summarized as follows. In the initial oxidation, a dense oxide layer (barrier layer) is formed while maintaining the same thickness under certain conditions. Then, anodic oxidation depends on ion diffusion in the electric field. Because the volume of alumina is greater than the volume of aluminum, a thick oxide layer may expand inside and produce large stress. In the common effect of electric field, stress, and acidic medium, weak spots of oxide film may experience destruction of selective dissolution to form a porous structure. Uniformity of the role of stress results in the lowest energy because the holes are self-assembled in a hexagonal close-packed ordered array.

With the template method, it is easy to control the scale and uniformity of nanowires to facilitate preparation of nanoarrays. Nonetheless, nanowires prepared in this way mostly have polycrystalline structures. Furthermore, nanowires are prone to damage when the template is removed. In addition, several of the aforementioned manufacturing methods are apparently difficult to apply in industrial production. Designing a new preparation method is an important creation itself. Many important materials are actually discovered when researchers try to devise new preparation methods. The description in this section is only an introduction and aims to initiate work that should be based on open ideas and innovations.

5.5 GMR Materials

5.5.1 GMR Effect and Applications [9]

While moving in a material, electrons may be affected by the cations brought about by the lattice atoms or the impure atoms, so their motion path may change and therefore they may collide with these lattice atoms, producing some heat; such a process is called the "resistance effect." In summary, the resistance effect is the result of the lattice hindering the motion of electrons. If the movement of electrons in materials is applied with a magnetic field, it will increase the probability of electron collisions with the cations, that is, it will lead to increased resistance. Because resistance is changed as a result of the impact of the magnetic field or a magnetic role, it is called *magnetoresistance* to differentiate from general resistance properties. In other words, magnetoresistance refers to the change in a particular magnetic field. Generally, both magnetic metal and alloy may accompany the magnetoresistance phenomenon, usually described by the rate of resistance change $\Delta\rho/\rho$ (or $\Delta R/R$). Common metal conductors have a very small $\Delta\rho/\rho$, which is only approximately 10^{-5}%. For a magnetic metal or alloy material (for example, Permalloy), $\Delta\rho/\rho$ may be up to 3–5%.

The so-called GRM effect manifests itself as a significant decrease in electrical resistance in the presence of a magnetic field. Such a decrease is usually 10 times that of the magnetoresistance values that may occur in magnetic metals and alloys.

In 1986, the German science team of Grunberg made an important discovery in Fe/Cr/Fe three-layer film: the chromium layer between two iron layers can produce coupling.

In 1988, Kent's research team at Paris University, France, first observed the GMR effect in [Fe/Cr] periodic multilayer films. When applied with an external magnetic field, its resistance decreased, presenting a rate of change up to 50% (Figure 5.19), so it is called the GMR effect. This discovery elicited great interest in the international science community.

In the 1990s, Fe/Ag, Fe/Al, Fe/Au, Co/Cu, Co/Ag, and Co/Au nanostructured multilayer films were found to have characteristics of GMR. In 1995, researchers used an Al_2O_3 insulating layer instead of the conductor Cr and observed the phenomena of large TMR.

Based on the discovery of GMR and TMR, a new branch of the discipline—magnetic electronics—is gradually taking shape.

From then on, scientists have been engaged in persistent efforts to transform the discovery into industrialization of the innovative information technology. IBM, in 1994, developed a

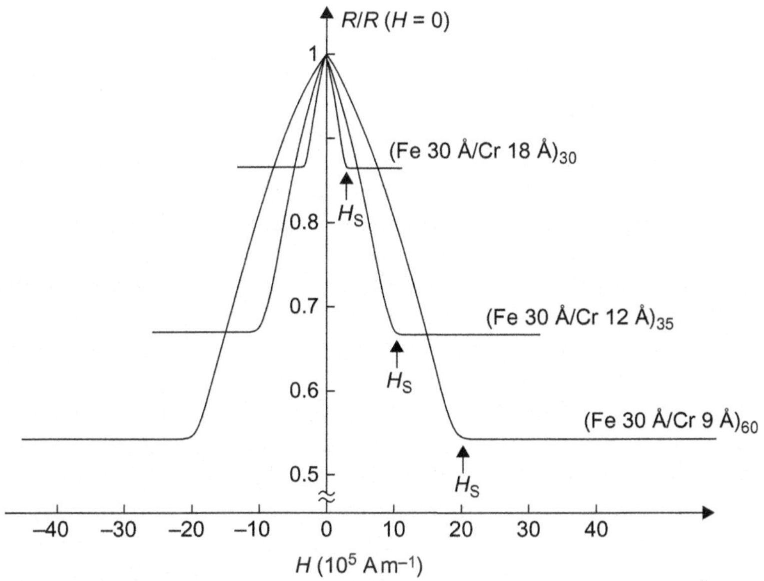

Figure 5.19 Fe/Cr multilayers with GMR (4.2 K) effect; H_s—intensity of saturation magnetic field.

read head with a giant magnetoresistive effect, increasing the disk recording density by 17-fold, and magnetic disks are again leading in the competition of recording density with CD-ROMs. The application of GMR heads led to rapid developments in the computer industry, breaking the bottleneck in the transmission and storage of images on the information superhighway. At present, the storage density has been as high as 56 GB/in.^{2}. GMR heads help generate a total of US$40 billion annually in the world market. Moreover, Motorola announced that it had successfully developed a GMR magnetic random access memory in 2001, which implies a market capacity of US$100 billion.

By using the GMR effect with different resistance characteristics in different magnetic states, magnetic random access memory (MRAM) can be made to retain information in the absence of power supply.

In 1999, a hard disk drive that used GMR heads (HDD) began to be sold, with a storage density up to 11 Gbits/in.^{2}. In 1990, this figure was merely 0.1 Gbits/in.^{2}, so the increase in one decade is 100-fold.

At present, research and development of GMR are booming, but the study commenced with respect to the aforementioned TMR multilayer films applied in the novel random access memory (MRAM). Based on the spin polarization effect, the spin transistor has also been suggested.

5.5.2 Classification and Comparison of Magnetic Resistance

In terms of the magnetic materials that have been studied, the magnetoresistance effect can be divided into ordinary magnetoresistance (OMR), technical magnetization-associated anisotropy magnetoresistance (AMR), GMR specific to magnetic multilayers and particulate film, colossal magnetoresistance (CMR) doped with rare earth oxides, and TMR. To understand the nature of the magnetoresistance, we must first understand the principle of electron scattering.

Electron scattering plays a fundamental role in all transport processes. In an ideal cycle field where rules of atom array are completely ordered, the electrons will be in a determining k-state and transition does not occur. But in reality, atoms do not statically remain in the grid points. Because of thermal vibrations, atoms often deviate from the grid points, which can be regarded as a perturbation of periodic potential field to cause electronic transitions known as lattice scattering. The size of the lattice scattering is proportional to the electron density of energy states on the Fermi side. In addition, the material impurities and defects also undermine the periodic potential field, resulting in electron scattering.

For a general nonmagnetic metal, the electron spin is degenerated without any net magnetic moment. And the electronic states in the vicinity of the Fermi surface, of course, are exactly the same as for the spin-up and spin-down. Thus, in transport processes, the electron flow is nonspin-polarized.

With a common nonmagnetic metal, electron scattering is mainly in the state of spin degeneracy between s electrons, showing a larger mean free path of electrons. With the Drude theorem $\sigma = ne2\tau/m$, it can be very easy to estimate that good metal conductors may have a mean free path of approximately 100 Å.

Unlike s electron scattering in ordinary metals, ferromagnetic metals, such as iron, cobalt, and nickel, feature the existence of s electrons and d electrons with great density of states at the Fermi surface. In the transport process, conduction electrons are expected to endure s–d scattering that is much higher than the s electron scattering, so the conduction electron here has a much smaller mean free path.

Because the spin-up 3d subband (more in spin) and spin-down 3d subband (few in spin) are not equal in density of states on the Fermi surface, the size of scattering will not be the same as that of the conduction electrons in a different spin mode. Therefore, the spin-up electrons may have a different mean free path ($\lambda_\uparrow$) from that of the spin-down electrons ($\lambda_\downarrow$).

Based on the theoretical and experimental evidence, the transport process of ferromagnetic metals or alloys can be decomposed into spin-up and spin-down channels of electronic conductivity, which have parallel connection and are almost independent of each other. This is a two-fluid model of spin-dependent scattering.

A normal magnetoresistance effect is common to all magnetic and nonmagnetic materials. Lorentz force from the magnetic field on the carrier can cause deviation from the motion of conduction electrons in space or the occurrence of spiral motion to increase the probability of electron impact and resistance; therefore, MR is always positive.

MR changes in magnetic conductors are associated with magnetic fields and the angle between the directions of current in the conductor, for example anisotropy, known as anisotropic magnetoresistance. When the external magnetic field is applied in a direction either parallel or perpendicular to the direction of the current, the resistivities $\rho_{//}$ and $\rho_{\perp}$ of anisotropic magnetic material are different from each other. Usually, $\Delta\rho/\rho_0 = (\rho_{//} - \rho_{\perp})/\rho_0$, which is used to evaluate the size of the anisotropic magnetoresistance. Here, ρ_0 is the magnetic resistivity in the ideal annealing state. A commonly used AMR material highly sensitive to magnetic fields is Permalloy ($Ni_{80}Fe_{20}$). At room temperature, its magnetoresistance change rate is 5%. In ferromagnetic metals, the AMR effect is strongly dependent on the direction of spontaneous magnetization. It is caused by the motion from anisotropy of the ferromagnetic domain in an outer magnetic field. In fact, this magnetoresistance corresponds to the technical magnetization, that is, the corresponding resistance changes from the demagnetization state to the process toward magnetic saturation. The zero-field resistivity is also associated with its history. In ferromagnetic materials, the anisotropic magnetoresistance effect comes from the anisotropic scattering, that is, it is derived from the spin-orbit coupling, which may reduce the symmetry of electron wave functions so that electron spin and its track are associated with each other.

In the past it was difficult to produce high-quality nanoscale samples. By the 1980s, however, because of the liberation from these restrictions, metallic superlattices have become a research frontier that has attracted the interest of many researchers. The researchers of condensed matter physics have been performing a wide range of basic research on the nature of such artificial materials, including magnetic ordering, interlayer coupling, electron transport, and quantum confinement.

Magnetic multilayers are a multilayer film system resulting from the alternately repeated growth of nanoscale ferromagnetic films (Fe, Co, Ni, and its alloys) and nonmagnetic films (including 3d, 4d, and 5d nonmagnetic metal). A combination of contemporary mesoscopic magnetism, nanomagnetic materials, and nanotechnology, it is developed as a kind of artificial superlattice magnetic material with unique magnetic properties. In 1988, when a single crystal (100) Fe/Cr/Fe three-layer film grew from MBE with a Cr layer thickness of 9 Å (at 4.2 K or below) and a 20 kOe external magnetic field, it showed a value of magnetoresistance of MR (%) $= \Delta\rho/\rho_s = (\rho(H) - \rho_s)/\rho_s$, up to 100%, and it is called the GMR effect.

With the sputtering process, Parkin and colleagues prepared multicrystalline Fe/Cr/Fe three-layer film and (Fe/Cr) multilayers also displayed the GMR effect. The MR values of the latter are 25% and 110% at room temperature and the low temperature of 4.2 K respectively.

Subsequently, research discovered that many magnetic multilayers, which constitute the alternative growth of a variety of iron magnetospheres (Fe, Ni, Co, and its alloys) and nonmagnetic layers, had a GMR effect, especially the polycrystalline (Co/Cu) multilayer films. Their GMR values were 130% and 70% at the low temperature (4.2 K) and room temperature respectively, and the added saturation field was approximately 10 kOe.

In artificial magnetic metal films with nanostructures, there is another essential material in addition to superlattices and multilayers: the granular film. It is a composite film formed by microparticles embedded in mutually nonsolutionable films. It has the double features of microparticles and thin film and their interaction effects. In principle, if any two elements are not soluble in equilibrium conditions, then either the co-sputtering or co-evaporation process can be used to prepare granular films. Therefore, research on the magnetic GMR effect is necessary in terms of its extension from magnetic multilayer films to granular films.

In 1992, Co–Cu and Co–Ag granular films were found to have a GMR effect similar to that in multilayer films. Subsequent work focused on two types of granular films comprising Fe, Co, and Ni metals and alloys based on Cu and Ag. Among the granule films that may produce the greatest GMR effect, the share of iron group elements in the volume has a percentage of approximately 15–25%, which is below the percolation threshold to form a network structure. At this time, iron group elements were embedded in thin films mainly in the form of microparticles, and the best size of microparticles is in the range of several nanometers to 10 nm, so that ferromagnetic particles of such sizes are usually in a superparamagnetic state at room temperature.

As with multilayer films, theoretically the GMR effect of granular films is associated with spin-dependent scattering and particularly with interface scattering. Theory suggests that the GMR effect of granular films is inversely proportional to the diameter of magnetic particles; for example, it is proportional to the specific surface area of the particles. Particles are usually randomly distributed in the films, and most of the conduction electrons are transmitted through the particles, similar to the case of multilayer film CPP (current perpendicular to it). So, a larger GMR effect can be easily obtained in granular films. However, there is a problem. Because of the ferromagnetic particles in a superparamagnetic state, the access of GMR usually requires a very high saturation field.

A magnetic tunnel junction has a sandwich structure, which is constituted by a ferromagnetic film, an insulator film, and a ferromagnetic thin film. The magnetoresistance phenomenon that corresponds to the tunnel junction is known as the TMR.

The TMR comes from the asymmetry between two ferromagnetic layers on the density of states of the spin-up and spin-down electrons. Assuming there is no spin-flip scattering (the spin conservation of electrons in tunneling process), the probability of spin polarization depends on the relative orientation of magnetization in two ferromagnetic layers. When the magnetizations are parallel to each other, the majority of electrons share the same spin orientation in the energy band of two ferromagnetic materials. The number of states that can be filled is maximized in the matching degree in the vicinity of a Fermi surface, creating the largest tunneling current and the smallest tunneling resistance. In contrast, in antiparallel magnetization, electron tunneling occurs between the majority of electronic states in a ferromagnetic layer and a few electronic states in the other ferromagnetic layer. This mismatch between densities of states results in the smallest tunneling current and the largest resistance. Because the external magnetic field is applied so that magnetization changes from parallel to antiparallel, or vice versa, the magnetoresistance phenomenon may occur. Because electron transport in a magnetic tunnel junction is based on the mechanism of the tunnel effect, it is called TMR.

TMR has the advantages of a small saturation field and high field sensitivity. In the magnetic tunnel valve, the magnetic field that has overcome two-layer ferromagnetic coercivity can subsequently have its magnetization directions transferred to be parallel with the direction of the magnetic field. At this time, tunnel resistance tends to have a very small value. If the magnetic field is reduced to a negative value, then the ferromagnetic field with the smaller coercivity will first reverse the magnetization direction. The two ferromagnetic layers are placed in the opposite direction of magnetization, making the tunnel resistance very high. Thus, only a very small external magnetic field can lead to a maximum value of TMR, so TMR is very high in magnetic field sensitivity. Miyazaki produced $Fe/Al_2O_3/Fe$ magnetic field sensitivity of 8%/Oe; Moodera and associates obtained $CoFe/Al_2O_3/Co$ sensitivity of 5%/Oe. Both showed results that are well beyond those of multilayer GMR and the oxides CMR. For resistance in the magnetic tunnel junction under zero field, the thickness of the insulating layer can be adjusted to change the number of orders of magnitude, but the TMR value is not changed. Meanwhile, changing their own resistances of multilayer films and metal oxides under the zero field is not as easy as it looks. Therefore, TMR can be used to make controllable high-resistance junctions.

5.5.3 Physical Mechanism of GMR

5.5.3.1 Magnetic Exchange Coupling

Two layers of ferromagnetic thin films separated by a layer of nonmagnetic metal thin film may have magnetic exchange coupling in between, from which the GMR effect occurs. Magnetic exchange coupling and GMR effect may have periodic oscillations depending on the nonmagnetic film thickness.

In the Fe/Cr/Fe system, the coupling state exists between adjacent iron layers. With the increase of the Cr layer thickness, coupled oscillation will be attenuated in the form of alternating positive and negative states. This makes the magnetic moments of adjacent Fe layers alternate from an antiparallel orientation with each other to a parallel orientation (Figure 5.20). In the outer magnetic field, the iron magnetosphere in multilayer films will also change in the antiparallel magnetization state. When electrified with current, this change in magnetic state is reflected in changes in the form of resistance. This is the physical mechanism of the GMR phenomenon.

The coupling between iron layers with the Cr electrons as an intermediary may have an oscillation attenuation as the Cr layer thickness increases. The average scope is 1–3 nm, which limits the Cr layer thickness. The electronic mean free path (10–20 nm) and the spin diffusion length (30–60 nm) are both

(A)	(B)
Ferromagnetic layer ←	Ferromagnetic layer →
Non-magnetic isolation layer	Non-magnetic isolation layer
Ferromagnetic layer →	Ferromagnetic layer →
Non-magnetic isolation layer	Non-magnetic isolation layer
Ferromagnetic layer ←	Ferromagnetic layer →
Non-magnetic isolation layer	Non-magnetic isolation layer
Ferromagnetic layer →	Ferromagnetic layer →
Non-magnetic isolation layer	Non-magnetic isolation layer
Ferromagnetic layer ← *Ma*	Ferromagnetic layer → *Ma*
(A) In the state of zero magnetic field	(B) In the state of over saturation field H

Figure 5.20 Spin configuration of the GMR multilayer film.

very short for metals, especially magnetic metals. This constitutes a further restriction on the individual sublayer thickness of multilayer films.

For these reasons, we can say that GMR and TMR studies depend entirely on the progress of nanomaterials science. No innovation or transformation can be made without this basis. However, the nanoscale is so tiny that it brings a challenge in the preparation of multilayer films and their microstructural characterization.

5.5.3.2 GMR Effects of Metal Superlattice

At present, it is well known that metallic superlattices to produce the GMR effect are required to satisfy the following three conditions. First, the system can change the magnetization state in the outer magnetic field. For example, in ferromagnetic conductor/nonmagnetic conductor superlattices, an antiparallel spin structure can constitute a zero magnetic field (Figure 5.20A) and can become a parallel spin structure when the external magnetic field is higher than magnetic saturation (Figure 5.20B). Second, metal superlattices should have a shorter cycle than the mean free path of the carrier electron. Third, the two kinds of electronics that differ in spin orientation (up or down) must be strikingly different in scattering on magnetic atoms.

Magnetoresistance phenomena were discovered as early as 1856, but an explanation was not available until the establishment of quantum theory. To tackle the problem of noninteger magnetic distance in ferromagnetic transition metals, Stoner proposed the band-splitting exchange model. Because of the exchange interaction, the d electronic band that contributes to the magnetic distance may split, causing the spin-up electronic energy bands to come down below the Fermi level. Therefore, the number of spin-down electrons is less than the number of spin-up electrons. The difference between the two caused the problem of noninteger magnetic distance in ferromagnetic transition metal elements (Figure 5.21). Inspired by this model, Mott proposed his famous two-fluid model. He indicated that in ferromagnetic metals, the conductive s electron is affected by the scattering from the magnetic distance of magnetic atoms (i.e., the role of local d electrons), and the scattering probability depends on the relative orientation of the spin direction of conductive electrons and the direction of the magnetic distance of magnetic atoms in solids. Based on this, a theoretical model of the GMR effect can be established. The following is a brief

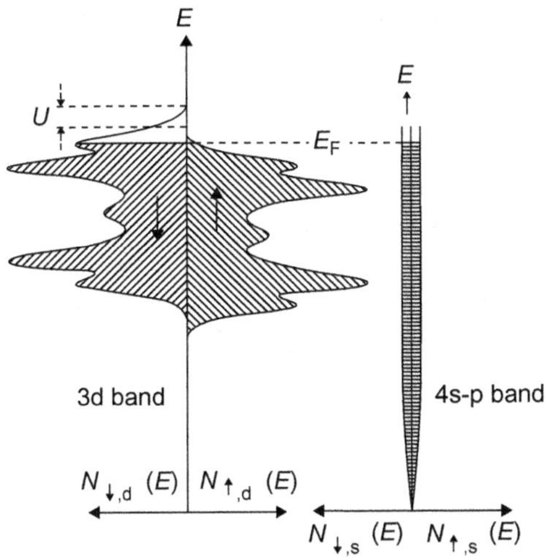

Figure 5.21 Schematic diagram of transition metal DOS.

introduction of how the two-fluid model is used to explain the GMR effect.

Magnetic metallic multilayers have nothing to do with the direction of the magnetic field in terms of the GMR effect, which only depends on the relative orientation of magnetic moments in adjacent ferromagnetic layers. The role of the external magnetic field is to change the relative orientation. This shows that electron transport is associated with electron spin scattering.

In s–d scattering associated with the spin, when the electron spin is parallel to the spin-up 3d subband (i.e., the majority in spin) in ferromagnetic metals, the average free path is longer and combined with a lower corresponding resistivity; when the electron spin is parallel to the spin-down 3d subband (i.e., antiparallel to the majority in spin) in ferromagnetic metals, the average free path is shorter and the corresponding resistivity is higher. Therefore, as antiferromagnetic coupling occurs in magnetic moments between neighboring ferromagnetic layers, electrons subject to weak scattering in a ferromagnetic layer must face strong scattering after entering into another ferromagnetic layer. As a whole, all electrons have strong scattering. If and when the magnetic moments of the adjacent ferromagnetic layer tend to be parallel under the influence of the magnetic field, the spin-up electrons will suffer weaker scattering in all the magnetospheres, equivalent to the spin-up electrons that have a short-circuit state. This is a simple interpretation based

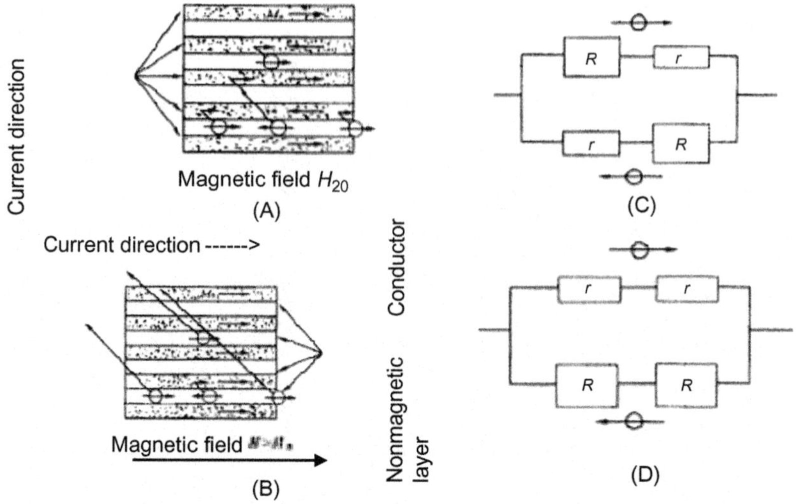

Figure 5.22 Network model of GMR effect: (A) antiparallel arrangement of the adjacent magnetic moments, (B) parallel arrangement of the adjacent magnetic moments, (C) resistor network diagram in the antiparallel arrangement of the magnetic moments, and (D) resistor network diagram in the parallel arrangement of the magnetic moments. Taken from Ref. [5].

on Mott's two-fluid model of the GMR effect. This process can be demonstrated with a simple resistor network (Figure 5.22). That is, if the magnetic moments in adjacent magnetic layers are in an antiparallel arrangement, then the resistor network is in a high-impedance state, and if a magnetic field is larger, with magnetic moments in adjacent magnetic layers in a parallel arrangement, then the resistor network would be in a low-impedance state.

5.5.4 GMR Biosensors [10–13]

The GMR effect has great potential. With external magnetic fields, the GMR effect may lead to reversal of the magnetic moment to trigger resistance mutations and, further, to the formation of a switching effect on magnetic power conversion. This makes GMR a possible candidate for computer read heads and, thus, a candidate to greatly improve the storage density of computers. Magnetoresistance and GMR read heads are designed to meet the needs of computers, multimedia, and the information superhighway for the development of high density, large capacity, and miniaturization. It is the key technology and an effective way to achieve new ultrahigh-density magnetic recordings. At present, a breakthrough has been achieved in improving the computer's storage density. GMR multilayers have been reported to be able to achieve readout heads of 22 GB/in.^{2}. GMR heads for hard disks have already been applied and 10–100 Mb of 3-in. disks will become a reality within a few years.

GMR can also be applied to high-sensitivity magnetic sensors, various types of motion sensors, and giant MRAM. In mechanical and electrical automation, automotive, and aerospace industries, GMR has also shown its wide range of application. Highlighted here is the description of GMR biosensors.

5.5.4.1 Introduction of Biosensors

Biosensors comprise a very active research and engineering field. Together with bioinformatics, biochips, biocybernetics, bionics, biology, and computer disciplines, it is a subject that combines life science and information science. Their common aims and features are: to explore and reveal the basic rules of production, storage, transmission, processing, conversion, and control of information in living systems; and to explore the application of the basic methods for human economic activity. Priorities in the research on biosensor technology include extensive application of bioactive materials and sensor integration and research and development of the transducers with recognition functions. In addition, biosensor technology is used to develop itself into the original technology in the manufacture of novel analytical instruments and analysis methods, with a particular focus on the research and development of their applications. A wide range of material objects with biological activity are applicable in biosensors, including biological macromolecules, cells, organelles, tissues, and organs, as well as synthetic molecularly imprinted polymers. Because the study of molecular DNA or protein molecule identification technology has become an independent field (biochips, i.e., DNA chips, protein chips), we do not discuss these areas in detail in this book.

Biosensor research originated in the 1960s. In 1967, Updike and Hicks made the first biological sensor by assembling the immobilized film and oxygen electrode of glucose oxidase, namely the glucose enzyme electrode. By the 1980s, the field of biosensor research had been basically formed. Since then, biosensors, including enzyme sensors, have gradually flourished. Biosensors, starting with one or more enzymes as molecular recognition elements, have been developed with the design of other biological molecules for identification components, for example enzyme–substrate, enzyme–coenzyme, antigen–antibodies, hormone–receptor, and molecules with DNA double-helix structures. One side is immobilized and may be used as a molecular recognition element to choose and measure the other. Apart from biological macromolecules, components with the recognition of certain environmental components, such as organelles, cells, tissues, and microorganisms, can also be candidates for the recognition element. Even synthetic receptor molecules and sensors can be combined to determine microorganisms, cells, and associated biological molecules.

Sensors combined with bioactive materials can be various types of physical or chemical sensors, such as electrochemical (potential measurement, conductivity measurement, impedance measurement), optical (photoluminescence, surface plasmon resonance), machinery (lever, pressure electric reaction), heat (thermal resistance), or electricity (ion or enzyme field-effect transistor). All of these combinations with biometric identification features are known as biosensors.

The 9/11 terrorist attacks in the United States were the beginning of mass terrorist activities by international terrorists. However, the aftermath of this human tragedy caused by terrorism is far from over. Biological weapons might be the most dangerous means of terrorist attacks. The danger lies not only in its immediate effect but also in the long-lasting potential threat it poses. Because there is an incubation period from committing crime involving biological weapons to the early symptoms of diseases that will have been detected, a serious pandemic disease caused during a short latent period may pose a great threat to social stability and security. Therefore, bioterrorism defense has been given the top priority in today's fight against terrorism by the international community. It is a very complex task that requires long term and arduous efforts. The need to prevent bioterrorism attacks has stimulated to some extent the research and development of biological sensors and detectors.

5.5.4.2 GMR Sensor Chip

The GMR sensor chip is a class of devices that mainly make use of magnetic nanometallic multilayer thin films with the GMR effect, which is compatible with semiconductor integrated circuits through the integration process. Since the discovery of the GMR effect, manufacturers and academia in various countries have attached great importance to it, making it quite popular in condensed matter physics.

The in-depth research and development and use of the GMR effect have opened a window to the gradual emergence of the spin-up subdiscipline. This is a novel discipline mainly aimed at research on the role of electron spin and the development of special purpose devices at the same time. The National Science Foundation claimed that the development and application of a spin subdiscipline will herald the arrival of the fourth industrial revolution.

Giant magneto-impedance materials are highly sensitive to magnetic fields (in the case of low-frequency magnetic induction and high-frequency magneto-impedance, a value up to 1,000%/Oe has been reported, and only a few Oe changes in external magnetic field are required). In addition, these materials feature high-temperature stability, small size, and low power consumption. In 1994, the United States first realized NVE's GMR effect in mass production of giant magnetoresistive magnetic field sensors. Because of its high sensitivity, thermal stability, and GMR, sensor chips can easily supersede Hall and the magnetoresistance (AMR) components, it is widely used in information, electrical, electronic power, energy management, automotive, magnetic information read–write, and industrial automation fields, among others.

5.5.4.3 GMR Biosensors

Considerable progress has been made regarding the current research on using magnetic particles as magnetic markers for detection of biological samples. Compared with other markers, such as conventional fluorescent markers, magnetic markers have many potential advantages. First, magnetic bead drops have very stable magnetic properties and can be used for repeated measurements because their magnetic properties will not be affected by chemical reagents and light. From the detection point of view, normally there is no strong magnetic background signal in samples of biological molecules. More importantly, the magnetic field is not shielded by water-based reagents or biological materials. Applied with a magnetic field, magnetic markers can achieve on-chip manipulation for molecular orientation or identification. Also, a strong magnetic field can be removed by analyzers with magnetic marker material to ensure the reuse of biochips.

GMR biosensors are commonly used in the detection of magnetic markers of biological samples. The new GMR biochip has several advantages. It is compatible with the IC process, biological information can be directly converted into electrical signals and subject to detection, and it is suitable for automated analysis. In addition, expensive high-precision optical measurement systems are not required. It can easily reduce or even eliminate background noise. It has high sensitivity and is expected to extract biological information from samples without undergoing PCR amplification. It is portable—a central laboratory is not required. Also, single-chip testing can be performed for multiflux.

GMR biosensors with magnetic markers work to identify biological molecules. First, the sample solution for testing and magnetic microspheres solution are mixed for reaction, whereby analyte molecules will complete the process of magnetic marking. Then, the marked sample solution is transported into the biomolecular recognition domain on the GMR biosensors, which are bound with a known specific receptor that is able to complete the specific combination with the corresponding target molecules. If the marked sample solution contains the target molecules that can be specifically combined with the known receptor bound on GMR biosensors, then a combination of both will remove the sample uncombined. GMR biosensors will detect the magnetic field signal produced by the microtouch balls marked on target molecules to achieve the purpose of biomolecular recognition.

Until now, there have been reports of a variety of GMR biosensors used to perform testing on biological samples with magnetic markers. Romanian scientists have also developed a GMR biosensor that can detect biological samples based on the magnetic markers. Under the effect of the outer magnetic field, magnetic particles may come close to the giant magneto-impedance microfilaments, causing changes in magneto-impedance of microfilaments. By measuring this change, we can understand the distribution of magnetic particles in solution. Because they are adhesive with objective molecules, the distribution of target molecules is available in solution. The range of application of this method can be further expanded with its sensing function in biological or medical testing of antibodies and biological specimens.

Researchers at Stanford University collaborated with the Genome Technology Center in the development of a magnetic biochip through integrated GMR of a high-density biosensor array with the standard CMOS chip. They used a "multipartition" array structure, whereby each biological sample point was consolidated with several sensor pixels to improve the sensitivity of low-concentration samples. Marked with labels to magnetic bead for biological molecules, HD-based technologies, such as GMR, can produce gene chips without using conventional fluorescence detection. This makes it possible to detect DNA hybridization at low cost. One way is to separate the DNA family to match the target DNA attached onto the chip surface of GMR sensors. Target DNA in solution, which is surrounded by the GMR sensors, hybridizes with matching DNA on the chip. Then, DNA with nanomagnetic labels is partially hybridized with the naked target DNA. As a result, magnetic labels are selectively attached to the chip surface. The GMR sensors can detect the amount of nanomagnetic tags attached, which combines biological molecules, cells, and bacteria. Unlike the complex and expensive optical detection systems, GMR biochip measurement of electrical signals comes directly from the sensors, making it a realistic candidate for a highly portable DNA detecting device.

Researchers have designed and manufactured biochips that contain 1,008 sensors within 1 mm^2. Such a device also has low noise and high-throughput readout channels for high-sensitivity DNA detection.

Another type of GMR biosensor is the so-called *bead array counter* (BARC). The development of GMR materials creates high-quality materials for highly sensitive magnetic sensors. This is the fundamental reason why BARC has boomed in recent years. However, the development of BARC is inseparable from the development of magnetic bead drops. They must be made with superparamagnetic materials—that is, the magnetic features are evident only when there is a magnetic field. The general magnetic bead droplets comprise a nucleus and a cladding layer, where the nucleus is iron oxide and the cladding layer is generally made of high polymers attached to proteins or antibodies.

As a DNA probe is based on the magnetic field rather than the optical and fluorescence features, BARC is composed of a linear magnetic field sensor array (Figure 5.23). Each microsensor is coated with a special DNA test agent that matches with a microbial genome in biological weapons. When the microbial DNA samples encounter the testing DNA probe, a single drop of the magnetic beads will be formed. When a sensor encounters a magnetic bead, the sensor's resistance will decrease. Although the decrease is very small, it can be detected. The more bead drops comprising a biosensor, the greater the resistance variation will be.

At present, biochips have found their largest application in the areas of biological research and disease diagnosis. Although they have enormous potential, there are some key factors that

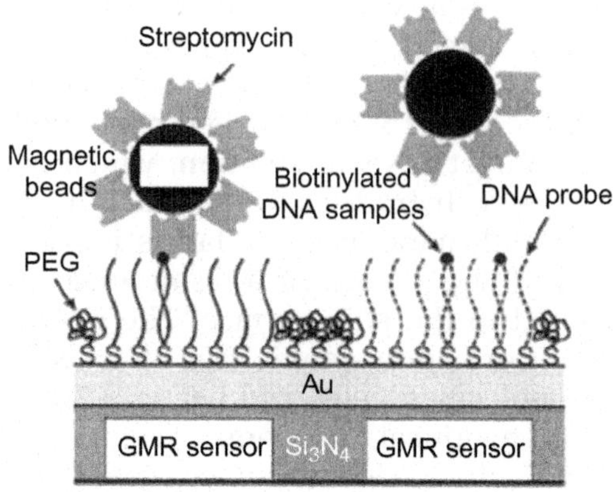

Figure 5.23 Principle of biosensors based on giant magneto-impedance field.

prevent them from large-scale application. First, biochip analysis involves too many steps that require well-trained professionals. Second, the biochip testing process depends on very expensive biochemical and optical testing equipment. Such devices are not only costly but also need to be implemented in a large central laboratory. Despite these disadvantages, this novel device, by replacing traditional fluorescent markers with magnetic nanoparticles and using GMR sensors to detect biological information, has a bright future.

References

[1] Lecture notes on nanomagnetic materials by State Key Laboratory of Magnetism under Chinese Academy of Sciences Institute of Physics. See <http://maglab.iphy.ac.cn/1-shouye.htmhttp://maglab.iphy.ac.cn/1-shouye.htm>.

[2] A.-H. Lu, E.L. Salabas, F. Schüth, Magnetic nanoparticles: synthesis, protection, functionalization, and application, Angew. Chem. Int. Ed. 46 (2007) 1222–1244.

[3] Y.W. Du, Recent progress in magnetic materials, Physics 9 (2006) 21–30.

[4] S.J. Jiang, W. Li, Magnetism in Condensed Matter Physics, first ed., Science Press, Beijing, 2003.

[5] Y.W. Du, Nano-magnetic materials and their applications, Chin. Univ. Technol. Transfer. 7 (2002) 38–41.

[6] S.Y. Chou, P.R. Krauss, L. Kong, Nanolithographically defined magnetic structures and quantum magnetic disk, J. Appl. Phys. 79 (1996) 6101–6104.

[7] C.A. Ross, I. Smith Henry, T. Savas, M. Schattenburg, M. Farhoud, M. Hwang, et al., Fabrication of patterned media for high density magnetic storage, J. Vac. Sci. Technol. B 17 (1999) 3168–3176.

[8] T. Nakamura, Snoek's limit in high-frequency permeability of polycrystalline Ni–Zn, Mg–Zn, and Ni–Zn–Cu spinel ferrites, J. Appl. Phys. 88 (2000) 348.

[9] M.B. Tian, Magnetic Materials, Tsinghua University Press, Beijing, 2001, pp. 319–343.
[10] R.L. Edelstein, C.R. Tamanaha, P.E. Sheehan, M.M. Miller, D.R. Baselt, L.J. Whitman, et al., The BARC biosensor applied to the detection of biological warfare agents, Biosens. Bioelectron. 14 (2000) 805–813.
[11] M.M. Miller, P.E. Sheehan, R.L. Edelstein, C.R. Tamanaha, L. Zhong, S. Bounnak, et al., A DNA array sensor utilizing magnetic microbeads and magnetoelectronic detection, J. Magn. Magn. Mater. 225 (2001) 138–144.
[12] F. Beaulieux, D.M. See, I. Leparc-Goffart, M. Aymard, B. Lina, Use of magnetic beads versus guanidium thiocyanate-phenol-chloroform RNA extraction followed by polymerase chain reaction for the rapid, sensitive detection of enterovirus, RNA, Institut Pasteur/Elsevier, Res. Virol. 148 (1997) 11–15.
[13] D.L. Graham, H.A. Ferreira, P.P. Freitas, Magnetoresistive-based biosensors and biochips, Trends Biotechnol. 22 (9) (2004) 455–462.

6

NANOTITANIUM OXIDE AS A PHOTOCATALYTIC MATERIAL AND ITS APPLICATION

CHAPTER OUTLINE

In recent years, the word "photocatalyst" has appeared frequently in the literature. What is it exactly? Strictly speaking, a photocatalyst is a substance that can promote chemical reactions through the conversion of light irradiation but the photocatalyst itself does not change. The promise of photocatalysis is in its ability to use solar energy to convert high-density chemical or electrical energy. Similarly, it is possible to use photoactived chemical species directly in the degradation, mineralization, and hydration of various pollutants in the air. As such, photocatalysis has great potential in environment remediation and renewal energy applications.

The most commonly used photocatalysts are TiO_2, CdS, WO_3, ZnO, ZnS, Fe_2O_3, and SnO_2. TiO_2 catalysts have been the most studied and used photocatalytic materials because of their high catalytic activity, chemical stability, low cost, and nontoxicity.

This chapter is an overview of the development of photocatalytic technology and focuses mainly on TiO_2 materials.

Nanomaterials and Devices. DOI: http://dx.doi.org/10.1016/B978-1-4557-7754-9.00006-8

6.1 Principle of TiO_2 Photocatalysis [1]

6.1.1 Development of Photocatalytic Technology

In 1972, Fujishima and Honda [2] discovered the photocatalytic decomposition of water on n-type semiconductor TiO_2 electrodes, thus initiating the study of photocatalysis. This photocatalytic effect is sometimes referred to as the Honda–Fujishima effect. In 1976, Garey removed the chlorine in polychlorinated biphenyls (PCBs) using TiO_2 photocatalyst. In the following year, Frank and Bard [3] achieved the photocatalytic oxidation of CN^- into OCN^-, symbolizing the start of the research of photocatalytic technology being applied in environmental remediation. Over the past decade, semiconductor photocatalysis technology has been developing rapidly in applied research on environmental remediation and solar fuel production.

With highly prospective applications, the current TiO_2 photocatalytic technology may cover the following main research areas:

1. Preparation of nanocatalysts for efficient wastewater treatment and air purification;
2. Applications of TiO_2 photocatalyst in coatings aiming to achieve application and industrialization in "self-cleaning" external and internal surfaces;
3. Research and application of TiO_2 photocatalyst in ceramic sanitary ware aiming to realize the industrialization of antibacterial sanitary ware;
4. TiO_2 photocatalyst applications in household appliances to achieve the application and industrialization of a self-cleaning series of appliances;
5. Use of TiO_2 photocatalyst featured with the surface superhydrophilicity and photocatalytic activity for research and development of nanophotocatalytic thin film materials with self-cleaning and fog-resistance performance.

6.1.2 Principles of Semiconductor (TiO_2) Photocatalysis [4]

The basic physical properties of titanium dioxide crystals are as shown in Table 6.1. Titanium dioxide crystals may exist in three phases, anatase, rutile, and brookite, in which TiO_2 with the anatase and rutile phases may have energy band structures as shown in Figure 6.1.

From Figure 6.1 we can see that the anatase and rutile phases share the same position of the valence band, with a

Table 6.1 Basic Properties of Titanium Dioxide Crystals

Structure of Titanium Dioxide	Relative Density	Crystal System	Lattice Constant (nm)			Band Gap (eV)
			a	*b*	*c*	
Anatase	3.84	Tetragonal	0.536		0.953	3.2
Rutile	4.26	Tetragonal	0.459		0.296	3.0
Brookite	4.17	Orthorhombic	0.915	0.544	0.514	

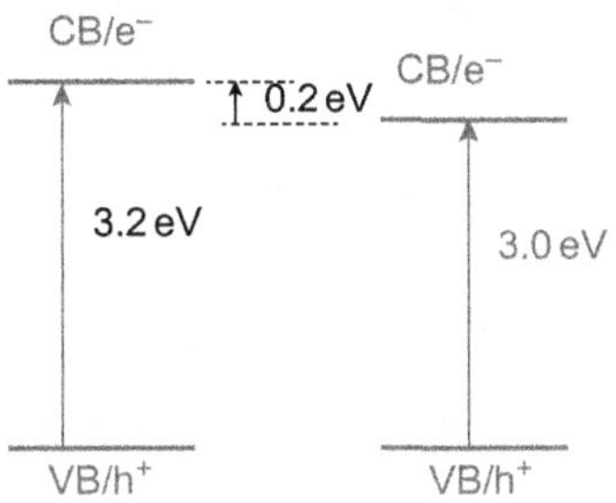

Figure 6.1 Band structures of anatase and rutile titanium dioxide.

photo-generated hole of the same oxidation capacity. However, the anatase conduction band has more negative potential and more capability in photo-electron reduction. After being irradiated with photons with a wavelength of less than 387.5 nm on the surface, the electrons in the valence band will be excited onto the conduction band, resulting in highly active photo-generated holes (h^+) and photo-generated electrons (e^-) respectively on the valence band and the conduction band. Under different conditions, h^+ and e^- will be isolated from each other and migrate to the surface of particles at different locations. Through a series of reactions, they will generate oxygen-containing small-molecule active species (O^{2-}, $\cdot OH$, H_2O_2, etc.). The TiO_2 catalytic reaction equation is as follows:

$$TiO_2 + h\nu \rightarrow e^- + h^+$$

$$h^+ + H_2O \rightarrow \cdot OH + H^+$$

$$e^- + O_2 \rightarrow \cdot O^{2-} \rightarrow HO_2\cdot$$

$$2HO_2\cdot \rightarrow O_2 \rightarrow H_2O_2$$

$$H_2O_2 + \cdot O^{2-} \rightarrow \cdot OH + OH^- + O_2$$

In the degradation of water and organic pollutants in the air, the active photocatalytic oxidation of TiO_2 has been proven to show remarkable results. Organic matter and bacteria can be decomposed and oxidized to CO_2 and H_2O_2. Because the photocatalytic reaction will generate OH radicals (i.e., $\cdot OH$), and because their energy is usually larger than the bonding energy from the CH, OH, CC, and C–Cl compositions of general organic matter, they can easily cut off the bond between molecules.

Practical application often involves the use of a mixed crystal effect based on these two-phase materials. Anatase and rutile mixed crystals may present a higher photocatalytic activity. This is because in the mixed crystal, a rutile thin layer will be formed on the surface of titanium dioxide anatase. This is a kind of coated composite structure effective in improving the separation of electron–hole pairs. The P25 variant of TiO_2 is a popular example of an effective photocatalyst with mixed anatase/rutile phases.

Photocatalyst particles can effectively enhance the quantum yield, which favors the catalytic reaction. Photocatalysts can show the following nanoeffects:

1. Quantum Effects

   When the semiconductor particle size is reduced to a particular critical value, the energy gap between the conduction band and valence band will be widened, and the light electrons and holes may have higher energies, and the oxidation–reduction capacity will increase accordingly.

2. Carrier Diffusion Effect

   The smaller the particle size, the less time required for photo-induced electrons to spread from the crystal to the surface. This reduces the chances for recombination of electrons and holes, increasing photocatalytic efficiency.

3. Effect of Surface Area Increase

   The factors of reduced particle size and increased surface area will enhance photocatalytic reactivity by exposing more reactive surfaces and allowing for more substrate adsorption.

TiO_2 has the following characteristics: high photocatalytic activity, strong absorption of ultraviolet light (due to the large band gap), strong reduction and oxidation of photo-generated electrons and holes, chemical stability (against acid and alkali and photochemical corrosion), biological nontoxicity, no absorption in the visible area, availability for the production of white or transparent films, and a rich supply of raw materials.

Nano-sized titanium dioxide TiO_2 surfaces contain a number of dangling bonds, which could constitute defect levels in the energy gap, enabling a very high surface activity of nano-TiO_2. This can have a great impact on the optical properties of nano-TiO_2.

Studies have found that nano-TiO_2 can be equally effective in photocatalytic reactions with considerable pollutants in water, such as halogenated aliphatic hydrocarbons, dyes, nitroaromatics, substituted anilines, polycyclic aromatic hydrocarbons, heterocyclic aromatic compounds, hydrocarbons, phenols, surface-active agents, and pesticides. Therefore, TiO_2 can be used for the detoxification, decolorization, and mineralization of water.

General bacteria (fungi, viruses) and odor are mainly composed of organic compounds, so the functions of deodorizers, antibacterials, and antivirals can be achieved through photocatalysis. Professor A. Fujishima and associates at the Faculty of Engineering at Tokyo University conducted some experiments and proved that nano-sized TiO_2 has a strong bactericidal effect against *Pseudomonas aeroginosa*, coliform, and *Staphylococcus aureus* and can be used in decontaminating hospital operating tables, walls, jar tiles, and so on.

Nano-TiO_2 can generate strong reducing capacity to restore heavy metal ions in water through light-generated electrons. Cr^{6+} in wastewater has strong carcinogenicity, with its toxicity being 100 times higher than that of Cr^{3+}. TiO_2 photocatalysis can restore 85% of Cr^{6+}. This is of great significance in practical applications dealing with wastewater containing Cr.

Currently, nano-TiO_2 is potentially one of the most effective optical catalysts. However, due to its wide band gap, TiO_2 can only work in the ultraviolet region. As such, there is active research to find new photocatalytic materials that are active under visible/solar light, with much of the work focused on the modification of titanium dioxide [5–7]. In the early stage of studies, the idea of getting the TiO_2 redshift to the visible light is concentrated in doping with metal ions or other metal-oxide semiconductor compounds. After 20 years of research, it was finally proved that while the modified cation is able to reduce the TiO_2 band gap, it also brings about significant reduction of the optical quantum efficiency. This is because the doped metal ions themselves induce electron–hole recombination. Researchers at Toyota Motor Corporation proposed that doping with anions can improve the photocatalytic properties of titanium dioxide in 2001. They found that by replacing oxygen with nitrogen, the light absorption capability of the obtained $TiO_{2-x}N_y$ photocatalytic materials was substantially more increased than that of pure titanium dioxide [5]. In September 2002, researchers from the United States found that carbon may be used to replace a portion of oxygen to obtain C-doped TiO_2, which also experienced a substantial increase of its optical absorption in the visible region [6]. Although these studies have shown some increase in the absorption of TiO_2 in the visible region with good photocatalytic efficiency, such methods are not suitable for practical purposes because the injected modifier N or C is unstable and easily breaks down under the illumination. So far, researchers have focused on the design of narrow-band TiO_2 while improving its catalytic capacity [7].

6.2 Preparation of TiO_2 Materials

TiO_2 powder can be prepared through physical or chemical methods. The physical method includes the building method (e.g., gas condensate) and the crushing method (high-energy ball milling method). Chemical preparation of TiO_2 materials and the comparison effects are shown in Table 6.2.

TiO_2 thin films can also be prepared in different ways, as shown in Table 6.3.

Table 6.2 Comparison of Titanium Dioxide Nanomaterials Prepared Using Chemical Methods

Method	Precursor	Features
Precipitation	$Ti(OBu)_4$ $TiCl_4$	Small size, evenly dispersed
Hydrolysis	$TiCl_4$	By control of reaction conditions, mixed crystalline and amorphous porous titania are available
Spray pyrolysis	$TiCl_4$	Electrolytes may affect the morphology and the size of aggregates
Sol–gel method	$Ti(OBu)_4$	Doping can be easily achieved, but the particle size is widely distributed
Oxidation–reduction	$Ti + H_2O_2$	The mixed crystal with anatase and rutile can be obtained by calcining at 200°C
Hydrothermal method	$TiCl_4$	Titanium dioxide in different crystalline phases can be prepared with a good degree of crystallinity

Table 6.3 Comparison of Preparation Methods of TiO_2 Thin Films

Method	Precursor	Features
Liquid deposition	Ammonium titanium fluoride	Transparent, uniform particle distribution
Sol–gel method	Ti $(OBu)_4$	Coating thickness is closely related to the number of coatings, plus dense thin film, small holes, particle of uniformity
Chemical vapor deposition	Titanium isopropoxide	All in anatase, without the rutile phase structure
Pyrolysis	Titanium isopropoxide	Film thickness can be effectively increased under control, with smooth film surface and uniform particles
Magnetron sputtering	Pure Ti	DC magnetron sputtering, with thickness of approximately 200 nm; high quality, high density, good combination, and strength

TiO_2 materials can also be prepared by modification of titanium dioxide and porous titanium in mesoporous materials:

1. Assembly of titanium oxide in zeolite (molecular sieves)

   TiO_2 is a polar molecule with surface polarization. This is not conducive to contact with nonpolar organic molecules. Vector may cause an increase of the catalytic activity of TiO_2 due to its acidic surface or its participation in the intermediate process of the catalytic reaction. It is prepared by impregnation.

2. Pore surface modification of mesoporous titanium oxide materials

   Mesoporous molecular sieves have a high specific surface area (>200 m^2/g), in which the surface area in contact with titanium dioxide and the substrate molecule can be increased to improve its catalytic activity, *in situ* synthesis, or secondary synthesis.

3. Preparation of porous titania

   Because the aperture is in the mesoporous range, substrate and resultants may rapidly proliferate in heterogeneous catalysis; titanium oxide nanocrystals in the powder can constitute micron-level aggregates with mesoporous holes inside, which not only have a higher activity but also are easy to filter and are quite significant for recovery and regeneration of photocatalysts. With high hydrothermal stability and thermal stability, its specific application can achieve further modification of its surface. Sol–gel synthesis preparation comprises two steps. The first is to prepare Ti–Si mixed oxides with controlled structure, and the second is to dissolve the silica contained within.

Light absorption of TiO_2 is limited to the ultraviolet range, limiting its use of sunlight. In addition, it is easy to recombine the photo-carrier to bring about some impact on photocatalytic efficiency. To address these problems, some methods such as surface modification can be used to prepare composite TiO_2 materials such as photocatalytic materials. In recent years, studies have shown that TiO_2 optical absorption can benefit by the introduction of impurities or defects from deposition of precious metals, doping transition metal ions, organic dye photosensitization, as well as compound semiconductors and other methods. Meanwhile, quantum efficiency and photocatalytic reaction rate can also be increased accordingly [8].

1. Compound Semiconductors

   Compound semiconductors can effectively improve the charge separation efficiency and extend the spectral response range. The semiconductor compound methods for the modification of TiO_2 include a simple combination of doping and multilayer structure, and the combination of different phases, as well as ion implantation and other methods. In recent years, $CdS–TiO_2$, $SnO_2–TiO_2$, $WO_3–TiO_2$, $ZrO_2–TiO_2$, and $V_2O_5–TiO_2$ have been reported to have shown photocatalytic properties that are superior to that of a single semiconductor [9,10] (Figure 6.2).

2. Metal Deposition

   Pt, Pd, Au, Ag, and Ru are all commonly used inert metals, of which the most commonly used is Pt. The deposition of a layer of precious metals on the TiO_2 surface is like

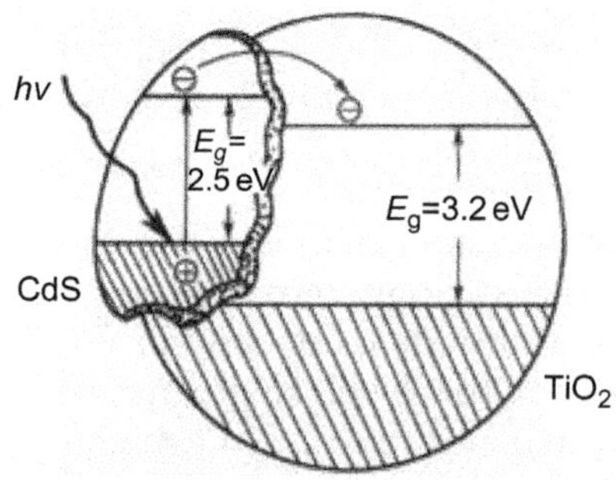

Figure 6.2 Optical excitation in compound semiconductor $CdS–TiO_2$ as photocatalyst.

building a short-circuit micro-cell with TiO_2 and inert metal micro-electrodes, so that photo-generated electrons and photo-generated holes can be effectively separated to improve the photocatalytic properties of the catalyst. Studies show that the surface of TiO_2 thin films deposited with silver, platinum, gold, palladium, ruthenium, and other metals can effectively improve its photocatalytic properties.

3. Ion Modification

The introduction of an appropriate number of metal ions can inhibit the recombination of photo-electronics and photo-holes to improve photocatalytic efficiency. Because metal ions can serve as an effective acceptor of electrons, they can capture the electrons in the conduction band. Competition of metal ions on electrons may reduce the raw carrier recombination, thereby enhancing the activity of the catalyst. Transition metal elements exist with multiple valencies. A small amount of transition metal ions (e.g., Fe^{3+}, Cu^{2+}), when doped into TiO_2, can result in the change of TiO_2 crystal lattice to introduce the lattice defects. Furthermore, this will form shallow potential wells to capture the photo-generated electron–hole pairs and lower the recombination risk of cavities and electrons to extend the recombination time, thereby enhancing the photocatalytic activity of TiO_2. For the improvement of the quantum efficiency of TiO_2 and the expansion of the light response range, a variety of metal ions can be chosen to conduct single-doping or multi-doping processes. Moreover, changing the concentration of the doped ions is combined with the use of different doping methods in an attempt to obtain the best possible doping effect.

6.3 Application of TiO_2 as Photocatalytic Material

TiO_2 photocatalytic material is applicable in many areas, such as sewage treatment devices, air purifiers, anti-fog and self-cleaning coatings, antibacterial materials, and photocatalytic decomposition of water [11]. Figure 6.3 illustrates the basic principles for photocatalytic purification.

1. Sewage Treatment: Solar Reactor for Wastewater Treatment

Traditional water treatment methods are constrained by limitations like low efficiency and high cost, in addition to secondary pollution and other problems. The development and application of nanotechnology may solve this problem.

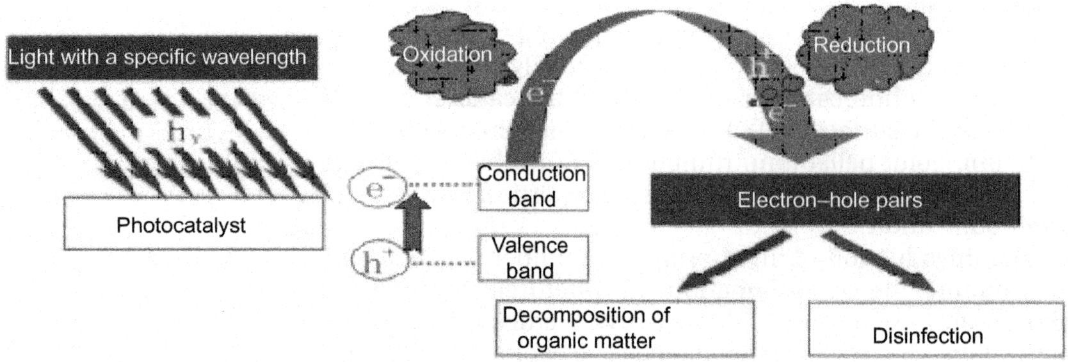

Figure 6.3 The basic principle of photocatalytic purification.

In addition to the aforementioned use of TiO_2 photocatalytic reduction in dealing with wastewater containing Cr^{6+}, solar reactors for wastewater treatment are also considered an important element.

Solar reactors use sunlight as the radiation source to conduct the photocatalytic reaction, which can degrade the organic pollutants in wastewater to reduce water pollution levels. Table 6.4 lists the various types of solar photoreactors with respect to their characteristics. The main organic photocatalytic degradation reactions are shown in Table 6.5.

2. Air Purifiers

At room temperature, photocatalytic oxidation technology can be used to remove air pollutants with the help of air, water vapor, and oxygen. TiO_2 as an air purification material is effective in degrading indoor and outdoor organic pollutants, and its oxidation functions can remove chlorine oxide, sulfide, and a variety of middle-atmospheric ozone. Also, TiO_2 is effective in degrading indoor harmful gases, such as formaldehyde released from decorative materials, as well as methyl mercaptan, hydrogen sulfide, and ammonia arising from the environment. Figure 6.4 shows the basic purification process of a photocatalytic air purifier.

3. Applications in Anti-Fogging and Self-Cleaning Coatings

Under normal circumstances, the TiO_2-coated surface has a larger contact angle with water and oily liquids such as ethylene glycol, *n*-hexadecane, and triolein. However, after UV irradiation, the water contact angle may decrease to less than 5°, or even down to 0°. This means they can fully seep into the surface of TiO_2. Therefore, the surface of TiO_2 shows affinity to both water and oil, and is a so-called superamphiphilic interface

Table 6.4 Solar Reactors Used for Wastewater Treatment

Nature of the Sun's Rays	Photocatalyst Status	
	Suspended	*Fixed*
Enhanced	Parabolic trough photoreactor Drop-down film photoreactor	Parabolic trough photoreactor
Nonenhanced, with a reflective device	Compound parabolic concentration photoreactor	Tubular photoreactor Optical fiber photoreactor
Nonenhanced, without a reflective device	Solar pool Tubular photoreactor Plate photoreactor Drip-style flat photoreactor Double-shell flat-plate photoreactor Drop-down film photoreactor	Plate photoreactor Drip-style flat photoreactor Double-shell flat-plate photoreactor

Table 6.5 Photocatalytic Degradation Reactions of the Major Organics

Organic	Catalyst	Light	Photolysis Products
Hydrocarbons	TiO_2	UV	CO_2, H_2O
Halohydrocarbons	TiO_2	UV	HCl, CO_2, H_2O
RCOOH	TiO_2	UV light, xenon lamp	CO, H_2, alkanes, alcohols, ketones, acid
Surfactants	TiO_2	Fluorescent	CO_2, SO_3^{2-}
Dyes	TiO_2	UV	CO_2, H_2O, inorganic ions, intermediates
Oxygenated organics	TiO_2	UV	CO_3^{2-}, NO_3^{2-}, NH_4^+, PO_4^{5-}, F^-, etc.
Organophosphorus pesticides	TiO_2	UV light, sunlight	Cl^-, PO_4^{5-}, CO_2

material. Therefore, bathroom mirrors, auto glass, and mirror surfaces, once coated with a layer of titanium oxide thin films, will have an anti-fogging effect. Because titanium oxide film may generate strong oxidation capacity and superhydrophilicity in sunlight, window glass, outer walls of buildings, highway guardrails, street lamps, and other surfaces coated with it will also have a self-cleaning effect (Figure 6.5).

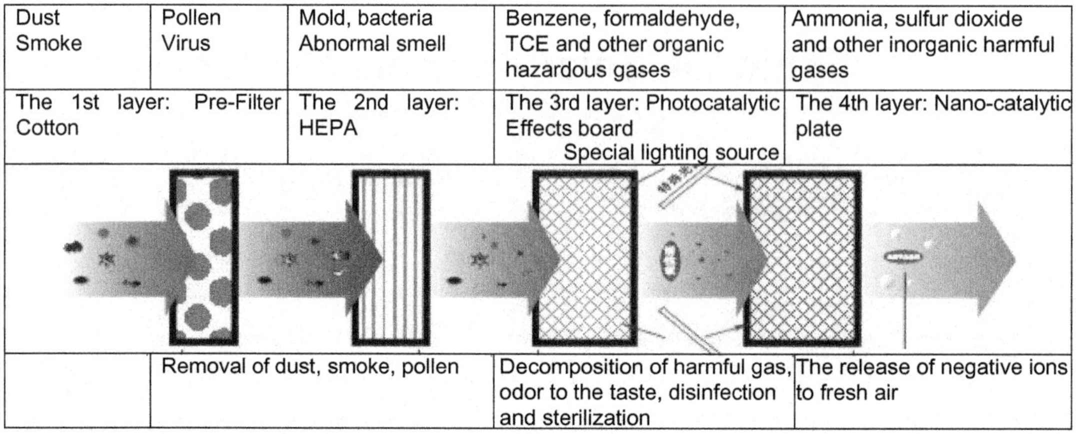

Figure 6.4 The basic purification process of a photocatalytic air purifier.

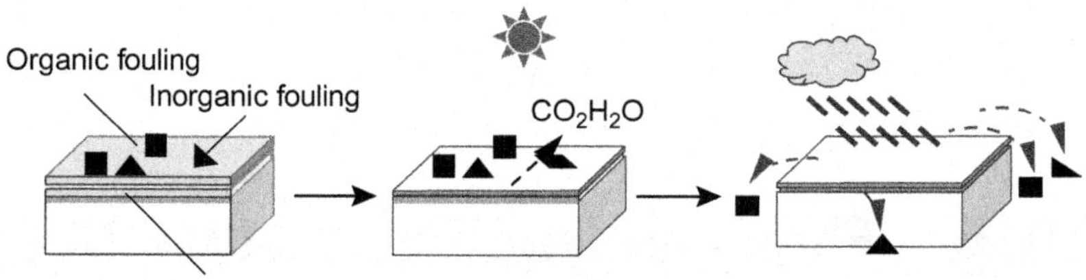

Figure 6.5 Diagram of the use of titanium dioxide thin films to achieve surface self-cleaning.

4. Antibacterial Materials

Active superoxide ion radicals and hydroxyl radicals are able to penetrate the bacterial cell wall to enter bacteria by degrading cellulose. In this way, they may prevent the transmission of film-forming material through blocking their respiratory system and electronic transmission system, thus killing bacteria effectively. The present study of antibacterial materials includes TiO_2 photocatalytic effects on bacteria, viruses, fungi, algae, and cancer cells. For example, there are antibacterial ceramic tiles and sanitary ceramics coated with nano-TiO_2 films that are used in hospitals, food processing sites, and other sites; the walls of some hospitals and public places are coated with antibacterial materials containing TiO_2 photocatalyst to inhibit the spread of germs. Some other applications include dentures containing nano-oxide titanium material, new types of dental bleaching agents containing nanometer titanium dioxide powder, and antibacterial fiber containing nano-titania powder.

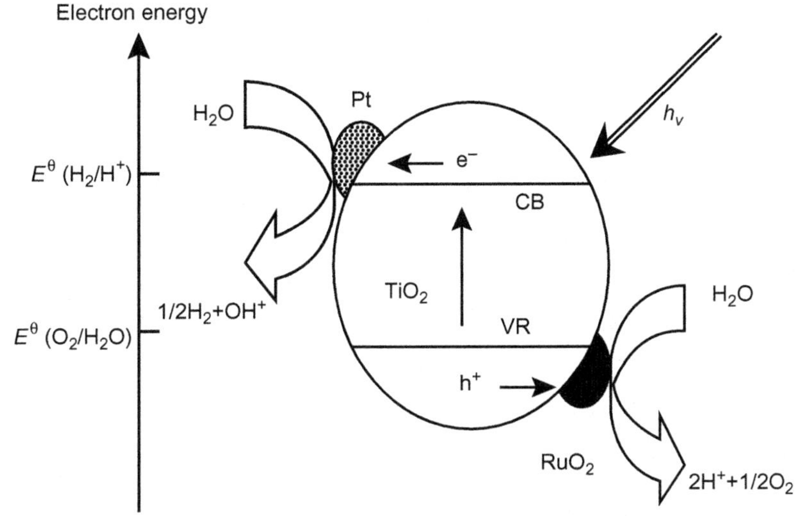

Figure 6.6 Reaction mechanism of titanium dioxide for water photolysis.

5. Photocatalytic Decomposition of Water

Titanium dioxide can be used for photocatalytic decomposition of water to produce hydrogen and oxygen to provide clean, efficient, and safe energy (Figure 6.6).

References

[1] L. Gao, S. Zheng, Q.H. Zhang, Titanium Dioxide Photocatalytic Nano-Materials and Their Applications, Chemical Industry Press, Beijing, 2002.

[2] A. Fujishima, K. Honda, Electrochemical photolysis of water at a semiconductor electrode, Nature 238 (1972) 37–38.

[3] S.N. Frank, A.J. Bard, Heterogeneous photocatalytic oxidation of cyanide and sulfite in aqueous solutions at semiconductor powders, J. Phys. Chem. 81 (1977) 1484–1488.

[4] P.W. Shen, Y.X. Che, Inorganic Chemistry Series (Vol. VIII, Sub-Family of Titanium), Science Press, Beijing, 1998.

[5] R. Asahi, T. Morikawa, T. Ohwaki, K. Aoki, Y. Taga, Visible-light photocatalysis in nitrogen-doped titanium oxides, Science 293 (2001) 269–271.

[6] S.U.M. Khan, M. Al-Shahry, W.B. Ingler Jr., Efficient photochemical water splitting by a chemically modified n-TiO_2, Science 297 (2002) 2243–2245.

[7] Y. Gai, J. Li, S.-S. Li, J.-B. Xia, S.-H. Wei, Design of narrow-gap TiO_2: a passivated codoping approach for enhanced photoelectrochecmical activity, Phys. Rev. Lett. 102 (2009) 036402.

[8] J.T. Zhou, H. Wang, B.B. Huang, X.H. Xu, W.F. Yao, Y. Zhang, et al., The latest research advances in photocatalytic materials, Funct. Mater. 35 (Suppl) (2004) 1964–1969.

[9] S.S. Srinivasan, J. Wade, E.K. Stefanakos, Visible light photocatalysis via CdS/TiO_2 nanocomposite materials, J. Nanomater. 2006 (2006) 7 [Article ID 87326].

[10] J.C. Yu, L. Wu, J. Lin, P. Li, Q. Li, Microemulsion-mediated solvothermal synthesis of nanosized CdS-sensitized TiO_2 crystalline photocatalyst, Chem. Commun. 9 (13) (2003) 1552–1553.

[11] M.R. Hoffmann, S.T. Martin, W. Choi, D.W. Bahnemann, Environmental applications of semiconductor photocatalysis, Chem. Rev. 95 (1) (1995) 69–96.

7

ELECTRO-OPTICAL AND PIEZOELECTRIC APPLICATIONS OF ZINC OXIDE

CHAPTER OUTLINE

7.1 Optoelectronic Applications

7.1.1 Optical Properties of Zinc Oxide [1]

Apart from III–V nitrides and II–VIII–VI selenium compounds, the unique nature of zinc oxide (ZnO) makes it a much anticipated material for the next generation of wide band-gap semiconductor laser devices. Although the methods for preparing ZnO of higher quality have yet to be further explored, ZnO has shown great superiority over other materials. Zinc oxide is

Nanomaterials and Devices. DOI: http://dx.doi.org/10.1016/B978-1-4557-7754-9.00007-X

considered to be an important direct wide band-gap semiconductor material. At room temperature, it can have a band-gap width of 3.37 eV and exciton binding energy of up to 60 meV, with rotation properties from the blue to ultraviolet bands. ZnO has high chemical and thermal stability and is not easily oxidized in the atmosphere. Compared with III–V nitrides and II–VIII–VI selenium compounds, it has unmatched material stability.

Because the ZnO grain size goes down to the nanoscale, significant changes will occur on its surface electronic structure and crystal structure, and thus it will show nanoscale size effect, surface effect, dielectric confinement effect, and quantum size effect that macro-block materials do not have. Meanwhile, it may have characteristics unparalleled in normal ZnO materials. Many of its special features can be applied in a variety of areas, including chemistry, electronics, biology, and medicine.

Zinc oxide-based semiconductor optoelectronic devices, including UV detectors, light-emitting diodes (LEDs), and semiconductor laser diodes (LDs), are widely used in optical communication networks, optical display and storage, conversion, and optical detection. In the field of optoelectronic displays, for example, the technology for flat panel displays is a basic requirement for the terminal display in the information age. Currently, it is available with key technologies, including liquid crystal display, cathode ray tube, plasma display panel, vacuum fluorescent display tube, and LED. The LED for flat panel displays exhibits the advantage of low drive voltage. Combined with IC design, it is secure in use, featuring a multicolor display, long life, and quick response. As an ideal flat panel display device, it has a wide range of potential applications for scientific instruments, portable computers, and aerospace and military fields. Commercialized red and green LED monochrome display devices have been available, yet color displays are not commercialized. The main reason is that the brightness and color purity of blue electroluminescence is lower than required, making it unable to display true colors with three primary colors, whereas white electroluminescent materials are mostly in the broadband radiation, which cannot satisfy the practical requirements because of lower effective radiation intensity of three primary colors from the filter chips. This is known as the blue electroluminescence problem in the field of electroluminescence. The development of the zinc oxide-based LED makes it possible to produce full-color LED displays, launching a blue electroluminescence boom worldwide.

The key technology for ZnO-based optoelectronic devices is the preparation of high-quality p-type ZnO thin films. Intrinsic ZnO is an n-type semiconductor, which must be doped with the acceptor before p-type transformation is achieved. However, there are more donor intrinsic defects in zinc oxide; they may produce a high degree of self-compensation effects on acceptor doping. Also, acceptor impurities in solid solution are very low; it is difficult to achieve p-type transformation, resulting in failure in making the core of semiconductor devices with a zinc oxide p–n junction structure. This greatly limits the development and application of zinc oxide-based optoelectronic devices. At present, p-type ZnO research has become an international focus, and a series of breakthroughs have been made accordingly. A ZnO homogeneous p–n junction with a two-tier structure of p-ZnO/n-ZnO has been created. In addition, some proposals like the donor–acceptor co-doping method that can be used to reduce acceptor ionization energy, resulting in the formation of shallow acceptors in ZnO, have been proven to be feasible [2]. These results have great significance for new attempts at making zinc oxide short-wavelength light-emitting devices, performing in-depth studies, and applying the doped ZnO thin-film crystal growth mechanism to broaden the application areas of zinc oxide thin films.

In recent years, blue–green lasers have become popular in semiconductor laser research. As early as 1991, the US-based 3M Company achieved the first II–VI ZnSe-based quantum-well laser (77 K), which was breaking news in the field of science. After that, the joint study group of Brown at Purdue University in the United States, Philips in the Netherlands, and Sony Corporation in Japan made lasers that could work continuously at room temperature. By 1996, Sony Corporation showed its II–VI blue–green laser, with ZnSe/ZnCdSe as quantum wells and ZnMgSSe as the restriction layer, had a continuous working life at room temperature up to 100 h. However, the development of ZnSe-based lasers has not been as fast as many had expected. A full 5 years were spent improving the laser's working life from 1 to 100 h, but further enhancements might lead to even greater difficulties. The main reason lies in the fact that ZnSe crystals have strong ionic and weak covalent characteristics. It is very "soft" and easily damaged. Under stimulated emission, it can easily have a large number of defects in proliferation due to an increase in temperature. It generated many doubts among researchers, who wondered if ZnSe, with its inherent shortcomings, is suitable for making a blue–green laser. In particular, after 1996, with the emergence of GaN blue LEDs and LDs, the ZnSe material series was facing bleaker prospects. Since the success of GaN lasers in continuous operation at room temperature, negative views of the blue–green laser in the ZnSe system have prevailed.

Are the II–VI semiconductors in dire straits regarding creating a blue–green laser? Some research groups still continue to improve the quality of ZnSe epitaxial material (such as the use of a ZnSe single crystal as the substrate, reducing the misfit dislocation of the ZnSe/GaAs interface) and increase the concentration of p-type doping and perfection of the ohmic contacts of the electrode. In addition, there are still two ways for new exploration. One is the use of the beryllium compound ZnMgBeSe. However, many people might hold back considering its highly toxic nature. The second way is to use ZnO materials.

Currently, zinc oxide at room temperature can be observed to have stimulated emission by optical pumping. Yet the observation of stimulated emission at low temperatures by using electron beams in ZnO bulk materials has been seen for many years. However, because of the fact that stimulated emission intensity is rapidly reduced with an increase in temperature, ZnO optoelectronic material has long been neglected by researchers. Because the GaN system has provoked more attention in recent years, ZnO material has been used as a buffer layer between the GaN epitaxial layer and the SiC substrate. The reasons are attributed to the following basic properties. First, ZnO shares the same wurtzite-type lattice structure as that of GaN. Its a-axis lattice constant is 0.325 nm, which is less than 2% of the lattice mismatch with GaN, and its c-axis lattice constant is 0.527 nm. Second, ZnO has an n-type doping concentration up to $10^{19}/cm^3$ and mobility of $260\ cm^2/Vs$, showing excellent electrical properties. Third, according to the common anion rule of electron affinity and energy band offsets, ZnO has a conduction band minimum that is lower by 0.7 and 0.4 eV respectively than those of GaN and SiC. Therefore, its use as a buffer layer between GaN and SiC will not cause a barrier to block electron motion. In the usual GaN/AlN/SiC structure, the buffer layer of AlN has a conduction band offset of 2.1 and 2.4 eV from GaN and SiC respectively, resulting in the buffer layer AlN becoming a potentially high electronic barrier. Finally, compared with the nitride semiconductor, ZnO material is quite soft and has a relatively small shear modulus. Its use as a buffer layer can prevent the dislocation caused by the lattice mismatch extending to the GaN active layer.

7.1.2 Epitaxial Growth of ZnO

Because ZnO bulk material tends to have some flaws, stimulated emission can only be achieved under low-temperature conditions. Therefore, it is not appropriate to use it for making different devices. The use of epitaxial thin films or a low-dimensional quantum structure is thus necessary to improve the crystal quality and the bound exciton energy. So far, high-quality growth of ZnO is accomplished by the use of molecular beam epitaxy (MBE) technology, with (0001) Al_2O_3 as the substrate. Although ZnO and Al_2O_3 may have a lattice mismatch of up to 16%, strain release could lead to a three-dimensional growth of ZnO quantum dots. In the process of growing ZnO in MBE, one of the greatest problems is the O_2 that may find a way into the growth chamber. The following is a brief overview of two kinds of technologies for MBE growth of ZnO.

7.1.2.1 MBE Technique with Microwave

Because the binding energy of oxygen molecules is as high as 5.16 eV, the traditional methods for material growth, for example MBE and chemical vapor deposition, are unlikely to decompose oxygen molecules into atoms. Here, one of the simplest ways is to decompose oxygen molecules into atoms before they react with Zn and, in this way, the flow of O_2 will also be reduced. A research team in Japan at the Tohoku Institute of Metallic Materials made a clever use of microwave methods that allowed oxygen molecules to be ionized into the plasma before passing into the growth chamber, where the microwave source power was 120 W, gas flow was controlled by a drain valve, the selected substrate was (0001) sapphire, and growth temperature was maintained at 500–550°C.

7.1.2.2 L-MBE Growth Technique

Another growth technique is the high-power laser ablation of the ZnO target with a purity of 99.999%, so that ZnO can be sputtered onto (0001) sapphire substrates at a temperature of 500°C. In this growth process, ZnO decomposition is inevitable under high-power laser and high local temperature. To ensure the quality of crystals, the growth chamber requires an oxygen-rich environment. In general, we can use the beam pressure of oxygen to increase the level of oxygen.

These two methods for epitaxial ZnO on sapphire are executed by using the Stranski–Krastanov growth mode. The whole

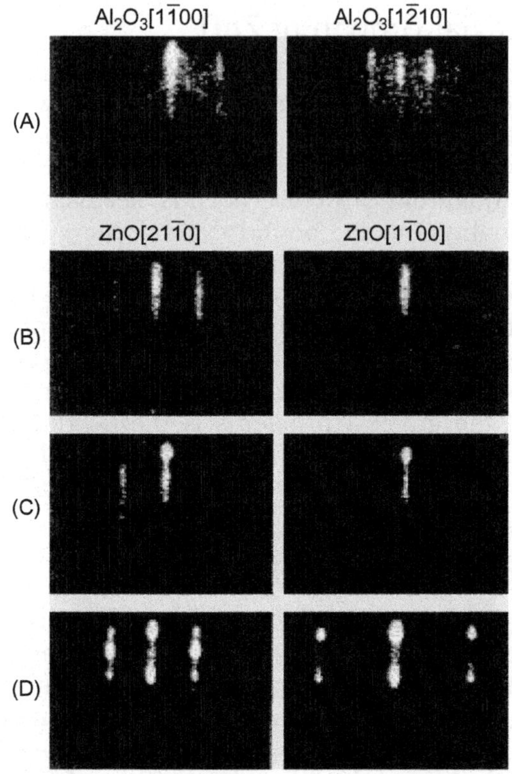

Figure 7.1 RHEED patterns of ZnO epitaxial layer: (A) (0001) sapphire substrates after annealing treatment; (B) the growth after 10 min; (C) the growth after 20 min; (D) growth after 1 h (growth rate: 1.5 nm/min). Taken from Ref. [3].

process is monitored by high-energy electron diffraction. In the initial stage of epitaxial ZnO, the thin stripe-like RHEED pattern shows that the growth is in a two-dimensional layer. As the ZnO thickness increases, strain energy in the epitaxial layer accumulates to a certain degree and will be released by way of dislocation, thus forming a three-dimensional growth nucleus. At this point, some dots in the RHEED pattern will be distributed on the thin line as shown in Figure 7.1. Epitaxial ZnO may have the $[21\bar{1}0]$ and $[1\bar{1}00]$ crystal orientations that are, respectively, parallel to the $[1\bar{1}00]$ and $[1\bar{2}10]$ of Al_2O_3. Because ZnO thickness is 55 nm, images under the atomic force microscope show the regularly ordered microcrystalline islets formed in three-dimensional growth with a diameter of approximately 50 nm and fluctuation of 10%. Compared with the bulk material, these two methods can reduce the concentration of impurities and defects so that stimulated emission may occur at room temperature. It should be made clear that in the growth of InAs on GaAs substrates, the lattice constants of both have a difference of 7%. Because the InAs layer thickness is approximately 0.5 nm, the formation of islets starts. The mismatch between the ZnO and Al_2O_3 may be 16%, whereas the critical thickness of the formed islets is 55 nm. This might be determined by three factors: ZnO surface energy, Al_2O_3 substrate surface energy, and the interfacial energy between the two.

7.1.3 Optical Properties of ZnO Quantum Dots

With the ZnO band gap of 3.37 eV, stimulated emission may be achieved at low temperatures. Nonetheless, such stimulated emission has never been observed at room temperature, and thus it is of little scientific interest to the research community. By the time ZnO quantum dots were successfully grown, the optically pumped stimulated emission observed at room temperature may achieve an exciton binding energy of 60 meV. Such high energy at low temperatures makes it possible to significantly reduce the lasing threshold. The following is a brief description of optically pumped stimulated emission of ZnO quantum dots.

With the cwHe–Cd (355 nm, 15 ps) laser as the excitation light source, a different composite process can be observed at different excitation intensity, as shown in Figure 7.2.

The free exciton peak A in Figure 7.2 corresponds to the stimulated emission peak E_{ex} at low-intensity excitation. In the case of higher excitation intensity, the free exciton peak may be gradually replaced by a new P_2 peak of approximately 70 meV, which is lower than E_{ex}. This is a spontaneous emission peak in the collision process of two excitons, the intensity of which presents a linear increase as the optical pump intensity increases. In the collision process of the two excitons, one of the excitons will absorb energy from another exciton and complete the transition to the high-energy state P_n. The exciton will release a photon in the transition from an unstable high-energy state back to the ground state.

$$P_n = E_{ex} - E_b^{ex}\left(1 - \frac{1}{n^2}\right) - \frac{3}{2}kT(n = 2, 3, 4, \ldots, \infty) \qquad (7.1)$$

Here, $E_b^{ex} \approx 60$ meV is the binding energy of the ZnO exciton, k is the Boltzmann constant, and T is the temperature. When the optical pump intensity reaches the threshold $I_{th}^{ex} \approx 24\ \text{kW/cm}^2$, the P_2 peak is gradually replaced by the new stimulated emission peak P, with the optical pump intensity

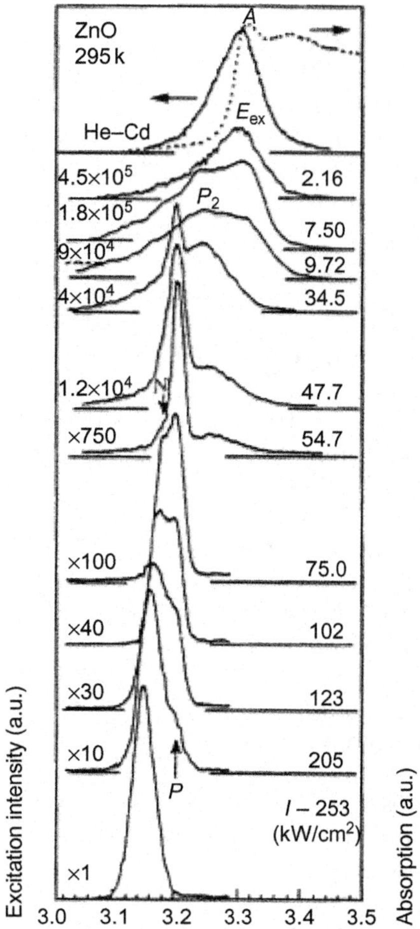

Figure 7.2 Stimulated emission peak at different excitation intensities. Taken from Ref. [3].

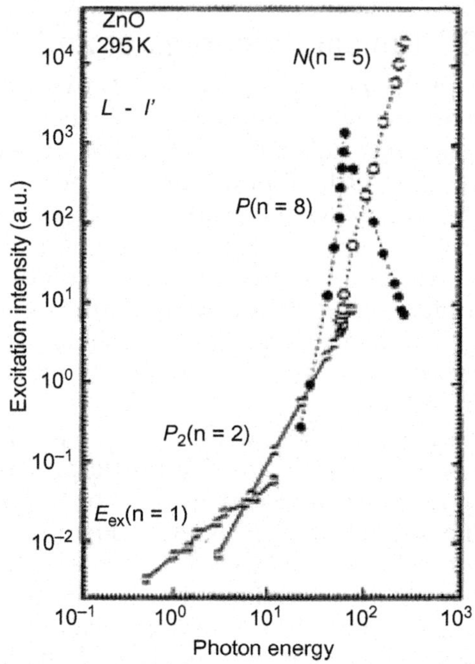

Figure 7.3 Peak intensity of the stimulated emission on changing the optical pump intensity. Taken from Ref. [3].

increased further to the threshold $I_{th}^{EHP} = 50\ \mathrm{kW/cm^2}$, and a new stimulated emission peak N will appear. For the relationship between stimulated emission intensity L and optical pump intensity I, see Figure 7.3. As the excitation intensity exceeds the threshold I_{th}^{EHP}, the peak P intensity may decrease with the increase of excitation intensity. This is because the excitation light source is too strong, thus resulting in high concentrations of plasma.

7.1.4 Controlled Synthesis of the Ordered ZnO Nanowire Arrays

As one of the one-dimensional ZnO nanomaterials, ZnO nanowire plays a key role in the development of devices using ZnO-based nanomaterials. The process of producing ZnO nanowires is a key issue worldwide. To date, scientists have invented many new techniques and new methods for preparing ZnO nanowires. The growth mechanism of ZnO nanowires can be categorized in the following ways: (1) vapor–liquid–solid (VLS) growth method, (2) vapor–solid (VS) growth method, and (3) hydrothermal method. These methods can be classified as follows.

7.1.4.1 VLS Growth

In the 1960s, R. S. Wagner and associates first proposed the VLS growth method in their research on the growth process of silicon whiskers. This is the traditional method for the growth of one-dimensional material. It is based on the catalyst of liquid metal clusters as the active sites of gas-phase reactant. The source material for preparation of the one-dimensional material is heated into vapor form. After the vapor is spread onto the surface of the catalyst to form saturated clusters, one-dimensional nanostructures will grow and take shape on the catalyst surface. The VLS method has been widely used for the preparation of nanowires of various inorganic materials, including ZnO nanowires. In the thermal evaporation VLS growth method for preparation of one-dimensional zinc oxide nanomaterials, catalysts are often made from copper or gold, or nanoparticles or nanofilms prepared with other transition metals.

Based on the catalytic VLS growth mechanism, ZnO nanowires can be synthesized using a simple chemical vapor transport and condensation method. Here, the catalyst plays a vital role in the growth of ZnO nanowires. By controlling the starting position and size of the catalyst, control of the location and diameter, respectively, of ZnO nanowires can be achieved, whereas the density of nanowires can be controlled by the density of catalyst particles. However, using the current catalyst VLS method, we cannot control the size and density distribution; therefore, we cannot control the diameter, location, and density distribution of ZnO nanowires.

7.1.4.2 VS Growth

The VS method includes one or several reactants heated to form vapor at high temperatures. The vapor in an inert gas flow is transported to the low-temperature zone in the reactor or vapor is rapidly cooled to achieve growth in one-dimensional structural materials. In 1997, Yang and colleagues used the improved VS method and successfully prepared an MgO nanorod array with a diameter of 7–40 nm and height of 1–3 μm. The preparation of one-dimensional zinc oxide nanomaterials using the VS growth mechanism is also very common. Lee and colleagues, by controlling the atmosphere, prepared ZnO nanowires by way of thermal evaporation, where zinc oxide powder was used as raw material with no catalysts and thermal evaporation temperature was 1,380°C. Nanowire growth substrate was placed at the export end of an alumina tube at a temperature of approximately 450°C. Xu and associates prepared ZnO nanowires and nanowhiskers with calcined ZnC_2O_4 nanoparticles and powders. Surface-active agent was used in the calcination process. In a tube-type furnace, the quartz boat had a temperature of 920°C and access to ammonia. The growth of ZnO nanowires is also interpreted as the VS mechanism in this preparation method.

The VS method has the advantage of requiring no catalyst; however, considering the complexity of the equipment for the synthetic process, the higher reaction temperature, very limited choice of substrate, the expense, and difficulty in controlling the size and shape of one-dimensional material, it is greatly limited in applications in the manufacture of devices.

7.1.4.3 The Hydrothermal Method

The hydrothermal method is in the liquid-phase synthesis method category.

The liquid-phase synthesis method, compared with VLS and VS, which require high temperatures (800–1,400°C), is a low-cost, low-temperature synthetic method. The hydrothermal method was first used for synthesis of ZnO nanorods. Zn nitrate solution ($Zn(NO_3)_2 \cdot 4H_2O$), mixed with hexamethylene tetramine, is placed inside a hydrothermal reactor at 95°C and heated for 1–10 h. From sediment on the F–SnO_2 glass after cleaning and air-drying, a large number of aligned ZnO nanorods can be obtained; however, the size of the product is a bit too large. There are improved methods, such as the two-step hydrothermal method. The two-step hydrothermal method is as follows. First, ZnO grains with a diameter of 5–10 nm are evenly scattered on an Si (100) chip as a crystal seed. When the silicon chip is uniformly covered, ZnO nanorods can be synthesized at 90°C by shaking the chip suspension for 0.5–6 h in an open crystalline plate of zinc-filled nitrate solution. After the chip is removed with salt ions, cleaned, and dried, we can get ZnO nanorods with a diameter of approximately 90 nm. In addition, the hydrothermal synthesis method can also be used to prepare ZnO nanowires. Currently reported array films of single-crystal ZnO nanorods created by a hydrothermal preparation have a highly consistent orientation, with an average diameter of approximately 40 nm. The film of approximately 390 nm emits strong fluorescence. A ZnO nanocolumn array created by hydrothermal synthesis has a nanocolumn length of 230 nm and a diameter of 38 nm. ZnO nanowire created by hydrothermal synthesis has a diameter of 50–80 nm and a length of more than 6 mm.

Compared with the preparation method of ZnO nanowires as described in Sections 7.1.4.1 and 7.1.4.2, the hydrothermal method has the unique advantages of being simple and economical and requiring no special large-scale equipment or high reaction temperature. If the size and location of ZnO grains (precursor of ZnO nanowires) can be controlled, then the ZnO nanowire size, location, and orientation, which are ultimately obtained with the hydrothermal method, can be controlled by the homoepitaxy principle.

7.2 Piezoelectric Applications of Zinc Oxide

7.2.1 Piezoelectric Effect

In physics, a number of dielectric ionic crystals (e.g., quartz, Rochelle salt, barium titanate) will have polarization not only under an electric field but also under mechanical stress.

In a certain direction, as the dielectric experiences a mechanical force to deformation, it will cause the relative transfer of its internal positive and negative charge centers to polarize. As a result, two of its relative surfaces (polarization surfaces) will be found with the bound charge Q with the opposite symbols, and its electric displacement D (the charge density σ in the MKS system of units) is proportional to the external stress tensor T:

$$D = dT \tag{7.2}$$

In this equation, d is the piezoelectric constant matrix. This process indicates a conversion from mechanical energy to electrical energy, called the positive piezoelectric effect.

If the aforementioned dielectric is applied with an electric field, then the relative displacement of the positive and negative charge centers inside the dielectric will also lead to dielectric deformation, with its strain S being proportional to the external electric field strength E:

$$S = Dt \cdot E \tag{7.3}$$

Dt is the converse piezoelectric constant matrix.

This phenomenon is known as the converse piezoelectric effect, also known as electrostriction, a process in which electrical energy is converted to mechanical energy. The piezoelectric effect is the effect of a charge polarization caused by the mechanical deformation of materials; it is an important physical process to achieve mechanical and electrical coupling and sensing.

Piezoelectric materials were discovered by the Curie brothers (Pierre Curie and Jacques Curie). They first demonstrated how to generate electricity through movement in 1880, and they constructed the world's first piezoelectric sensor. In the following year, Jacques Curie discovered the function that converted the electric energy created by movement to another type, called the converse effect. The piezoelectric technology demonstration led to the building of the world's first ultrasonic submarine detector by French military scientists in 1917. This was the prototype of sonar.

At that time, piezoelectric material had to be extracted from quartz; these quartz bodies may act as a frequency reference and flexibility tester and so on. In acoustic holographic technology, they can be used to detect small structural defects. Today, however, the piezoelectric material can be manufactured in the laboratory and can also be made into various synthetic compounds, which include ceramic converters, microphones, accelerometers, starters to the filter of surface acoustic waves, and others. The synthesized piezoelectric materials also can be used as solid-state devices to replace the solenoid and the static "muscle" in computer motor systems.

The piezoelectric effect is one of the most important physical effects that has been applied to the sensing and control of subjects. When applied with stress, some of the crystals with the special structure can generate a voltage difference. Conversely, under the influence of an external electric field, the crystal can produce elastic deformation. The piezoelectric effect demonstrates an important physical process through which stress is converted to electrical signals.

7.2.2 Piezoelectric Application of Zinc Oxide: Nanogenerators

7.2.2.1 Why Do We Need Nanogenerators?

A nanodevice is small, uses less power, has high sensitivity in reflecting, and has all the unique advantages of macroscale devices. Nanodevices have been at the forefront of research in academia. To make these tiny devices work properly, they must be driven with electricity. However, only self-powered nanodevices can be regarded as a true nanosystem. Because a nanosystem must be very tiny (so it can be implanted into the human body, for example), its power supply system must be miniaturized. Nonetheless, the current study is only focused on the nanoscale device itself, without considering the issues of power-inputting systems.

Energy is needed for generating power. For example, wireless sensors that are planted into organisms require a battery to supply the power directly or indirectly. Many nanodevices are used for health care and, in such circumstances, conventional batteries that contain lithium and cadmium and other toxic substances cannot be candidates for biomedical devices to be implanted in the human body.

Alternatively, people are aware of recycling and reuse. In fact, there is much wasted energy around us, such as the vibration of passing vehicles, electric heat, a variety of acoustic and electromagnetic waves, and so on. If we can develop a technology to recycle that wasted energy, then it will be a truly remarkable contribution to improving energy efficiency.

In fact, people generate energy when walking and breathing. So, can we convert the energy in the human body into the electric power for nanoscale devices? In other words, can the tiny vibrating mechanical energy in the surrounding environment be collected and transformed into electricity to provide energy for nanodevices such as sensors and detectors? The vibration of mechanical energy exists in the natural world and in people's daily lives, such as the various frequency noises caused by air or water flow, engine rotation, air-conditioning, or the functioning of other machines, as well as energy used to stretch the muscles in people when walking or the compression energy from the feet stepping on the ground. Even in the human body, subtle changes in pressure caused by breathing, the heart beating, or blood flow somewhere in the body can also be used to drive nanogenerators. A tiny device that could transform the body's bioenergy into electricity for nanodevices so we can achieve the miniaturization of nanodevices and its power supply system would be most ideal.

The concept of the nanogenerator is proposed just in this context. By using the unique nature of zinc oxide nanowires with the vertical structure, nanoscale mechanical energy can be successfully converted into electrical energy. The development of nanogenerators of this kind is based on zinc oxide nanowires with a vertical structure. By using a conductive atomic force microscope probe, the vertical zinc oxide nanowires can be bent and input with mechanical energy, whereas the piezoelectric effect in the zinc oxide nanowires may result in the polarization of charge. Thus, mechanical energy is transformed into electrical energy. Because of the semiconductor features of zinc oxide, the semiconductor metal Schottky barrier can be used for the temporary storage of electrical energy in nanowires. The conductive atomic force microscope probe is applied to access the power for transmission to the outside. Thus, a perfect nanoscale-generating function is made ready. Presently, the power-generating efficiency of nanogenerators can reach 17–30%. The invention of nanogenerators opened a new chapter in nanoscience and technology. Principally and theoretically, it laid the foundation for self-generating nanodevices.

7.2.2.2 Principle of Piezoelectric Nanogenerators [4–6]

The principle of the nanogenerator is based on the piezoelectric effect by the use of one-dimensional nanomaterials (nanowires or nanobelts); that is, a longer (micron level) nanowire is pushed using an electron microscopy (e.g., atomic force microscopy, ATM) probe to produce a charge. Also, as the probe is moved away, the charge is released as current. Thus, mechanical energy is converted into electrical energy to enable the normal operation of a nanogenerator. Here, the piezoelectric effect is a result of the bending of the nanowire, so it can be called the bending-electric effect. Macrobending is quite difficult, and the bending-electric effect is not so significant. But nanoscale one-dimensional bending is relatively easier, so the bending-electric effect is manifested more easily. Of the known one-dimensional materials, ZnO is the most important. The main advantages of ZnO are that it is a semiconductor with a piezoelectric effect, a basic feature that allows it to be used as electrical and mechanical coupling sensors and inverters; in addition, in the current nanostructures that have been invented, ZnO has the most varieties, such as nanowires, nanobelts, nanosections, nanorings, nano-arched structures, and nanohelical structures [3,7,8].

People have gained experience in how to control and grow high-quality ZnO nanostructures. As mentioned, a nanogenerator may find its major application in biological systems, which require relatively higher biosafety and biocompatiblity of the materials involved. ZnO meets these requirements.

It is vital that zinc oxide is featured in semiconductors with a piezoelectric effect. This guarantees that zinc oxide nanowires are able to complete the cost-effective transition from mechanical energy to electrical energy. Conventional piezoelectric materials, such as PZT, are usually insulators. Their bending or compression can also generate electric potential changes. Nevertheless, because they cannot work with metal to form a Schottky barrier with a one-way conductive feature, the transition from charge accumulation to release cannot be achieved. Thus, while there are some studies that tend to use conventional piezoelectric materials as a power source, they require complex external circuitry to achieve the accumulation of charge, making it really difficult to miniaturize the devices. More importantly, conventional piezoelectric materials, due to

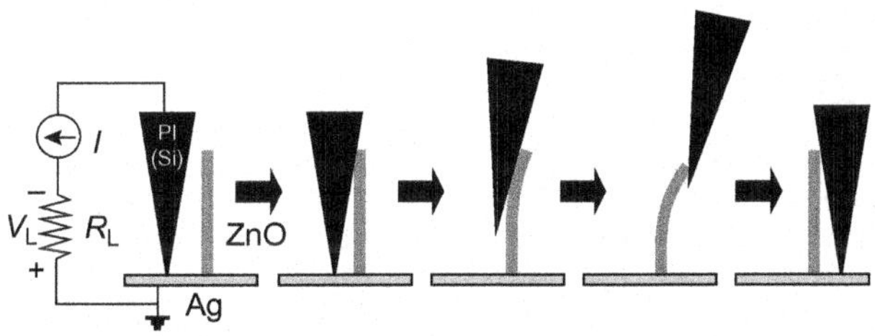

Figure 7.4 Experimental device and the process using a conductive AFM tip to bend a piezoelectric nanowire to generate electricity. From Ref. [9].

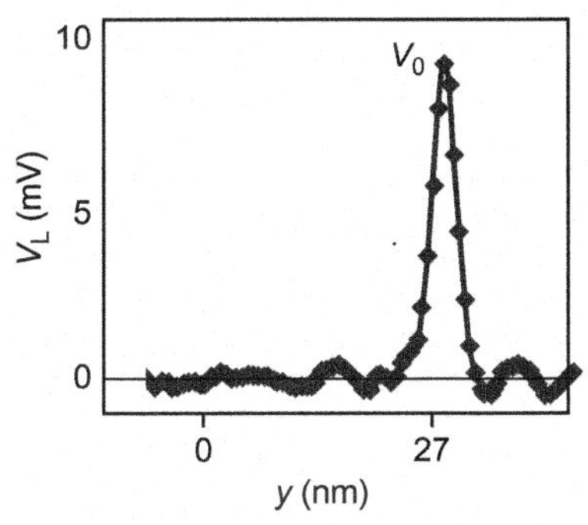

Figure 7.5 Life profile of voltage output signal obtained by AFM tip scanning along a vertical nanowire. From Ref. [9].

the higher complexity of their chemical composition and crystal structure, are difficult to synthesize into high-quality nanoscale structures. Therefore, the high output power can only be achieved by increasing the size and external forces, thus making it impossible to achieve small size and high power at the same time. On the contrary, as a new type of semiconductor piezoelectric material, zinc oxide has a relatively simple chemical composition and crystal structure. Nanowire produced from this has a small diameter, usually less than 100 nm, but its length can reach a few microns. This favorable aspect ratio makes it able to be bent and able to produce electrical potential difference with very little force. As long as we can perform this process, for example bend a large number of vertical nanowires continuously, then nanogenerators with high output power are likely to be made.

Experimental devices and processes using a conductive AFM tip to bend a piezoelectric nanowire to generate electricity are shown in Figure 7.4. The corresponding voltage signals are shown in Figure 7.5, where y is the bending amplitude of the nanowire from the vertical.

References

[1] L.S. Yu, Semiconductor Heterostructures Physics, second ed., Science Press, Beijing, 2006.

[2] L. Esaki, R. Tsu, Superlattice and negative differential conductivity in semiconductors, IBM J. Res. Dev. 14 (1) (1970) 61–65.

[3] L. Ke, X.Y. Miao, Y.F. Wei, J. Wang, X. Wang, New materials for II–VI semiconductor lasers—ZnO quantum dots, Physics 28 (1999) 30–34.

[4] M.C. Tamargo (Ed.), II–VI Semiconductor Materials and Their Applications [M], Taylor & Francis Inc., New York, NY, 2002, 132.

[5] X.W. Fan, Y.M. Lv, Research progress of ZnSe semiconductor blue light-emitting and laser devices, Physics 23 (1994) 393.

[6] N.T. Pelekanos, J. Ding, M. Hagerott, A.V. Nurmikko, H. Luo, N. Samarth, et al., Quasi-two-dimensional excitons in (Zn,Cd)Se/ZnSe quantum wells: Reduced exciton–LO–phonon coupling due to confinement effects, Phys. Rev. B 45 (11) (1992) 6037.

[7] T.N. Oder, J. Li, J.Y. Lin, H.X. Jiang, Photoresponsivity of ultraviolet detectors based on $In_xAl_yGa_{1-x-y}N$ quaternary alloys, Appl. Phys. Lett. 77 (6) (2000) 791.

[8] J. Li, K.B. Nam, K.H. Kim, J.Y. Lin, H.X. Jiang, Growth and optical properties of $In_xAl_yGa_{1-x-y}N$ quaternary alloys, Appl. Phys. Lett. 78 (1) (2001) 62.

[9] Z. Wang, The principle of the piezoelectric nano generator and potential applications, Physics 35 (2006) 897.

8

SUPERCONDUCTING NANOMATERIALS

CHAPTER OUTLINE

8.1 Superconductivity [1]

Superconductivity is defined as electrical flow through a material with zero electrical resistance. In metals it occurs at very low temperatures, but higher temperature superconductivity occurs in some ceramic materials. Materials with superconductivity are called superconducting materials. In 1911, Heike Kamerlingh Onnes, a physicist at Leiden University, first discovered the

Nanomaterials and Devices. DOI: http://dx.doi.org/10.1016/B978-1-4557-7754-9.00008-1

phenomenon that at superconductivity of 4.2 K (i.e., −268.8°C, known as the transition temperature or critical temperature, T_c), the electrical resistance of mercury suddenly disappears. In 1933, Meissner found that when a substance in the superconducting state was applied with a magnetic field, magnetic field lines could not penetrate the sample while maintaining the magnetic flux inside the superconductor as zero. This feature is called the *Meissner effect* and is the second characteristic of superconducting materials. The third characteristic is the *Josephson effect.* Superconducting electrons may be in the formation of Cooper pairs. Cooper pairs can come through the barrier at a certain probability, known as the tunneling effect of Cooper pairs. In other words, the Josephson effect is a phenomenon whereby current comes through the two kinds of superconducting materials separated by a thin insulating layer in between. It was theoretically predicted by Josephson in 1962. Josephson was awarded the Nobel Prize for this work in 1973.

Superconducting materials have significant applications as low-loss current conductors, creating ultrahigh magnetic fields, and in microwave applications. Because of its enormous economic potential, it has been given increasing attention. There have been important research results and discoveries over the years, perhaps most noted by Nobel Prize awards to scientists in superconductivity research in 1913, 1972, 1973, 1987, and 2003.

Considerable progress has been made in superconducting materials research. Nonetheless, the current T_c of superconducting materials remains very low. The highest transition temperature ($T_c \sim 130$ K) is obtained by the use of copper oxide high-temperature superconducting materials. This temperature is higher than the boiling point of liquid nitrogen, so it can be used in many applications. At present, all the world's applications of superconducting devices cannot operate without liquid nitrogen or liquid helium cooling equipment, which is expensive and inconvenient. It goes without saying that if superconductivity can be achieved at room temperature, then modern science and technology will certainly undergo profound changes (Figure 8.1).

Over the past 20 years, the exploration and discovery of new types of superconductors have changed our understanding of superconductivity. Researchers used to think that superconductivity was a rare phenomenon that would appear only at ultralow temperatures. Now, we have realized that superconductivity is a basic state of many materials, even those compounds that have little metallic character.

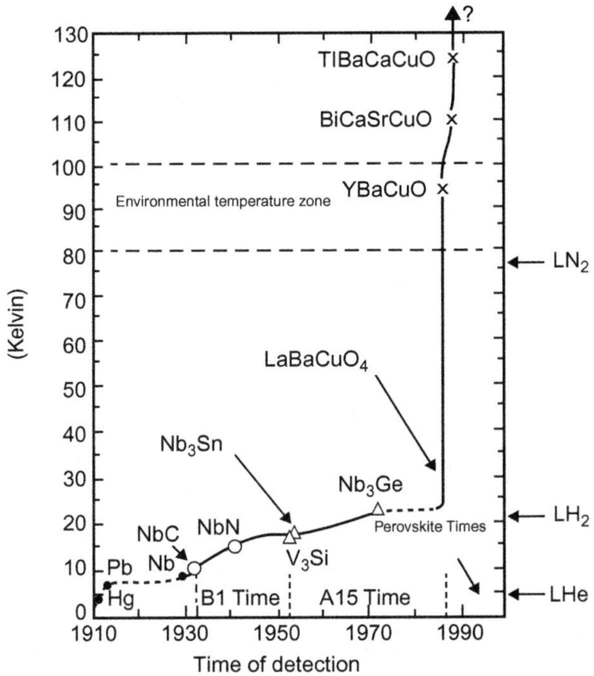

Figure 8.1 Development of superconductor materials.

8.2 The Physical Principles of Superconductivity [1]

Ever since the phenomenon of superconductivity was discovered, researchers have been eager to reveal the mystery of its physical mechanism. After years of studies, Bardeen, Cooper, and Schrieffer believed that superconductivity originated in the electron and lattice interactions. This theory is known as the *BCS theory*. As shown in Figure 8.2, when an electron A comes through the crystal lattice, due to the mutual attraction of the charges with different signs, positive charges in the positive ion dot-matrix of a crystal lattice will be close to electron A, resulting in a partial density increase of the positive charges. This may lead to a local excess of positive charges. This disturbance of the partial positive charge may spread in the form of a crystalline lattice wave, which will, in turn, affect the second electron, namely electron A may further attract the other electron B in an indirect way. Of course, there is Coulomb repulsion

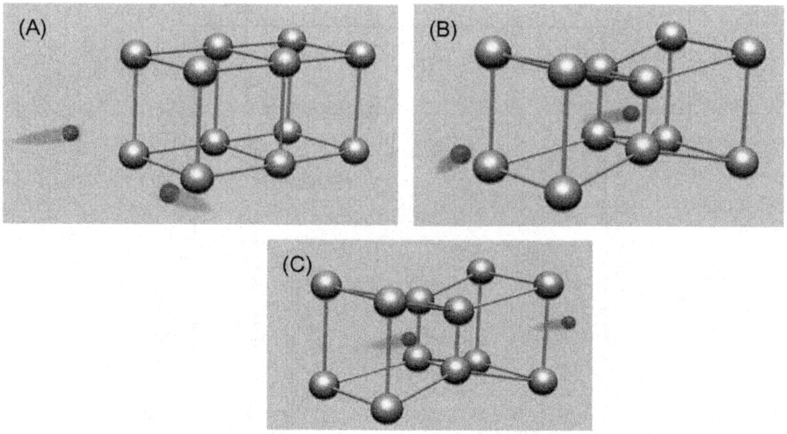

Figure 8.2 Diagram of the BCS theory: (A) as electrons get close to the crystal lattice, they will attract the positive charge on the lattice because of their negative charge; (B) the attracted cation may cause lattice deformation, while part of the electronic energy will be transmitted onto the lattice; (C) when the electrons pass through the lattice, the lattice deformation still remains. This creates a positively charged area to attract another electron, which may receive energy from the crystal lattice.

between the electrons. However, as long as the interaction of the electron and lattice is strong enough, the indirect effect of attraction between the electrons is likely to exceed the Coulomb repulsion, so that the electron pairs may have a net amount of interattraction. The result is that a bound state can be formed between them. This bound state is an electronic pair formed by two pairs of electrons, called Cooper pairs. From the momentum space, the two electrons are set with a total momentum of K. Cooper's work showed that when $K=0$, the binding energy reaches its maximum, and electron pairing has the lowest energy. From the momentum space, the two electrons in Cooper pairs may involve a quantum state with a momentum that is equal in size and opposite in direction, and also include those with an opposite spin. In this case, the Cooper pairs have an energy that is less than that of the two electrons in "their own action" and thus are more stable.

In the Cooper electron pairs, momentum (speed) of a single electron can be different. However, the total momentum of each Cooper pair is constant. Metallic resistance in the normal state is caused by the electrons under the scattering of crystal lattice waves. However, in the superconducting state, the electrons form Cooper pairs, conducting continuous scattering. During this process, the total momentum is conserved, and thus the current also remains unchanged. This is the reason why superconducting current is unfettered. Positive ions in the crystal lattice have thermal vibration that makes these "*lone warriors*" in difficulties, resulting in energy consumption during the electronic movement. This explains the resistance existing in normal metals.

8.3 The Classification of Superconductors [2]

8.3.1 Low-Temperature Superconductors

Superconductors with the critical temperature below the liquid helium temperature are known as low-temperature superconductors. After discovering properties of mercury, researchers discovered superconducting characteristics in tin, lead, and many other metallic elements and alloys, as well as compounds, but their critical temperatures have remained very low (below the temperature of liquid helium). Through years of efforts, researchers now have discovered most metallic elements exhibit superconductivity. By the introduction of special techniques (such as high-pressure technology, technology in precipitation into a thin film at low temperature, and very rapid cooling), those metals that were once believed to be unable to turn into superconductors have achieved the state of superconductivity under certain circumstances.

8.3.2 High-Temperature Superconductors

Low-temperature superconductors below the liquid helium temperature have limited industrial applications. For this reason, the practical application of superconductors had been stagnant for quite some time before the discovery of high-temperature superconductors (HTSs). Finally, in 1987, superconducting technology achieved a decisive breakthrough. Based on iridium, barium, and copper oxide, American scholars Mow-Kuen Wu and Paul Chu made HTSs ($YIBa_2Cu_3O_7$) with T_c in the range of 90–100 K. These temperatures are above the boiling point of nitrogen (77 K). We call such materials HTSs. The discovery of cuprate HTSs has had a great impact on science and technology. This finding is a significant indicator of the development of superconducting materials over the past 10 years, not only because of their many applications already known but also because they revolutionized the basis of condensed matter physics. The reconstruction of a new theoretical framework still continues but is controversial. Many researchers believed that "understanding of the high-temperature superconducting mechanism" is an open issue that is considered to be the most critical issue in condensed matter physics. The discovery of cuprate HTSs illustrates some of the facts. Superconductivity is no longer regarded as a phenomenon that occurs at low temperatures only. Room-temperature superconductivity is now considered possible.

Almost all of the HTSs that have been discovered since 1987 are copper-containing ceramics. Although the critical temperature has been greatly improved, the HTSs have not yet attained the necessary stability. Also, the carrying capacity has also slightly decreased. All materials with the highest critical temperature are complexes with low-dimensional structure, and each of them contains light elements in the chemical formula. The high phonon density of states and high energy are very attractive. However, before a widely accepted theory of HTSs is established, the real key factor must be determined. The application of superconductors seems to have a long way to go.

8.3.3 Other Novel Superconductors

The newly discovered superconductors are found widely throughout the periodic table, from lighter elements boron and lithium to the transition metal uranium series, and many complexes of transfer salts. Here, C_{60} may have a greater potential for development because of its greater elasticity, making it easier to be molded than the oxide ceramic, which has a hard and crisp texture. Moreover, it has a larger critical current, critical magnetic field, and coherence length. These characteristics make C_{60} superconductors more likely to be put into practical use. C_{60} is hailed as a "star" of new materials in the twenty-first century, for it has demonstrated a wide range of novel features and application prospects in machinery, light, electricity, magnetism, and chemistry. Some researchers predicted that the giant C_{240} and C_{540}, once synthesized successfully, are likely to become superconductors at room temperature.

Some superconductors have a higher T_c, for example the alkali metal "bucky ball" A_xC_{60} has a critical temperature at 33 K, and RNi_2B_2C and YPd_2B_2C have a critical temperature of 23 K. Some superconductors are very unusual, such as the heavy fermion superconductors $CeMIn_5$ (M = Co, Rh, Ir), $CePt_3Si$, and $PuCoGa_5$, which have a critical temperature of 18 K. Sr_2RuO_4 has its parameters in p-wave symmetry. Ruthenium copper oxide is not limited to HTSs in its antiferromagnetic order, even if they are only at a distance of a few angstroms. Magnesium diboride (MgB_2) has dual-band superconductivity and a superconducting transition temperature of 39 K. Discovery of magnesium diboride provided new ways to study the new class of HTSs with simple compositions and structures. Magnesium diboride can be easily synthesized for processing and can be easily made into films or wires. It can be used in power transmission, supercomputer devices, as well as CT scan imaging instruments. Its discovery brought excitement to the world of condensed matter physics. Superconducting boron-doped diamond is an interesting research topic in the field of electronics. $Na_xCoO_2 \cdot H_2O$ is associated with another very interesting new material system. In the case of high pressure, researchers have also found that many elements may have a higher T_c. Some elements, under a pressure of 1 million pascals, may have a T_c of more than 10 K, such as Li and S ($T_c = 17$ K), Ca ($T_c = 15$ K), and B ($T_c = 11$ K). The recently discovered CaC_6 superconductor has T_c of 11.5 K, which is twice the value of the critical temperature of graphite electrode compounds.

8.4 Nanosuperconductors [3]

8.4.1 Research Progress

Researchers are interested in the superconductivity of nanometer-sized materials for the following reasons. First, nanomaterials can be used to make ultrasmall superconducting devices such as quantum bits (to be introduced later). Second, when some or all of the material sizes of a system are equal to or smaller than the feature sizes of length, such as the penetration depth, coherence length, grain size, and cell size, new quantum effects will be seen. Superconductivity may occur in the state of ultrathin film, wires, and quantum dots, which correspond to two-dimensional, one-dimensional, and zero-dimensional (2D, 1D, 0D) systems, including human-made structures such as knots, mesh, or multilayer heterostructures. Two-dimensional superconductors have an ultrathin film nanostructure. In the superconducting state, properties of the material largely depend on the spatial dimensions. Three-dimensional materials have the strongest superconductivity. In superconducting wire and superconducting thin films, thermal fluctuations and quantum fluctuations play an important role and are able to change the transport properties of electrons through the weak link. When the superconductors become small or thin, they often lose their superconducting properties. Superconducting properties have not been found to exist in ultrathin films, but the study results from one-dimensional superconducting coil are encouraging. As mentioned, in the superconductor, the current is hosted by the electrons in a configuration of Cooper pairs. Typically, the pairing of electrons has opposite spin and momentum; at any one time, an electron spins up and shifts to the left, then the other electron spins down and moves to the right. The force applied by the magnetic field will try to undermine the balance of the electron movement while keeping the spin directions of both of the two electrons arranged toward the magnetic field. Of course, due to the need for enough energy to undermine the Cooper pair, superconductors can withstand a certain intensity of magnetic field before being converted into the state of a normal conductor. This field is called the critical magnetic field. It may limit the application of superconducting coils and disable the superconducting coil to transfer high current, because the magnetic field generated by currents could weaken or destroy the coil superconductivity.

Phil Adams and his colleagues at Louisiana State University produced superconducting thin beryllium with a thickness of 2–30 nm in glass substrates [4], on which a 0.5-nm thin film of gold was precipitated. They found that when the magnetic field was applied in the direction parallel to the film, this gold–beryllium membrane could withstand a magnetic field much higher than that of pure beryllium film. In the thinnest films of 2 nm, the critical magnetic field was increased 10-fold. Adams and his collaborators believed that, in their case, the most critical factor was the interaction between the electron spin and the large positive charge in gold nuclei. Because of the relative movement between the electrons and nuclei, the action of a charge on another charge is equivalent to the occurrence of current, and the current would produce a magnetic field. The magnetic field produced by gold nuclei is much larger than that generated by the beryllium nuclei, so this magnetic field that comes from gold nuclei would cause each pair of Cooper electrons to be arranged in slightly the opposite direction. These distorted electrons are more easily adaptable to an external magnetic field because their spin in the rearranged direction is closer to the direction of the magnetic field. However, if the magnetic field is not parallel to the film, then the magnetic field would destroy superconductivity by interfering with the balance of the electronic movement rather than the spin. Although practical applications are still some way off, Adams saw the potential in the form of multilayer (sandwich) structures in the design of ultra-wires. These ultra-wire coils are likely to produce a stronger magnetic field than the presently used magnet, which is normally used in nuclear magnetic resonance instruments and physics research fields. Singular effects, demonstrated by Adams and his colleagues, provided the potential to help researchers in their study of Cooper pairs in HTSs, whose superconductivity is still not fully understood.

We know that superconductivity comes from "Cooper pairs" formed by spin-pairing electrons. When a superconductor is placed in a strong magnetic field, the electronic Cooper pairs will be destroyed. Meanwhile, the spin of electrons will also be affected, and the superconductivity will be weakened or completely disappear. But when the size of superconductors is reduced, the destructive effects of the magnetic field will also be weaker; when the superconductors become the size of the nanometer scale, the magnetic field will be unable to undermine the Cooper pairs. Bezlia Kim, a professor of physics at the University of Illinois, and associates confirmed this in their experiments. They placed a single layer of carbon nanotubes on the silicon wafer and these were etched with a "ditch" approximately 100 nm wide; the surface of carbon nanotubes was coated with a layer of Mo–Ge superconducting material. The temperature decreased below the critical temperature and the response of the nanoscale superconducting material in a strong magnetic field was observed. It was found that the impact of the strong magnetic field was significantly weakened on the nanoscale superconducting materials. Bezlia Kim and associates hypothesized that because of the very small ultra-wire diameter, only approximately 10 nm, interaction between the electronic Cooper pairs offsets the effects of the magnetic field on superconductivity.

This feature of nanoscale materials will broaden the application prospects for superconductivity. For example, in the past, superconducting coils could not transport high current because the magnetic field generated by currents could weaken or destroy the superconductivity of the coil. However, if the general nanoscale superconducting coil is mixed with the superconducting filaments, then transmission of strong current can be easily accomplished. In addition, the nanoscale superconducting material can also be used for magnetic resonance imaging and in other applications.

Certainly, the size of nanoscale superconducting materials cannot be infinitely reduced; if that were possible, then the mutual interference between the electronic Cooper pairs would undermine the superconductivity. In addition, the nanoscale superconducting materials cannot reach zero resistance consistently. This is similar to large superconducting materials.

In the past 20 years, researchers have made considerable progress in the preparation and characterization of nanostructured superconductors, including nanoparticles, quantum dots, nanowires, ultrathin film, and single-layer superlattice materials. In the field of high-temperature superconducting materials, many have lengths of only nanometers (e.g., single-crystal domain size, coherence length, etc.). Therefore, the experimental study of the nanometer-sized structures has considerable significance in helping researchers understand the mechanisms of the microworld.

Many experiments have been performed with regard to the zero-dimensional tunnel junction, such as the experiments with the so-called completely confined system. The capacity in preparation of low capacitance and small tunnel junction promoted research on the transition between the similar Josephson and charge performances in various Josephson junctions and junction arrays. The latter is the phase model of a physical instance of the shift from superconductivity to insulator, whereas the former shows the dissipation that reduced quantum fluctuations, thereby changing superconducting behavior. Therefore, through dissipation that reduces the quantum fluctuations, we can control the transition from superconducting to insulating. Technology in ultrasmall tunnel junction structures is essential to the study of the quantum bit (qubit). The qubit is one of the basic elements for building a quantum computer. Studies of the double tunneling junction have revealed the characteristics of the Coulomb blockade and odd–even effect that may be found in superconductivity.

Some features of one-dimensional nanowires have already been mentioned in the preceding chapters. The development of e-beam lithography makes it possible to perform structural design at an order of magnitude close to the superconducting coherence length. This has stimulated enthusiasm in physics research of superconducting nanowires, which are prepared by using growth approaches such as lithography and electrochemical templates with carbon nanotubes as template. A recent study reported the inverse proximity effect in superconducting wire: when the electrodes in contact with superconducting wire also show superconductivity, the superconductivity of the wire is then curbed. Perhaps this is an example of dissipation control of superconductivity. Another study showed that after the chaotic nature of the superconducting wire reaches a certain level, superconductivity completely disappears (the wire shifts from superconductor to insulator). This quantum shift is observed in the nanowire template constructed. The critical point can occur in the form of the critical resistance or the critical diameter, where the critical resistance is close to the quantum resistance

($h/4e^2$), whereas the experiment showed a critical diameter of approximately 6 nm.

The emergence of two-dimensional ultrathin superconducting films provided a research platform for the topological phase transitions. The Kosterlitz–Thouless–Berezinskii transition [5] is a typical case of contemporary condensed matter physics.

For disordered films, a further decrease of thickness leads to the superconductor–insulator transition (rather than the superconducting metal transition). This transition can be adjusted by changing the degree of disorder, magnetic field, or electrostatic charge. This phenomenon, as an example of a quantum phase transition, recently created great interest. However, the basic principle of this phenomenon is unclear.

Heterostructures and superlattices are layered structures that contain an interface between different materials. Their nature is strikingly different from that of those composing materials. Sandwich junctions have been used to study superconductivity and other competing order parameters with respect to the interactions between them. Particular attention has been given to the interface between superconductor (S) and ferromagnetic (F) or antiferromagnetic (AF) materials. In an FSF three-tier structure, the transport process that came through the iron magnetosphere led to the emergence of a π-junction. Many superlattice materials composed of nano-size thin layers have been synthesized and studied. Recently, research attention has shifted to SFSFS-type superlattice material. Superconducting thin films have been found to have considerable unconventional quantum phenomena, such as the oscillation of critical temperatures and critical current density acting as a function of the thickness of magnetic iron, which were predicted earlier and have now been observed.

8.4.2 The Main Difficulties

For copper salts with high critical temperature and Sr_2RuO_4 or heavy fermion superconductors, the fundamental knowledge needed to discover this type of unconventional novel superconductor in low-dimensional geometry is currently lacking.

8.4.2.1 Incredible Magnetic Nanoclusters

Recent theory proposed that globular clusters with a specific number of superconducting electrons will show superconductivity at relatively high temperatures. This theory may be established for some simple metal clusters such as aluminum, gallium, zinc, and cadmium. For example, Ga_{56} was predicted with a critical temperature of approximately 150 K (−123°C). Mesh material composed of such clusters can be formed on the surface and can carry the superconductor current at higher temperatures. In 1984, W. Knight discovered the existence of the so-called shell structure, which is the basis of this phenomenon. At that time, this event did not arouse the attention of the superconductivity research community. Rather, it was largely ignored because of research on the cuprate HTS discovered in 1986, a hot topic at the time. At present, research on nanoclusters is steadily developing. It is quite possible that superconductivity will be observed in these clusters. Experiments in this area require artificially measuring the selected cluster concerning its excitation spectrum at low temperatures (above the critical temperature). Superconducting pairing mechanisms will lead to significant differences in the excitation spectrum at temperatures such as the critical temperature. Relevant technologies have been developed in this respect, including mass spectrometry, generation of an energy beam at different temperatures, as well as photoelectron spectroscopy. Another noteworthy achievement is the growth of isolated clusters in the matrix and molecular crystals, which can form an orderly three-dimensional lattice.

8.4.2.2 Quantum Fluctuations and Strong Correlation in Nanowires

Researchers have not yet come to any conclusion regarding the exact role that quantum fluctuations play in superconducting nanowires. The issue has aroused the interest of the scientific community in low-temperature superconducting electronics technology, because it may limit the length of the nanowire dimensions.

With the one-dimensional limit, the traditional Fermi liquid theory is no longer applicable; it is replaced by the Tomonaga–Luttinger liquid theory [6]. In the latter theory, there is no quasi-particle, and the excitation system is based on completely decoupled and separately excitable spin and charge degrees of freedom. This has been observed in filaments based on GaAs heterostructures. The relationship between the superconductivity and Luttinger liquid and the possibility of the Luttinger liquid itself being a superconductor is an interesting open issue.

8.4.2.3 Ultrathin Film

The open issues in this area include the following: the existence of an intermediate metallic regime in two-dimensional systems with Cooper pairs like ultrathin films rather than the nonelectronic type of quasi-particles such as the charge carrier; whether the insulating state is a new state of matter of the Bose insulator; and whether there are similarities between ultrathin films of superconducting behavior and the pseudo-gap regime of underdoped cuprate superconductor.

8.4.2.4 Proximity Effect

In the synthesis of nanomaterials, boundaries, crystal defects, chemical defects, and inhomogeneous strain may frequently cause the system to have various forms of "disorder," thus affecting the properties of the materials. Physicists knew that when the superconducting material got close to nonsuperconducting materials such as excess oxides, pollution, impurities, and the substrates, a "proximity effect" might occur. This is because the Cooper pairs could get inside the normal material to a certain extent, whereas the normal atoms would have an impact on the superconductors. As a material gets smaller in physical size, its electronic wave functions, electromagnetic fields, and other properties will also make a difference. Therefore, in addition to the inherent nature of the material, we need to understand its nature in proximity to other materials, as well as the features it exhibits when reduced to nanoscale. As a result of the superconducting proximity effect, when a normal metal (N) touches superconducting materials, superconductivity will penetrate into the normal metal. The production of the proximity effect can be attributed to the coherence in electron transport passing through the metal/ superconductor interface, a process called *Andreev reflection*. Regarding the proximity effect, we know it is a local effect. In copper salts, however, we have observed a long-range proximity effect, and there has been no explanation for this.

8.4.2.5 Nanosuperconductors and Hybrid Structures

In this regard, research directions include the fluctuation in nanosuperconductors, Josephson array, shot noise and dynamics, Andreev state and related phenomena, small Josephson circuit and the vortex dynamics in the Josephson array, nonstationary effect in meso-scale superconductors, and ferromagnetic/superconducting heterogeneous structures. By using the method of atomic layer molecular beam epitaxy (MBE), multilayer types have been greatly expanded. This method can also be used in the design of novel functional heterogeneous structures, including spin effects. This heterogeneous structure would also provide the conditions for basic research into the interaction between the magnetic order and the superconducting regions.

8.4.2.6 Links Between Superconductors and Nanostructure

The proximity effect and nanotechnology together can upgrade the traditional transmission and magnetic properties of superconductors. This effect had been predicted and then observed a long time ago, but it was only in recent years that, because of continuous improvement of the advanced technology in preparation of nanostructured materials, its different parameters and scope of applicability were substantially broadened. For example, in the interaction between the nanostructured magnetic particle arrays and superconducting thin films, simultaneous interaction with the vortex lattice may lead to a pinning effect. Arrays of magnetic particles can be prepared with the help of a variety of electron beam lithography or self-assembly, combined with sputtering or MBE and other thin film deposition technologies. Due to the geometry of the pinning arrays, resistivity of magnetic particle arrays, their magnetization, and other physical properties will show cyclical changes. Asymmetric magnetic nanostructures will produce a ratchet effect. The interactions between vortex and nanostructured arrays are considered in a certain frequency range to enhance a similar Josephson effect so the frequency range can be adjusted by an external magnetic field.

Based on the proximity effects of the two materials, boundary conditions and geometry of the nanointerface connecting with the superconductors will lead to the emergence of a novel electronic structure. For example, the common electronic reflection on the superconducting interface may become one that contains Andrew reflections, resulting in novel standing-wave states in small samples.

8.5 Application of Nanosuperconductors

Nanometer-sized materials can be used to make ultra-small superconducting devices, such as quantum computing devices. Currently, use of the nanosuperconducting Josephson junction as a qubit is an ideal candidate among a number of quantum bits.

8.5.1 Quantum Computers

The quantum computer is a class of physical devices that can perform high-speed mathematical and logical operations and that can store and process quantum information following the laws of quantum mechanics. When a device's processing and calculation are based on quantum information and run with quantum algorithms, it is a quantum computer. The concept of the quantum computer stems from research on reversible computers. The purpose is to address the issues regarding energy consumption of computers.

In the 1960s and 1970s, researchers discovered that energy consumption would lead to overheating of computer chips, which greatly affected the integration scale, thus limiting the ultimate speed of computers. Studies found that energy consumption comes from the irreversible operations in calculations. So, must calculation involve irreversible operations to complete? The answer is, for all the classic computers there is the possibility of a corresponding reversible computer with unchanged computing power. Because every step of the operation can be converted into a reversible operation, in quantum mechanics it can be expressed as a unitary transformation. The early quantum computer was a classical computer described by quantum mechanics languages and did not involve the use of the essential features of quantum mechanics, such as quantum superposition and coherence. In a classical computer, the basic information unit is a bit; the computing object is various bit sequences. Similarly, in a quantum computer, the basic information unit is the quantum bit, with operation targeting qubit sequences. The difference is that the quantum bit sequences not only are located at all orthogonal superposition states but also can be at the entangled state. In addition to the possibility of quantum parallel computing, these particular quantum states also brought about many wonderful features. In contrast to classical computers, quantum computers can perform arbitrary unitary transformation. Obtaining the output state, measurement can lead to the computing results. Thus, quantum computing generated a tremendous expansion of classical calculations. In the form of mathematics, classical computing can be seen as a special class of quantum computation. Quantum computers may transform a stack of each component; all of these transformations are completed at the same time and would stack at a certain rate of probability to give the results. This calculation is called quantum parallel computation. In addition to parallel computing,

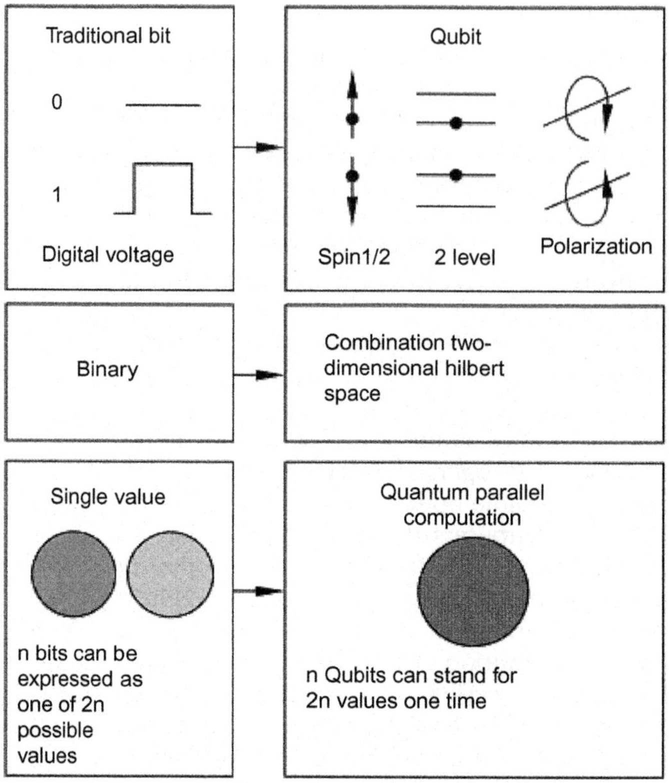

Figure 8.3 Comparison of traditional bits and qubits.

quantum computers have another important use, that is, to simulate quantum systems. This is a task beyond the competency of classical computers (Figure 8.3).

Whether using quantum parallel computing or quantum simulation computing, both are required to make use of quantum coherence. Unfortunately, it is difficult to maintain quantum coherence in a practical system. In quantum computers, quantum bits are not an isolated system. Rather, they will interact with the external environment, leading to the attenuation of quantum coherence, namely decoherence. So, the core issue in designing a quantum computer is to overcome decoherence. The discovery of quantum coding is by far the most effective way to overcome decoherence. The major quantum coding schemes include quantum error-correcting code, quantum error-avoiding code, and quantum error-prevention code. Quantum error-correcting code is the classic analog error-correcting code, which is a class of codes that attracted the most attention from researchers. Its merit lies in its wide application, but its drawback is its lower efficiency.

So far, there are still no quantum computers in the real sense in the world. However, many laboratories around the world are pursuing this goal with great enthusiasm. Many methods have been proposed regarding how to realize quantum computing. The problem lies in the fact that achieving the experimental manipulation of the micro-quantum state is very difficult. The currently available programs are mainly based on the interaction of atoms and cavity, as well as cold trap bound ions, electron or nuclear spin resonance, quantum dot manipulation, and superconducting quantum interference. It is difficult to say which option is more promising. From the perspective of integration and miniaturization, the quantum point program and superconducting Josephson junction program combined may be most appropriate. In the future, perhaps a novel design will emerge and make all the existing proposals obsolete; the new method might be based on some kind of novel material. Quantum computers are not intended to replace existing classical computers. The concept of quantum computers provides a new perspective for computation and is different from other new concepts of computers, such as optical computers and biocomputers. The role of quantum computers is far more than to solve some intractable problems that are possibly found in classical computers.

8.5.2 Nanosuperconductor Quantum Bits [7–10]

The production and control of the stable qubits system is a key issue in the practical use of current quantum information and the physical implementation of quantum computing. For nonsolid qubit systems, researchers have successfully demonstrated the operation of quantum logic gates and simple quantum algorithms. Nevertheless, practical quantum computing needs to overcome many known and unknown difficulties and to transcend some barriers of principles and nonprinciples.

To achieve practical use of the quantum computer in physics, we need to organically integrate universal quantum logic gates, to keep quantum entanglement between qubits, and to be able to manipulate them. With the increase in the number N of integration of qubits, the loss of quantum coherence (or quantum decoherence) will increase exponentially at the power of e. Recent studies have shown that the loss of quantum coherence of a single particle may decay at the index of e. But with the use of quantum entanglement with the characterization of complicated entanglement (concurrence), it will be almost lost

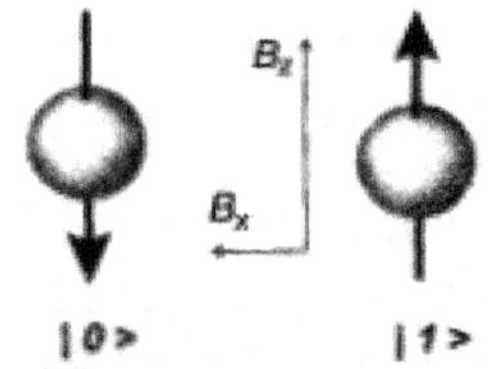

Figure 8.4 Diagram of a two-level system. It is constituted by the spin 1/2 particles in a magnetic field. Two basic quantum states correspond to spin orientations upward and downward.

in a limited period of time. These results suggest that a multi-qubit being simply integrated would have more vulnerable quantum coherence. Basically, this is a challenge to current theoretical and experimental subjects.

Microscopic quantum systems, such as atoms, spin, and photons, are easy to separate from the environment, thereby reducing decoherence, which is the advantage of the microscopic quantum system. However, the microscopic quantum system does not facilitate the integration to create the actual quantum computing devices. From this point of consideration, the macrosystem appears to be more flexible in the use of standard integrated circuit technologies. Macroscopic quantum bits currently proposed are based on nanostructured electronic circuits, or on the composition from the quantum dots, or by the superconducting Josephson junction. Josephson junction quantum bits are divided into three categories: charge qubits, flux qubits, and the corresponding hybrid qubits.

In fact, any physical system can be used as qubits, as long as it can be clearly defined with the two-level state. For example, the spin 1/2 particle in a magnetic field, with the spin only taken as +1/2 or −1/2 along the magnetic field direction, can constitute a two-level system (Figure 8.4).

A long time ago, researchers discovered that superconductors could exhibit the phenomenon of macroscopic quantum transport. Because of the superconductor coherence of Cooper electron pairs, all the microelectron pairs can be regarded as a single macrowave function. The tunneling nature shown by the macroscopic quantum wave function can be observed in the superconducting Josephson junction system. At present, the superconducting Josephson effect can be used to produce quantum devices that have advantages such as easy operation, longer coherence time, and ability to integrate for production as one of the potential quantum bit systems. Some superconducting quantum devices have been developed in the past,

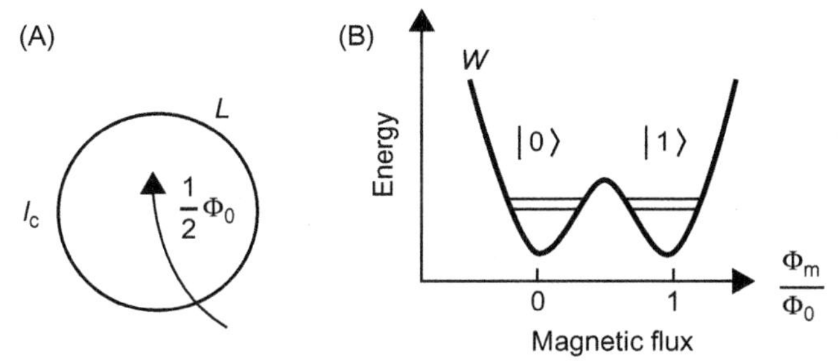

Figure 8.5 (A) A typical circuit diagram of magnetic flux quantum bits. (B) SQUID energy changes with the internal magnetic flux. External magnetic flux $\Phi_{ext} = \Phi_0/2$, and the horizontal level is representative of a mixed-energy state.

including quantum interference devices (SQUID) (flux qubit) and the single-electron device (charge qubit); all have been confirmed preliminarily to be excellent quantum bits. Here, we introduce their principles and characteristics.

The flux qubit is as shown in Figure 8.5A. The SQUID quantum interference device is a superconducting ring, on which there are one or several Josephson junctions. The SQUID level constitutes a two-level system (Figure 8.5B). Here, the role of the Josephson junction covers two aspects: one is to separate the superconducting phases on the right and left sides and the second is to allow superconducting Cooper pairs to come through in the form of electron tunneling. Superconducting current coming through the Josephson junction is determined by the superconducting phase difference between the left and right sides γ. When the phase difference is zero or an integer multiple of 2π, the current is zero and the system has the lowest energy. Generally, the Josephson junction energy is expressed as $-E_J \cos \gamma$. Interestingly, as an external magnetic field is applied perpendicular to it, the magnetic field will cause an increase of one phase in superconducting quantum wave functions along the direction of the ring. Mathematically, this increased phase is expressed as the line integral of vector spaces, equivalent to the magnetic flux coming through the superconducting rings. Due to the single-valued nature of the superconducting wave function, the phase added to an external magnetic field must be compensated by the phase difference between both sides of the Josephson junction. If the phase difference is not zero, and if the superconducting ring is to produce a superconducting current around it, similar to shielding current, also known as *persistent current,* then, in fact, this persistent current can produce a magnetic flux $\Phi_{eff} = LI$. Here, L is the self-inductance of

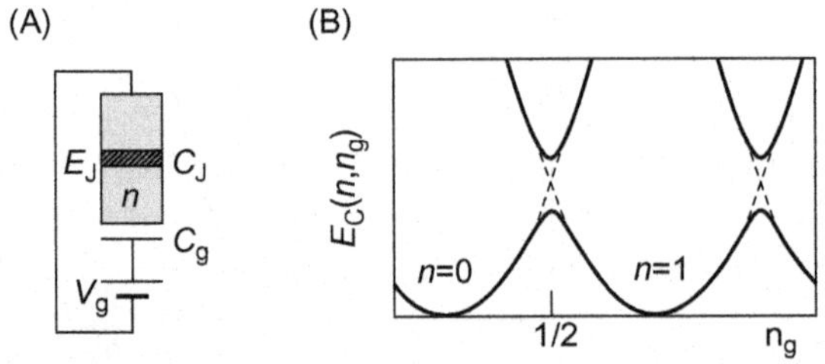

Figure 8.6 (A) A typical circuit diagram of charge qubits. (B) The dotted line denotes charging energy at $n = -1, 0, 1$. Solid line represents the energy curve of the charge qubit in quantum state at different n_g. Josephson coupling energy makes it possible to produce anti-crossing of energy levels at $n_g = 1/2$.

superconducting rings. The Josephson junction has a superconducting phase difference that satisfies the equation:

$$\sum_k \gamma_k = \frac{2\pi\Phi}{\Phi_0} + 2n\pi \quad (8.1)$$

Here, Φ is the magnetic flux contained in the ring, including the applied external field Φ_{ext} and the combined effect of continuous current. Φ_0 is a constant and $\Phi_0 = h/2e$ is magnetic flux in the magnetic field that causes an increase in phase of 2π.

Another type of superconducting quantum bits are charge qubits. We can use the phase conjugate physical quantities, that is, electric charge, as the two basic states of quantum bits. The principle of charge qubits is similar to that of single-electron devices. Figure 8.6A shows a typical superconducting single-electron device.

In addition, the smaller Josephson junction area of connected devices plus the lower volume of the central superconducting region (also known as the island, or box) may then reduce the equivalent capacitance of the central island. This creates Cooper pairs with a charging energy E for access to the central island greater than the disturbance of heat energy. Therefore, at low temperatures, we can see that the Cooper pairs are hindered in their access, with the whole line rendering a breaking circuit phenomenon, which is called the *charge effect*. Interestingly, when we use the gate voltage to change the potential energy of the central island, the electron energy for the access of a Cooper pair can be changed to zero. At this point, the line turns into an access but the charge transfer can only come through a central island one-by-one. When the charge effect arises, the amount of charge on the island becomes an important physical quantity. In general, we can write the following Hamiltonian to describe it:

$$H = 4E_C(n - n_g)^2 - E_J \cos\theta \tag{8.2}$$

Here, n represents the charge amount on the island described in the number of Cooper electron pairs. θ is the superconducting phase of the central island. n_g is the equivalent amount of charge caused by the gate voltage. Because θ and n are conjugates, and because the contribution that can be obtained from Josephson energy is actually $+1$ or -1 in the number of Cooper electron pairs on the island:

$$H = \sum_n 4E_C(n - n_g)^2 |n\rangle\langle n| - \sum_n E_J(|n+1\rangle\langle n| + |n-1\rangle\langle n|) \tag{8.3}$$

As n_g changes between 0 and 1, the lowest two energy states are $n = 0$ and $n = 1$. If we consider only two states, $n = 0$ and $n = 1$, then it can be viewed as a two-level state. Described in a spin 1/2 model, $4E_C(1 - 2n_g)$ is the magnetic field in the z direction, whereas E_J is that in the x direction. An ideal superconducting charge qubit can be the connection between the left and right sides to form a Josephson junction with two superconducting rings, similar to a quantum interference device, and may be called the controlled SQUID. The benefit is that as we change the magnetic flux coming through the superconducting rings, the equivalent of the Josephson coupling can be adjusted as $E_J = E_J^0 \cos|\pi\Phi/\Phi_0|$, the principle described previously. Therefore, we can regulate the size and direction of the equivalent magnetic field by changing the gate voltage and magnetic flux. Devices of this kind are not in a simple two-level state. To simplify the infinite charge states into two, we must widen the energy gap for each energy state. To achieve this, we require $E_C \gg E_J$.

In fact, more and more types of superconducting devices can be used as quantum bits. A suitable two-level state that can be found is not necessarily required to be the magnetic flux or charge. For example, a single Josephson junction can be used as qubits. In the Josephson junction system, there is a natural formation of energy levels with equal intervals of $\hbar\omega_p = \sqrt{2E_C E_J}$. A current coming through the Josephson junction tilts the wave-like potential energy wells equivalent to the Josephson effect. Under certain conditions, two more stable phase states can be acquired while the other higher energy states would decay quickly. The remaining two stable states become quantum bit systems that we need. Further manipulation of the quantum bits can be performed by changing the current size and the

Josephson coupling energy. But its impact is less straightforward, as for the magnetic flux and charge qubits described previously.

Another possible improvement is the use of a Josephson π junction to supersede the current 0 junction. General Josephson junction has a current of zero if the phase difference is zero. The π junction has a current of zero if the phase difference is π. This is equivalent to the potential energy curves of the 0 junction being translated for a distance of π. This makes the superconducting rings using a π junction in the absence of the applied field able to automatically achieve the two magnetic flux states from energy degeneracy. At present, there are three types of experimentally feasible π junctions. The first is the use of d-wave properties of HTSs, in which the superconductor crystal plane is rotated by 90° to complete the bonding. The second includes a magnetic metal layer being added to the middle of the superconductor junction. Magnetic material flipping its electron spin is used to change the phase of Cooper electron pairs in tunneling. The third involves a metal layer being added to the middle of the superconductor junction, which is injected with high-energy electrons. Inside the metal layers, the electronic distribution of a nonthermal equilibrium state may result in changes in the relationship between superconducting current and phase. Because the third type of junction is located in a high-energy excited state, one can imagine that its systems may have a very short coherence time. For this reason, research on quantum bits is more prone to obstacles. They can be fine-tuned from the 0 junction to the π junction for a single device. Theoretically, another particular Josephson junction was proposed, which is formed by the combination s-wave–d-wave–s-wave, with the received energy of the Josephson junction being $-E_J \cos(2\gamma)$. Linked with the general Josephson junction in the parallel state, it can also be a quantum bit device.

References

[1] K. Huang, R.Q. Han, Solid State Physics, Higher Education Press, Beijing, 2003.

[2] G.C. Che, A.F. Dong, Z.X. Zhao, Introduction to the direction, way and method for exploration of new superconductors, J. Low Temp. Phys. 27 (2005) 97–102.

[3] National Science Library of Chinese Academy of Sciences (Wuhan branch), Dynamical Monitoring Bull. Scient. Res. (New Mater. Adv. Manufacturing) 15 (2006) 1–10.

[4] X.S. Wu, P.W. Adams, Y. Yang, R.L. McCarley, Spin proximity effect in ultrathin superconducting Be–Au bilayers, Phys. Rev. Lett. 96 (2006) 127002.

[5] S.S. Yan, Basis of Solid State Physics, second ed., Tsinghua University Press, Beijing, 2000.
[6] J.M. Luttinger, An exactly soluble model of a many-fermion system, J. Math. Phys. 4 (1963) 1154.
[7] H.C. Guo, Q.D. Chen, Progress of quantum computing in small superconducting structures, Physics (Bimonthly) 25 (4) (2003) 510.
[8] R. Waser (Ed.), Nanoelectronics and Information Technology—Advanced Electronic Materials and Novel Devices, second, corrected ed., Wiley-VCH Verlag, Weinheim, 2005.
[9] Y. Makhlin, G. Schon, A. Shnirman, Quantum-state engineering with Josephson-junction devices, Rev. Mod. Phys. 73 (2001) 357–397.
[10] G.F. Mao, Y. Yu, Superconducting quantum bits based on Josephson devices, Prog. Phys. 26 (1) (2007) 34.

9

NANOBIOLOGICAL MATERIALS

CHAPTER OUTLINE

Nanomaterials and Devices. DOI: http://dx.doi.org/10.1016/B978-1-4557-7754-9.00009-3

Based on life science, modern biotechnology refers to the design and building of novel materials or systems with specific expected performances by using the features or functionality of organisms (or tissues, cells, and other components of organisms). It also integrates technologies with engineering principles to process goods or to provide services. The scope of this technology is very wide, including gene technology (e.g., "cloning"), which is used for reforming or restructuring the biological genes and enabling the recombinant gene to be expressed in cells to generate new materials that humans need: biological molecules engineering technology, which starts from the simple, common raw materials, is used to design the best route and select the appropriate enzymes in the synthesis of essential functional products; biological production technologies (e.g., fermentation) that use biological cells in a large number of processing, manufacturing products; biological coupling technology, by which the biological molecules and the electronic, optical, or mechanical systems can be linked together while the information captured by biological molecules can be amplified, transmitted, and converted into optical, electric, or mechanical information; nanobiotechnology, which can be used to study the fine structure of biological macromolecules and their functions in the nanoscale and to transform their structure to assemble molecular devices; and bionic technology that simulates biological objects or biological systems, organizations, and organ structure.

As an integration of nanotechnology and biotechnology, nanobiotechnology will become an important component of modern bioengineering. It has promising potential applications in the biomedical sector, electronics, materials, and environmental science.

Nanobiotechnology research is mainly focused on two aspects: first is the use of existing biological molecules such as DNA, proteins, and ion channels for the development of self-assembly nanotechnology and exploration of nanobiology; and the second is the use of processing nanomaterials and technologies, such as gold nanoparticles, branched molecules, quantum dots (QDs), and magnetic nanoparticles (magnetic fluid), and through appropriate chemical modification or in connection with biological molecules, making them applicable in detection, drug delivery or release treatment, biomimetic roles, gene transfer, cell separation, specific biological molecules, or separate calibration.

Nanoparticles for biological applications can be divided into two categories: organic and inorganic. Organic nanoparticles, such as the gelatin nanoparticles made from protein, can bond with DNA or peptides by modifying their surface amino acid content. Another example is the polymer nanoparticles from monomer polymerization, such as acrylic acid and styrene, in addition to the biomedical research purposes.

Inorganic nanoparticles, such as silica nanoparticles as a carrier of DNA and protein by chemical modification, are applicable to analysis detection and gene transfer purposes. Gold nanoparticles, magnetic nanomaterials, branch molecules, and QDs also fall in this category.

9.1 Nanobiological Materials [1,2]

9.1.1 Overview

Related to both life science and material science, biomaterials are natural or synthetic materials that are used for diagnosis, treatment, replacement, or repair of damaged tissues or to enhance function. Current applications cover a variety of biological materials, such as inert or active implant materials, drug delivery materials, dental materials, sutures, wound membranes, catheters, adhesives, and other materials that may have contact with body fluids. In addition, there are sensor materials, probing materials, and electrode materials that can be used in medical diagnostics and medical equipment.

In the context of this chapter, biological materials largely comprise biopolymers, with the rest including metallic materials for medical purposes, bioceramic materials, and biocomposite materials. Depending on function, biological materials can be divided into bioinert materials, biologically active materials, biodegradable materials, and bioabsorption materials.

Biocompatibility is the key to biological materials in research and applications. It refers to the chemical, physical, and mechanical reactions in the interaction of biological materials and organisms, including immune response, blood reaction, tissue reaction, and biochemical reaction.

There are two major types of applications of nanobiological materials. Novel nanomaterials that are developed with functional biological molecules are exploited for their self-assembly properties to create multifunctional materials. Other nanomaterials are suitable for biomedical applications and may or may not have biological activity without causing adverse reactions. The main materials are polymer nanoparticles, inorganic nanoparticles, and the nanostructured biomaterials in tissue engineering with specific identification and orientation-induced functions.

Polymer nanobiomaterials may include nanoparticles, nano-microcapsules, nanomicelles, nanofibers, and nanopore structured biological materials. Polymer nanoparticles have particle sizes in the range of 1–1,000 nm, with a large surface area and the emergence of new properties and new features different from common materials. Polymer nanobiological materials can be synthesized by using methods such as microemulsion. Polymer nanobiomaterials exhibit their applications mainly in immunoassays, drug and gene carriers, DNA nanotechnology, gene therapy, nanoliposomes (the drug delivery system for imitated biological cells), biomolecular adsorption separation, invasive diagnosis, and treatment. Immunoassay here refers to the quantitative analysis of protein, antigen, antibody, and the cell as a whole. It can be divided into fluorescence immunoassay, radioactive immunoassay, and enzyme-linked analysis by means of the markers used in the process.

Immunoassay operations can be performed through the following three steps: in a specific carrier, covalent bonding corresponds to the immune affinity molecules of the analysis object; solution containing the analysis object is cultured with the carrier; and by detecting the amount of free carriers through a microscope, accurate quantitative analysis can be performed on the analysis object. Immunoassay carriers are usually selected as the polymer nanoparticles, particularly some of the particles with a hydrophilic surface. They have a very small amount of adsorption of nonspecific protein and are widely used as a marker carrier.

9.1.2 Drug and Gene Carrier Nanomaterials

In nanobiomaterial research, the current hotspots are drug nanocarrier and nanoparticle gene delivery technology. This technique is based on nanoparticles as drug and gene transfer vectors, by which the molecules for gene therapy, such as drugs, DNA, and RNA, are wrapped in the nanoparticles or adsorbed on the surface. At the same time, they couple with specific target molecules on the surface, such as specific ligands and monoclonal antibodies. The target molecules are combined with specific cell surface receptors and can find a way into specific cells for safe and effective targeting of gene therapy.

Drug nanocarriers come with the advantages of a highly accurate degree of targeting and high level of drug release control and can improve the dissolution rate of insoluble drugs and absorption rate, drug efficacy, and lower toxicity. Nanoparticles as gene vector have the following significant advantages: they can be used for wrapping, concentrating, and protecting nucleotides from degradation via nucleases; they have a larger specific surface area plus biocompatibility; they easily couple to specific targeting molecules on the surface to achieve specificity of gene therapy; cycle time in the circulatory system is significantly prolonged as compared with ordinary particles and, in a certain period of time, it will not be rapidly cleared by phagocytic cells like ordinary particles; they allow a slow release of nucleotide and effectively extend the time of action and maintain an effective concentration of the product to enhance the transfection efficiency and bioavailability of the transfected products; they have fewer metabolites, fewer side effects, and no immune rejection reactions.

Biodegradability is one of the features that cannot be ignored for drug carriers or gene vectors. Through degradation, carriers and drug/gene fragments can be directed into the target cell, keeping the surface of the carrier biodegradable, while the drugs contained in the core are released to provide a curative effect. In this manner, drug release in other tissues can be avoided.

Drug nanoparticle carrier (nanoparticle drug delivery) technology, as one of the important development directions of nanobiotechnology, will bring great changes to the treatment of diseases like malignant tumors, diabetes, and Alzheimer's disease. Nanomaterials as drug carriers are largely divided into the following two categories: synthetic polymer materials, such as poly-alkyl cyanoacrylate ester (PACA, including methyl, ethyl, butyl, iso-ester, hexyl ester, as well as dissidents ester, hexadecyl ester) and polyester (mainly including polylactic acid (PLA), polylactide, polycaprolactone (PCL), PCL ester, poly-hydroxybutyric acid, glycolic acid, etc.) and their derivatives and copolymers. Drug-loaded nanoparticles prepared from these materials mainly include the following types.

9.1.2.1 Nanolilmsome

Nanolilmsome (lipid vesicles) has become one of the research hotspots in recent years. It is easy to prepare, convenient to use, and can be used for multipurpose administrations. It is a carrier with a phospholipid bilayer structure similar to that found in natural systems. Nanolilmsome as a drug carrier has unique advantages, including protecting the drug from degradation, helping drugs get to target sites, and reducing toxic side effects. Nanolilmsomes have had some issues, such as low encapsulation efficiency, rupture-prone liposome membrane, leakage, poor reproducibility, and *in vivo* instability and quick release. Preparation of nanoliposomes is mainly dependent on the ultrasonic dispersion method and reverse-phase evaporation method.

9.1.2.2 Solid Lipid Nanoparticles

Solid lipid nanoparticles (SLNs) are prepared by the use of carriers of a variety of lipid materials, such as fatty acids, fatty alcohols, and phospholipids; the medicines are wrapped in lipid materials to make solid particles. SLNs have a certain sustained-release effect, mainly suitable for the wrapping of insoluble drugs. They are used for intravenous injection or local administration to achieve target positioning and controlled release, thus avoiding drug degradation and leakage. SLNs are mainly applicable to lipophilic drugs; for hydrophilic drugs, they have the defect of low encapsulation efficiency. SLNs are commonly prepared by the high-pressure emulsion smoothing method and the microemulsion method.

9.1.2.3 Nanocapsules and Nanospheres

Nanocapsules (NCs) and nanospheres (NSs) are mainly prepared by using biodegradable polymer materials, such as PLA, polylactide/glycolide, chitosan, and gelatin. They can be used to wrap hydrophilic or hydrophobic drugs. The route of administration depends on the performance of different materials, such as their intravenous targeting effect and intramuscular or subcutaneous sustained-release and controlled-release effects. NCs and nanoparticles in oral delivery can also be nondegradable materials, such as ethyl cellulose and acrylic resin. Preparation of such carriers is mainly through precipitation, emulsification-solvent evaporation, and other methods.

9.1.2.4 Polymer micelles

Polymer micelles are novel nanocarriers that have been developed in recent years, with both hydrophilic groups and hydrophobic groups. Dissolved in water, they will spontaneously form polymer micelles and complete the solubilization and wrapping of drugs containing a hydrophilic shell and hydrophobic core suitable for carrying different types of drugs. In addition, they enable drugs to escape phagocytosis by mononuclear macrophages, that is, they have a "stealth" feature. The hydrophilic chain segment is commonly made from polyethylene glycol (PEG), polyoxyethylene (PEO), and polypropylene, whereas the hydrophobic chain segments are made from PLA (P), polylactide glycolide, chitosan, and so on. The current study highlights the block copolymer PLA–PEG of PLA and PEG, whereas chitosan and its derivatives are being closely studied because of their excellent biodegradable properties.

Another type of nanoscale drug carrier is prepared from natural materials such as lipids, sugars, DNA, proteins, and so on. DNA nanotechnology is designed on the physical and chemical principles of DNA mainly used in molecular assembly. Self-assembly of nanoparticles is conducted in the following ways: a single strand of the DNA fragments is connected to the surface A of nanogold particles with diameter of 13 nm, and then another single-stranded DNA fragment with a complementary sequence is connected to the surface B of nanogold particles. A and B are mixed, then, in the DNA heterozygous condition, A and B are automatically linked together. Finally, based on the complementary features of double-stranded DNA, the automatic assembly of nanoparticles is completed.

Plasmid DNA can be used for gene therapy. After entering the target cells, it can repair the genetic errors or produce therapeutic factors (e.g., peptides, proteins, antigens). Nanotechnology enables DNA targeting in cells through an targeting role, and plasmid DNA will be condensed to the size of 50–200 nm and applied with negative electricity, contributing to the effective invasion of the nucleus. The process depends on the nanoprocess techniques available.

9.1.3 Bioceramic Nanomaterials

Bioceramic nanomaterial was developed in the mid 1980s. This type of new materials is composed of microstructures at the nanoscale level, with its grain size, grain boundary width, second phase distribution, pore size, and defect size only limited to the order of 100 nm.

Bioceramics are nontoxic, free from side effects, and have good compatibility with biological tissues. They are applicable in the production of artificial bones, bone nails, artificial teeth, dental implants, and nails within bone marrow, ranging from short-term replacement and filling to permanent transplantation. They have been developed from bioinert materials in bioactive materials. Nanobioceramics can enhance biological properties and mechanical properties, showing a unique superplasticity, whereas the increased chemical activity leads to higher biological activity.

Bone-like nanoapatite crystals and polyamide synthetic polymer can be synthesized into a complex as artificial nanobone, which can bond with the natural bones and grow tightly together with human muscle and blood vessels. In addition, it can induce cartilage formation, showing a variety of features equivalent to those of human bones. Bioceramic nanomaterials can also be used in making artificial eyes; by using nanoceramic material in the making of the shell of the construct, the artificial eye can be controlled by the muscles to move around, or electrical pulses are applied to stimulate the nerves of the brain to achieve the visual effect.

Bioceramic nanomaterials can be used for nanocell separation technologies, cell staining, as well as antibacterial function and sterilization. In nanocell separation, the first step is to produce amorphous SiO_2 nanoparticles, with the size controlled within 15–20 nm. Its coated monomolecular layer works as the adhesion layer, and the finished size of coating is approximately 30 nm; then, in the second step, PVP colloidal solution containing multiple cells is produced by uniformly mixing the two and using centrifuge technology to separate out the desired cells. The intracellular staining technique requires selection of the type of antibody. Nanogold particles are mixed with prepurified antibodies or the monoclonal antibody to prepare a variety of (nanogold antibody) complexes; respectively, composite particles are combined with various intracellular organs and skeletal systems to form different complexes, by which certain featured colors will be rendered in the irradiation of white or monochromatic light, and thus different colored labels are affixed to various combinations.

Once biological materials are used in the human body, the surrounding tissue may be at risk of infection. Therefore, antibacterial operation and sterilization are especially important. In the reaction to synthesize hydroxyapatite (HA) nanoparticles, silver, copper, or another aqueous solution of soluble salts is added in the reaction, so that antibacterial metal ions are able to get into the apatite crystals. A typical representative of nanomaterials with bactericidal or antivirus features is known as photocatalyst titanium dioxide (TiO_2) and works only in ultraviolet radiation, as described in greater detail in Chapter 10.

9.1.4 Magnetic Nanoparticles

Magnetic nanoparticles have a good surface effect, which is characterized by surge of specific surface, larger functional group density, and selective adsorption capacity, so the capacity for carrying drugs or genes is increased. In the physical and biological sense, paramagnetic or superparamagnetic iron oxide nanoparticles may experience a temperature increase up to

40–45°C under the influence of an external magnetic field, which can kill tumor cells.

Practical application of magnetic nanoparticles for biomedicine covers the following aspects: coating magnetic nanoparticles with polymer materials and combining them with targeting moieties. Then, they can be used as drug carriers in the human body. Under the action of an external magnetic field, magnetic navigation is applied to move the carriers to lesions to complete the treatment; 10–50 nm of the Fe_2O_3 surface coated with methacrylic acid may have its size increased to 200 nm and is able to carry proteins, antibodies, or drugs for cancer diagnosis and treatment. Its local treatment effect is good, with few side effects.

9.1.5 Biocomposite Nanomaterials

Inorganic–organic nanocomposite materials are most commonly found in biological organisms. Biomineralization refers to the process of minerals (biominerals) being formed inside a living body. In the biomineralization process, through the interactions on the interface of organic macromolecules and inorganic ions, inorganic mineral precipitation is controlled at the molecular level, so that biominerals have a special multilevel structure and assembly modes. During the process of biological mineralization, cell secretion can complete the auto-assembly of organic matter, which acts as a template for the formation of inorganic compounds, so that inorganic mineral has a certain shape, size, orientation, and structure.

To a large extent, the manufacture of nanobiological materials is inspired by the biomineralization process. In nature, the cell membrane of some bacteria may be mineralized at different levels. Outside the cell membrane, the protein molecules in regular arrangements can be used as a template to induce the synthesis of microstructured nanomaterials. Biomineralization provides an effective means for the design and processing of nanobiomaterials. US scientists have found a gender-peptide molecule; one end is the hydrophilic arginine-glycine-aspartic acid (RGD), and the other end is amino acids containing phosphorylated residues. RGD helps the adhesion of materials and cells, whereas the phosphorylated amino acid residue can interact with calcium ions.

Tissue engineering is the research and development of organism alternatives aimed at restoring, maintaining, and improving organizational functions by the application of the basic principles and methods of engineering and life science. The idea is first to isolate cells *in vitro* for culturing, and then a certain amount of the cells are grown onto a three-dimensional biological material to be cultured further; eventually, some tissues and organs with a certain structure can be formed. Stent materials play an important role in tissue engineering, because adherent-dependent cells are only adhesive to the materials before they can grow and divide. By imitating the natural extracellular matrix-collagen structure, nanofibers containing biodegradable materials are used *in vitro* and in animal experiments in tissue engineering; they demonstrate an attractive application potential. In composition, the currently reported nano-HA/collagen composite is an imitation of inorganic and organic components in natural bone matrix, with a nanoscale microstructure similar to natural bone matrix. The 3-D stent formed from porous nano-HA/collagen composite provides a microenvironment similar to the body for osteoblasts. Cells are able to grow well and secrete the bone matrix on such stents. Experiments *in vitro* and animal experiments show that such an HA/collagen composite is a good nanobiological material for bone repair.

As the biodegradable polymer material for tissue regeneration template, nanostructured tissue engineering stent material has the following properties: (1) good biological compatibility; (2) cells having good adsorption and proliferation at the material surface; (3) capability of inducing cell growth by prefabricated patterns; and (4) after tissues are grown, the materials can be degraded and excreted through metabolism (Table 9.1).

9.2 Nanobiomedical Materials

There are a wide range of nanobiomedical materials, from nanobiomedical materials (including drug carrier, drug release

Table 9.1 Nanoscale Tectonic Units in the Living World

Erythrocytes	6000–9000 nm
General bacteria	2000–3000 nm
The size of virus particles	~50 nm
Protein	1–20 nm
RNA protein complex	15–20 nm
Diameter of DNA chain	~2 nm

control, composite materials) to the nanodrugs getting into cells, as well as nanotechnology that is conducive to drug absorption, targeting, and controlled release. The following is a detailed description of both the inorganic and organic aspects.

9.2.1 Nanobioinorganic Materials

Nanoinorganic biological materials mainly include ceramic oxides, alumina, titanium dioxide, porous glass granules, nanocarbon materials, nanosilicon oxide (SiO_2), and nanomagnetic particles.

1. Ceramic Oxides

   Traditional ceramics have long been used to make artificial bones in clinical practice, as well as bone screws, artificial teeth and dental implants, artificial sponge free from impact of bearing weight, and bone intramedullary fixation materials. As for nanoceramics, its strength, hardness, toughness, and plasticity are significantly increased, so it has great future potential for artificial organs and other clinical applications.

2. Alumina, Titanium Dioxide, and Porous Glass Granules

   The adhesion of alumina (Al_2O_3) and titanium dioxide (TiO_2) on bone cells has been greatly improved, so they can be better combined with living cells. Nanoporous glass powder is a novel inorganic nanobiological material. Its applications in the biomedical field include functional matrix materials, microreactors, or a biochemical separation matrix, bioenzyme catalyst carrier, or the carrier of the drug controlled release system.

3. Nanocarbon Materials

   Carbon is an essential element in human tissues. Studies of artificial blood vessels revealed that carbon has an excellent antithrombotic function. Nanocarbon materials are featured with fewer defects, large surface area, and compact structure. When it is used in artificial organs, artificial bones, artificial teeth, and artificial tendons, the strength, hardness, and toughness can be increased. It can also be used in blood purification systems to remove particular viruses or components.

4. Nanosilica

   Nanosilica (SiO_2) can be used to separate a very small amount of fetal cells from the blood samples of pregnant women at approximately 8 weeks of gestation and then accurately determine whether there are genetic defects in the fetus. It can also be used to detect cancer cells in the blood of a tumor in its early stages or check the muscle protein in blood to help treat heart disease.

5. Nanomagnetic Particles
   Iron oxide (Fe_3O_4) with a diameter of approximately 10–50 nm, after being coated on polymer material, followed by externally binding with protein, can be injected into organisms. Nanomagnetic particles with polymer and protein can be used as a drug carrier. Through intravenous injection, such drug carriers can enter organisms. Navigated by an external magnetic field, it can carry the drug to the lesions.

9.2.2 Nanoorganic Biological Material

9.2.2.1 Nanopolymeric Biological Materials

Nanopolymeric biological materials can be generated from microemulsion polymerization, with particle diameters of approximately 10 nm. As such nanoparticles are much smaller than red blood cells and can move freely in the blood, and they can be injected into all parts of organisms for inspection and treatment of diseases. Also, they are suitable for the production of drug carriers in different forms and the biofilters used to separate biological substances. Among this type of materials, lactic acid–acid polymers can be used as drug carriers. Animal experiment results show that carriers containing dexamethasone may be effective in the treatment of arterial stenosis by arterial administration. Polymers containing the antiproliferative drug are appropriate in the prevention of coronary artery stenosis by coronary delivery.

Polymer materials can also be used for drug delivery. Nanoparticles with inclusive or surface-combined vaccines can provide sustained release of the wrapped antigens. Polymethacrylate nanoparticle anticancer drugs in nanoparticle sustained release are able to extend the retention time of the drug in the tumor, preventing the growth of the tumor. As tumors normally have higher vascular permeability, nanodrugs administered by intravenous operation can be transmitted largely within tumors, thus improving efficacy and reducing dosage and toxicity.

Nanopolymer materials can be modified to increase the targeting specificity toward the tumor. For instance, PEG-modified nanoparticles can reduce the uptake by the reticuloendothelial system while increasing the uptake of tumor tissue to avoid the toxicity caused by nonspecific aggregation. By the use of antibiotic therapy of intracellular infection, a study has shown that nanoparticle-coated Ampicillin may have an effect 20 times higher than the dissociative Ampicillin.

Polymer materials can be used as gene delivery systems as well. The use of poly-cyanoacrylate alkyl to adsorb oligonucleotides has been proven to play a role against RNase in both the buffer solution and a cell culture medium to prevent nucleic acid degradation. Cholesterol can be adsorbed to poly-cyanoacrylate alkyl lipid nanoparticles and then transfected into human bladder cancer cells. This complex can complement the proto-oncogene mutation area, and thus inhibit the proliferation of bladder cancer cells in culture medium.

9.2.2.2 Nanobiocomposite Materials

Many nanobiocomposite materials are derived from natural tissues, such as the human body; most such tissues are composite materials, whereas teeth and bones are nanocomposite materials composed of nanoapatite crystals and polymers. Through the imitation of natural hard tissues, some medical composite materials have been developed.

In 1994, British scholar W. Bonfield made high-molecular-weight polyethylene and nano-HA. In the study, he indicated that this kind of nanocomposite had somewhat worse mechanical properties. Dutch scholars studied the composite material from nano-HA needle crystals and polyactive polymer. Because of the use of a dry HA needle crystal, this kind of composite material is easy to gather into clumps that may affect the dispersion in the composite materials.

In natural bones, nanoscale HA materials deposit in the collagen matrix. In 1992, Yu-bao Li and associates, by using the hydrothermal synthesis process, obtained nanoscale synthetic bone-like apatite crystals, which are very similar to natural bone apatite crystals in shape, size, composition, and structure.

A solution environment can be directly used to prepare the polyamide (PA66)/nanocrystalline nano-HA bioactive composite. Nano-HA can be uniformly dispersed in the composite, with a content up to 60%. This figure is close to the ratio of natural bone apatite, able to form a bond at the interface between the two phases. Composite material is close to human cortical bone in respect to compressive strength, flexural strength, and modulus of elasticity, and it has cartilage-inducing properties. Animal experiments demonstrated that nano-HA/PA66 has good biocompatibility and biological activity and is an ideal material for bone repair.

9.2.3 Nanotechnology in Drugs

Nanomedicine is essentially a nanocomposite material, which is the nanostructural system assembled and synthesized according to the design of the human body. They are made of nanowire or tubes as the basic unit that is assembled and arranged in one-dimensional, two-dimensional, or three-dimensional spaces, and are associated with stability, gastrointestinal irritation, toxic side effects, high drug utilization, targeted drug delivery, and sustained-release effect.

We know that smaller drug particles have a greater dissolution rate than larger drug particles. So, by controlling drug particle size, we can control the dissolution rate of particles. For water-soluble drugs as oral tablets made of granules, particle size is the key to controlling a drug's pharmacological effect. For nonwater-soluble drugs, stable water suspensions can be made for subcutaneous injection into the body and can travel with the blood circulation. In this case, the size of drug particles should be controlled very strictly to not block the blood circulatory system. For gas sol–spray pharmaceutical agents, particle size is the key factor determining its effectiveness. The current pharmaceutical particles can only reach the micron level.

In applying nanotechnology in pharmacy, QDs and magnetic nanomaterials have enjoyed a high priority in recent years. QDs are nanoscale metals and can be used as novel fluorescent labeling materials because of their light-emitting characteristics. Magnetic nanomaterials are mainly iron oxide nanomaterials with excellent biosecurity and biocompatibility. Because the nanoscale can lead to an improvement of magnetic properties, iron oxide nanomaterial has become an independent development in the field of biomedical testing and treatment.

9.2.4 Biochips

The biochip is an integrated assembly of one or a variety of biological activities on a small surface; by using tiny samples of physical or biological matter, it can simultaneously detect and study different biological cells, biomolecules, and DNA characteristics, as well as the interaction between them to acquire the life patterns of microactivity. The biochip can be approximately divided into a cell chip, protein chip (biological molecules chip), and gene chip (A-chip), all of which have the advantages of integration, parallel and rapid detection. It has become cutting-edge biomedical engineering technology in the twenty-first century. DNA chip, a representative of the biochip, has been developing rapidly recently and is typically symbolized by nanotechnology, integration, and multifunction. The scope of its use covers molecular biology, disease prevention, diagnosis and treatment, drug development, the development of biological weapons, judicial appraisal, and monitoring and supervision of environmental pollution and food hygiene.

The biochip in a narrow sense is a *microarray*, including DNA microarrays, oligonucleotide arrays, protein microarrays, and microarrays of small-molecule compounds. Addressable elements are fixed by way of a lattice to the surface of a certain size of substrate (silicon, glass, plastic, etc.). Each dot in the lattice can be regarded as a sensor probe. Biochip operation typically includes the following steps: first, the object of detection can be labeled using a chemical fluorescence method, enzyme label method, isotope marking method, or electrochemical method; second, the labeled object of detection getting into the biochip will respond at the appropriate address, and this specific address will be displayed; third, a scanner is used to search for biochip marks on the display for analysis with calculating software.

Biochips, in a general sense, can be used for rapid parallel processing and analysis of biological components or molecules. They are a kind of solid-thin device with a size of less than a cubic centimeter. Typical biochips include the microarray chip, filtration separation chip, dielectrophoresis separation chip, biochemical reaction chip, and capillary electrophoresis chip.

Instruments commonly used for observing and testing the biochip are the transmission electron microscope and scanning probe microscope. Transmission electron microscopy is required to be conducted in a high vacuum to observe dry samples. There are many types of scanning probe microscope, such as scanning tunneling microscope, atomic force microscope, magnetic force microscope, and scanning electrochemical microscope.

9.2.5 Future Development of Nanobiomedical Materials

The development of biomedical materials at the nanoscale is mainly focused on nanorobots, nanotargeted drugs, invasive diagnostic capabilities and intelligence, drug delivery systems, and medical complex material.

9.2.5.1 Nanorobots

Nanorobots are the most attractive field in nanobiology. Controlled by nanomicroelectronics, coupled with nanobiomedical materials and nanomedical inorganic materials and crystal structure, nanorobots will be made at a size even smaller than red blood cells of humans. This will certainly bring about a profound revolution in human health care. The first generation of nanorobot is an organic combination of biological and mechanical systems, such as a combination of enzymes and nanogears. Such nanorobots can be injected into veins to become molecular robots able to complete operations in blood vessels. These robots may get energy from glucose and oxygen dissolved in the blood, and they are able to explore any object they have encountered according to preprogrammed procedures. Molecular robots can perform a full medical check, clearing a thrombus and the fat deposits in heart arteries, eating bacteria, killing cancer cells, or monitoring the body's diseases. The second generation of nanorobots is nanoscale molecular devices, which are assembled directly from the original elements, presented with specific functions. The third-generation nanorobot is a nanocomputer. This is a device equipped with the function of performing human–machine dialog. Such nanorobots, once successfully developed, would be able to complete several billion calculations within 1 s, thus bringing about a revolutionary change to humanity.

9.2.5.2 Targeted Nanomedicine

Drugs made in nanosize are directly injected into the lesion and will greatly enhance the medical effect and reduce side effects. As such, drug targeting will be a hot topic.

9.2.5.3 Capabilities and Intelligence of Invasive Diagnosis

Nanobiomedical materials will enable interventional diagnosis and treatment to develop in the direction of microtype, trace, and mini-invasive or noninvasive, and will help develop specific and intelligent materials.

9.2.5.4 Drug Delivery Systems

Nanodelivery systems can have a slow-release effect and create a novel route of administration. Under the premise of ensuring drug action, the system is designed to reduce the dose to reduce or avoid toxicity. Furthermore, the system can improve the stability of drugs, making them easy to store; a nano-controlled release system, through proper modification, can also come through the blood–brain barrier to deliver drugs to the central nervous system. Pills, such as "mini-pills" or "drug-loaded nanoparticles," which contain sensors, storage capsules, and micro-pressure pumps, can be developed; "mini-pills" are pumped to the designated location inside the body to release drugs, which is a more effective method than the traditional ways of injecting or ingesting medicine. There are more advanced "nanomagnetic particles." The United States has successfully developed a targeting drug with nanomagnetic materials as a drug carrier, known as a "biomissile." Namely, drugs may be carried on the surface of the protein, wrapped with magnetic Fe_2O_3 nanoparticles, and injected into human blood vessels. Navigated through the magnetic field, it is transported to the lesions to be released. This can prevent the liver, spleen, and kidney from suffering side effects of drugs. Therefore, the nanodelivery system is a very promising form of drug dosage.

9.2.5.5 *Medical Composite Materials*

Nanobiocomposite material is widely present in the whole organism (e.g., bamboos, shells, bones, teeth), but real nanobiocompound synthetic materials are rare. With the simulation of human tissue composition, the structural and mechanical properties and bionic medical composite material of nanobioactivity appear to be very important in research.

9.3 Magnetic Particles in Medical Applications [3]

Miniaturization is the most significant advantage of nanomaterials in biomedical applications, but it also makes the separation of such materials more difficult. Although magnetic nanomaterials are also nanoscale particles but with magnetic properties, they are easily separated from other materials. This is the advantage of magnetic nanomaterials.

Magnetic nanomaterials are usually composed of iron oxide, e.g., Fe_2O_3 and Fe_3O_4. Among the current products, it is common to use embedded or coated magnetic particles in combination with antibodies in the cell separation and purification steps.

There are two ways to prepare magnetic nanomaterials: (1) if Fe^{2+} and Fe^{3+} are heated as a mixture, magnetic nanomaterials can be derived from the reaction precipitation, and (2) the mixture of Fe^{3+} and SO_3^{2-} under alkaline conditions can lead to the formation of Fe_3O_4 nanoparticles.

Magnetic nanoparticles for biomedical purposes must be dispersed in aqueous solution. However, magnetic nanoparticles are generally composed of a single crystal, such as Fe_3O_4, $MnFe_2O_4$, or $CoFe_2O_4$. These single crystals are not soluble in water, so magnetic nanoparticles need to be coated on the surface with a layer of hydrophilic surfactant, so that magnetic nanoparticles can be stably dispersed in water. The solution is the so-called magnetic fluid. In magnetic fluid, magnetic nanoparticles are subject to the effect from the thermal fluctuation of water molecules. In the absence of an external magnetic field, the fluid has no spontaneous magnetic dipole. But when a magnetic field is applied to the fluid, the magnetic moment of magnetic particles in the liquid tends to follow the direction of the external magnetic field, resulting in a magnetic dipole. When the applied field is removed due to the thermal fluctuation of water molecules, magnetic nanoparticles will again show a zero magnetic dipole. This phenomenon is called superparamagnetism. It enables magnetic particles to have a strong magnetic field in the presence of a magnetic field, but after the removal of the magnetic field, magnetic properties will also disappear accordingly. Because of this, magnetic nanoparticles are traceable, recyclable, and can also be analyzed quantitatively. Fe_3O_4 was also the first to be used in biomedical magnetic resonance imaging (MRI).

Magnetic nanoparticles have another important feature, i.e., their size is similar to that of biological molecules. Therefore, the proper transformation of the biochemical characteristics of the particle surface will enable the combination of particles with specific biological molecules, giving biological molecules a magnetic marker. At present, this characteristic is often used in biomedical technology for immunoassay, cell separation, and transgenic applications.

Regarding cells, particular cells may have particular biological molecules on the membrane. For example, the surface of stem cells has CD34, a unique biological molecule. Some antibodies corresponding to these particular biomolecules, therefore, can be coated on the magnetic nanoparticles and mixed with blood containing the separation cells. These specific cells will be coated with magnetic nanoparticles. As shown in Figure 9.1, the black circles represent the magnetic marked cells and the white circles indicate other kinds of cell. So, when a magnet passes near the blood, due to the magnetic attraction, magnetic nanoparticles will make that particular cell

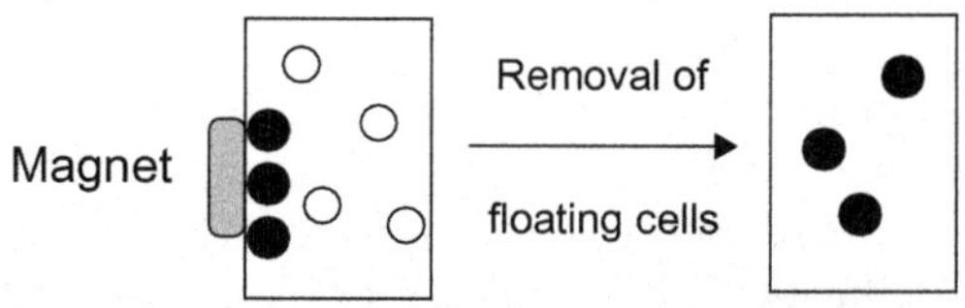

Figure 9.1 The use of magnetic nanoparticles in the separation of specific cells required. From Ref. [3].

(Figure 9.1, black circle) adsorbed by the magnet. Then, deionized water is used for washing the blood to remove the other kinds of cells adsorbed by the magnet, followed by removal of magnets; that is, the required specific cells will be separable out of blood.

The black circles are the separated cells that have been magnetically marked, and the white circles are the other kinds of cells.

Compared with traditional separation technology, magnetic separation technology is simple and requires less time. Moreover, it has higher separation efficiency and specificity. This technique can also be applied to the purification of proteins and other biological molecules, and it has a considerable impact on the biological and chemical industries, so it is highly valued by academia and industry.

Advantages in magnetic drug delivery are numerous. First, the dosage of drug is less. The drugs can be delivered to the lesions highly concentrated without the need of surgery. This will not affect the rest of the body with side effects. Its combination with MRI can be used for development of new drugs for the analysis of the disease on cell activity, side effects, and usage.

But magnetic nanoparticles also have some disadvantages. For example, the drape modification method used in preparing them will make the particle size increase from 5–10 to 5,000 nm. So, further development is needed for this method.

Magnetic nanomaterials have great development potential as immobilized carriers of enzymes and proteins, DNA and RNA used for the reagents and tracing agents of MRI, as well as purification and separation agents for biochemicals, cell separation, calibration and control, drug delivery or gene therapy, and biochips.

The following highlights the application of the surface nanomagnetic particles coated with bioactive agents in drug delivery and medical treatment.

We know that cancer patients after tumor excision still need to receive radiation therapy regularly to kill cancer cells in the body. However, radiotherapy will inevitably harm normal cells, especially the bone marrow cells in the body, which are responsible for hematopoietic function and the immune system. Therefore, radiation therapy requires the patient's bone marrow to be extracted first. After conducting radiation therapy on the cancer lesions, bone marrow will then be implanted. For terminal cancer patients, cancer-diseased cells often have spread to the bone marrow. After radiation treatment, cancer lesions will still be retained in the bone marrow that is to be replanted back into the matrix, so transplanting the matrix will just make the treatment procedures useless. So, before the bone marrow fluid is transplanted back to the matrix, the cancer cells must first be separated from the bone marrow fluid.

The following experiment is an example of separation with mouse bone marrow liquid to illustrate how magnetic nanoparticles work in cancer treatment processes. The experiment involved using Fe_3O_4 nanoparticles with particle size of approximately 50 nm and polystyrene with outer cladding diameter of 3 μm.

The experimental process of mouse bone marrow liquid separation can be summarized as follows:

1. Start by taking out antimouse antibody Fc, or immunoglobulin, from the sheep body.
2. Combine into complex molecules with magnetic nano-Fe_3O_4 particles coated with organic polymer.
3. Remove the bone marrow containing normal cells and cancer cells from the mouse body, and add in the hybrid antineuroblastoma antibodies, which will only combine with the cancer cells in bone marrow fluid.
4. Extract this antibody and adsorb it onto the magnetic nanoparticles covered with polymer. The obtained composite particles are injected into the bone marrow fluid containing the cancer cells, and then the antibodies on the magnetic particles will only combine with the cancer cells with antigen.
5. Using magnetic separation devices with nanomagnetic particles, cancer cells can be easily separated from the bone marrow fluid, with a differential rate of more than 99.9%.

In addition to cell separation, magnetic nanoparticles also have their application values in disease detection and drug treatment. Magnetic nanoparticles can be used as a drug carrier to enter into the animals through intravenous injection. Then, an appropriate magnetic field is applied *in vitro* in the animal to control the movement of nanomagnetic particles in the body. This is so-called magnetic navigation, which can guide the drugs to be released in a suitable location.

9.4 Nanoparticles in Bioanalysis

The rapid development of life science has highlighted many new issues of analytical chemistry currently focused on the analysis of peptides, proteins, nucleic acids, and other biological macromolecules, as well as biopharmaceutical analysis, ultra-trace analysis of bioactive substances, and even microbiological analysis. Therefore, biochemical analysis has become one of the most important frontiers in the development of modern analytical chemistry. Many researchers in analytical chemistry are trying their best to find new methods and techniques. The application of nanoparticles is one key initiative.

Fluorescence analysis is commonly used in the clinical determination of the content of certain elements in biological samples, of which the determination of RNA and DNA is of most importance. Fluorescent probes are commonly used in this method.

Normally, the fluorescent agent in such probes is an organic dye but, in most cases, their excitation spectra are narrow, so it is difficult to stimulate a variety of components at the same time, and the distribution of the spectra is asymmetric. Using nanoparticles as biological fluorescent probes can address these issues well. Compared with conventional fluorescent probes, nanocrystals have wider excitation spectra and a continuous distribution, whereas the emission spectrum is symmetrically distributed with narrow width and adjustable color; that is, nanocrystals of different sizes can be excited by a single wavelength of light to show different colors of light. Meanwhile, it has higher photochemical stability and is resistant to photolysis.

There are three types of nanoparticles that can be used as fluorescent tags:

1. Metal nanoparticles with optical activity.
2. Fluorescent nanoparticle emulsion (commercialized).
3. Light-emitting QDs.

The following is an overview of these three kinds of nanoparticles and their development prospects in bioanalysis.

1. Use of metal nanoparticles in bioanalysis

   When gold (Au) particles reach the scale below 10 nm, the surface activity will be much larger, making it qualified as a catalyst. With the change in the size of gold nanoparticles, the phenomenon of wavelength shift will arise from its optical color due to the differences in quantum effects, from large to small, including yellow, orange (100 nm), green (50 nm), and red (13 nm). The 13-nm gold nanosolution, after the inclusion of NaCl, will result in condensation effects that change it from the original dark red solution to a blue one.

   Preparation of Au nanoparticles can be achieved using the following three methods: (1) laser shot stripping method, which uses a laser to break gold blocks into nanoparticles; (2) gas phase synthesis, whereby gold atoms are changed into gas through gasification and condensed into nanoparticles during aggregation; and (3) chemical reduction, which uses hydrogen tetrachloroaurate and sodium citrate to produce the gold nanoparticles by mixing them in an appropriate proportion.

Gold nanoparticles are mainly used to probe the genetic labeled material and as a fluorescent quencher.

Recently, Elghanian and colleagues proposed a highly selective colorimetric method for polynucleotide detection. They used metal nanoparticles as the receptor, which was modified with polynucleotide alkanethiol. After hybridization, polynucleotide probes not only bind with the target polynucleotide by a specific sequence but also form a converged network with each receptor cell connected to a number of short double-helix fragments.

With the progress of hybrid, system colors will vary with the change in the optical properties of nanoparticles, because part of the optical properties of nanoparticles depends on their distance in the aggregation network. When this distance is much greater than the average diameter of particles, it appears to be red; and when they are approximately equal, it is blue.

This change is caused by the resonance of metal nanoparticles on the surface plasma. Hybridization can shorten the distance between particles, resulting in a corresponding color change and the formation of nanoparticle aggregates. Thus, by the color change, it can be determined whether hybridization has occurred.

Elghanian and colleagues also found that the DNA-marked gold nanoparticles, when placed at a high temperature (80°C) or in a higher concentration of salt solution (0.1 mol/L of NaCl solution), can remain stable for several days. This is mainly due to the surface of gold nanoparticles linked to DNA that prevents their integration with each other. This is very important for the hybrid, because DNA hybridization needs to be performed in salt solutions with high concentrations.

It was found in experiments that prehybridization solution was red; after hybridization, it is pink (or purple), and after being dried it is blue. If no hybridization occurs or if the temperature exceeds the thermal decomposition temperature, then it will show a pink color. Moreover, according to the change of the hybrid system color at approximately the thermal decomposition temperature, it is discernible whether a hybrid is an exact match.

This method is effective in detecting ultra-traces (10 fmol, i.e., 10–14 mol) of oligonucleotides, which can be widely used in designing a high-resolution nucleic acid detection system; moreover, because the equipment is low cost and easy to operate, it is particularly suitable for small laboratories.

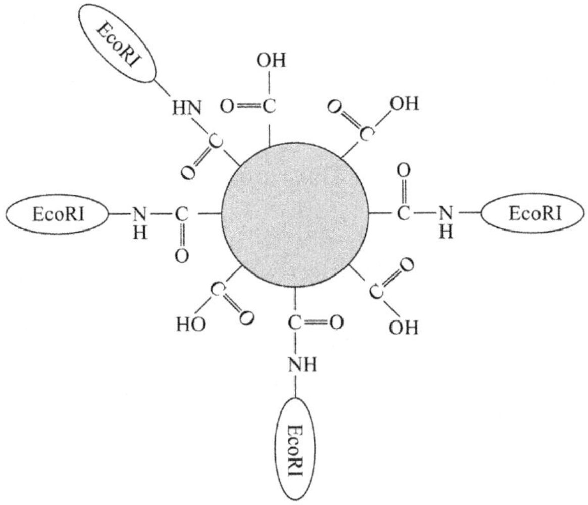

Figure 9.2 *Eco*RI enzyme molecules capable of covalent bonding with the 20 nm size of fluorescent nanoparticles.

2. Fluorescent latex NSs in bioanalysis

In contrast to single-dye molecules, each of the fluorescent latex nanoparticles (or nanoparticles) contains approximately 100–200 molecules, and each molecule also contains chromophores protected by the external environment. The protective effect of the emulsion arises from separation of different chromophores (as different chromophores generally contain conjugated ring structures, which, if not isolated, often resulting in $\pi-\pi$ stacking, so that peaks will be widened and a red shift will result).

These fluorescent nanoparticles are difficult to break down and can emit stable fluorescent light (no flickering).

Taylor and colleagues used proteins labeled with nanoparticles to determine the specific order of the individual DNA molecules straightened.

*Eco*RI enzyme can identify a specific sequence GAATTC of double-helix DNA molecule through 12 hydrogen bonds; it can combine with the 20-nm fluorescent nanoparticles through amide bonds (Figure 9.2).

This combination undergoes a reaction with the λ-DNA (there are five positions that can be combined), and then the DNA molecules after the reaction are straightened by applying the fluid mechanics theory and fixed onto the glass slides coated with polylysine. The experiment showed that *Eco*RI enzymes labeled by fluorescent nanoparticles can recognize and split a single λ-DNA molecule. Using a multicolor fluorescence microscope, we can see the green (530 nm) single DNA molecule and yellow orange (580–620 nm) NS. In this way, fluorescent images can reveal the location of connected nanoparticles. Experimental results showed that more specific orders on the same DNA molecule can be determined simultaneously.

Nanoparticles combined with organisms are very conducive to real-time observation and dynamic studies of the proteins and enzymes on a single DNA molecule.

3. Light-emitting QDs in bioanalysis

   In recent years, the potential value of nanocrystal methods has caused great concern in immune biology and clinical testing studies. Nanocrystals are a special kind of nanoparticle characterized by a neat arrangement of atoms and a bulk crystal structure. A QD is a kind of nanocrystal that is a class of semiconductor nanoparticle made from semiconductor material. It is stable and soluble in water, with a radius of less than or close to the exciton Bohr radius. The next section is devoted to presentation of the light-emitting QDs in biological analysis. It must be emphasized that the application of the nanoparticles in biochemistry is a new area that deserves high priority. The development of nanoparticles in biological markers has brought new opportunities for a large number of multicolor experiments and diagnostics; they have tunable optical characteristics that make them available to be directly used as a probe or as a traditional probe sensitizer. Of course, there are still some problems using nanoparticles as biological fluorescent markers. For example, stable and efficient light-emitting nanoparticles are difficult to prepare, and their biological compatibility and accessibility of large molecules have yet to be further improved.

9.5 QDs in Biological and Medical Analysis [4]

QDs have unique optical properties. A number of semiconductor QDs (nanocrystals) can emit laser-induced fluorescence, and fluorescent colors (peak position on the fluorescence spectroscopy) are controlled by the physical scale of QDs. They can be synthesized by using either a top-down crystal surface etching method or a bottom-up chemical method. In the 1970s, its application was mainly focused on the electronic and optical

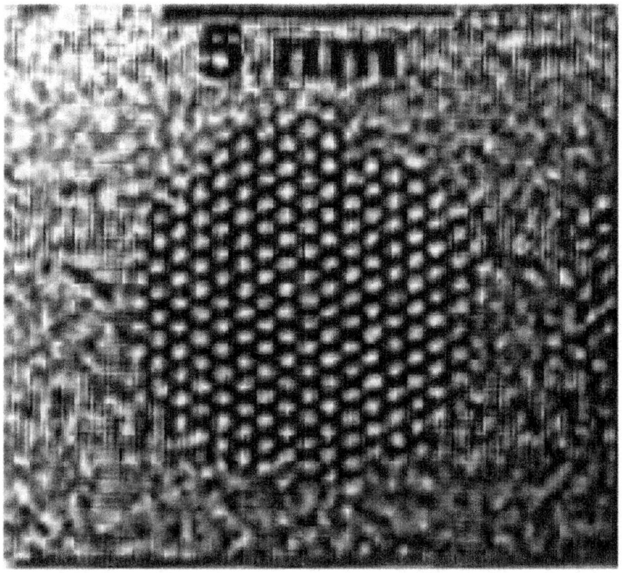

Figure 9.3 TEM diagram of atomic arrangement in CdSe QDs with a size of 5 nm.

sectors. In the 1980s, biologists began to have a strong interest in QDs. But because of its lower fluorescence quantum yield, the work was focused on the research of the fundamental characteristics of QDs. Since 1997, preparation technology of QDs continued to improve and their application in biological research became possible. QDs available as biological probes can be traced back to two research groups: Alivisatos et al. [5] and Chan et al. [6] in 1998. The functions of QDs were further discovered and promoted, making them a popular field of biological research. QD emission has a wide wavelength range and a sharp emission peak, and the emission wavelength can be adjusted through nanoparticle size. This is particularly suited for markers of biological systems (Figure 9.3).

9.5.1 QDs in Biological and Medical Analysis

Fluorescence is a widely used analytical method, particularly in the clinical determination of the content of certain elements in biological samples. The traditional fluorescent labels' absorption and emission wavelength are often restricted to a narrow range, with the intensity in continuous decline. If we use fluorescence spectroscopy directly to study them, then the bases and nucleic acids may have very low fluorescence quantum efficiency. Only tryptophan, tyrosine, and phenylalanine have natural fluorescence, so the best way to detect them is to use a variety of fluorescent probes.

At present, the fluorescent agent in such probes is an organic dye; however, in most cases, their excitation spectra are narrow, so it is difficult to stimulate a variety of components at the same time and the distribution of the spectra is asymmetric. The most serious flaw is the poor photochemical stability. QDs can absorb a wider range of wavelengths of light and emit a single specific wavelength. QDs have a structure of approximately hundreds to thousands of semiconductor atoms, with the size generally within 10 nm. This gives QDs atomic-like discrete energy levels. After QDs absorb the photon irradiation, an electron can skip the energy gap to an excited state. When falling from the excited state back to the ground state, the appropriate wavelength of photons can be launched. It is possible to design the QDs in different light colors. QDs can be created using group II–IV, CdSe, CdTe, CdS, and ZnSe materials, but also can be formed from group III–V, InP and InAs materials. In 1993, Bell Labs developed a highly efficient synthesis of light-emitting semiconductor QDs, thus beginning the utilization of QDs.

Compared with conventional fluorescent probes, the excitation spectra of QDs are wide and distributed continuously; that is, nanocrystals of different sizes can be excited by a single wavelength of light and emit the light in different colors, while the emission spectrum is narrow and distributed symmetrically. Meanwhile, it has higher photochemical stability and is resistant to photolysis. QDs as fluorescent probes have shown broad application prospects in the field of biomarkers and diagnostics.

QDs generally used in the biochemistry and medical fields function via QD bioconjugation. It is based on QDs as the core, followed by the formation of an organic outer layer by way of coating or chemical modification, which then combines with biological molecules to form a conjugate.

Bioconjugated QDs can be made in the following ways: the use of bifunctional base cross-linking agent for modification with a ZnS disulfide chain; hydrophilic adsorption; surface silylation; surface ionization; and micronanospheres.

QDs highlighted in the research are mainly composed of group II–VI or III–V elements (Table 9.2).

A QD is usually a core–shell-type nanobody, with CdSe as the nucleus and CdS or ZnS as the shell. Compared with traditional organic dyes, it has unique properties; QDs feature a

Table 9.2 QDs Composed of the Elements in Groups II–VI or III–V

Group	QDs								
	MgS	MgSe	MgTe	CaS	CaSe	CaTe	SrS	SrSe	SrTe
II–VI	BaS	BaSe	BaTe	ZnS	ZnSe	ZnTe	CdS	CdSe	CdTe
	HgS	HgSe							
III–V	GaAs	In GaAs	InP	InAs					

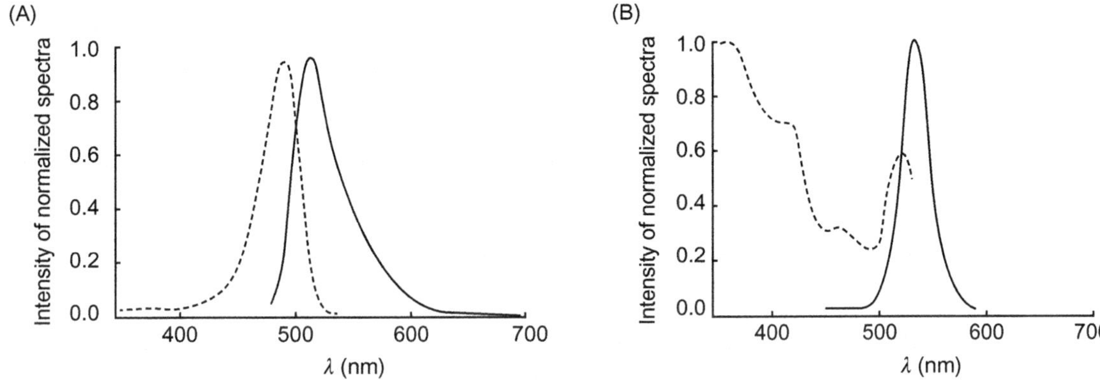

Figure 9.4 (A) Typical water-soluble nanocrystals of fluorescein in excitation: spectrum (dotted line) and fluorescence spectrum (solid line). (B) PBS. From Ref. [7].

large Stokes shift and a narrow, symmetrical fluorescence spectrum (Figure 9.4).

QDs have a broad range of applications, including a variety of fields and instruments, but their greatest application is for cell and disease detection. Compared with ordinary fluorescent substances, QDs are presented with a strong luminescent capacity and a longer light-emitting time. Nontoxic and easy to detect, they have a broad absorption wavelength range but only emit a single wavelength. They are valued as a future hot commodity to supersede fluorescent materials. QDs are available with a variety of colors in light emission. The color depends on the size of QDs. For the same excitation wavelength, there will be a range of excitation lights qualified for simultaneously detecting multiple targets. Other features of QDs include resistance against light-induced bleaching, safety, low toxicity to cells, suitable for living cells and *in vivo* studies, and long fluorescence (the fluorescence time is thousands of times longer

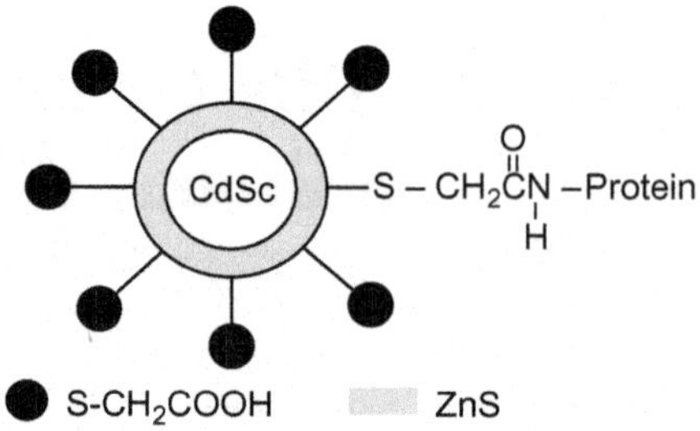

Figure 9.5 ZnS-coated CdSe QDs covalently coupled with proteins. From Ref. [7].

than that of ordinary fluorescent molecules, which facilitates long-term tracing and storage of the results).

QDs can be used for ultra-sensitive detection of nonisotope-labeled biological molecules. For example, the surface of QDs can be connected to mercaptoacetic acid ($HS{-}CH_2COOH$), so that QDs not only can be water soluble but also are able to be combined with biological molecules (e.g., proteins, multipeptides, nucleic acids). Then, QDs can be detected by photoluminescence, thus enabling biomolecular recognition of some specific substances (Figure 9.5).

QDs of different fluorescent characteristics can be combined into hollow polymer balls, creating fluorescent nanoparticles with different spectral characteristics and brightness features available to be labeled to biological macromolecules. Taylor and associates once used nanoparticle-labeled protein to determine straightened individual DNA molecules, where *Eco*RI enzyme could be combined with the fluorescent nanoparticles at a size of 20 nm through amide bonds and identify the specific sequence of double-helix DNA molecule through 12 hydrogen bonds.

Temperature may directly affect the size of QD particles. In general, the higher the T, the smaller the QD particles obtained, and the shorter the fluorescence wavelength. So, QD particles of different sizes can be displayed in different colors (Figure 9.6). This character can be used to track the activity and proliferation of amino acid receptors in nerve cell membranes.

The diversity of QD colors is able to meet the requirements of analyzing biological polymers (protein, DNA) that contain vast amounts of information. Polymers and QDs can be combined to form polymer beads, which are free to carry QDs of different sizes (colors) and begin to emit on irradiation. Transmitted through prism refraction, a variety of spectral lines of specified density (bar codes) can be formed. This type of bar code has a promising application potential in gene chip and protein chip technology.

Figure 9.6 Four kinds of CdSe with different particle sizes irradiated with 302 nm UV light. From right to left, the luminous colors change with wavelength (4.6, 5.0, 5.8, and 7.0 nm respectively).

QDs can also be used to detect the characteristics of DNA and proteins. Researchers have been able to bring a mixture of a variety of QDs encapsulated into rubber balls of millionths of a meter in diameter each, and they can radiate different colors of light. Researchers can use these rubber balls as different markers of gene sequences or antibodies to identify different DNA or antibody proteins. This provides a new approach to further probe the nature of DNA or antibody proteins. In addition, the hydrophobic modified polyacrylic acid used to coat QDs, combined with immunoglobulin G and streptavidin–biotin, can acquire accurate results for marking on the cell surface protein, cytoskeleton proteins, and the proteins in the nucleus. Its anti-bleaching performance is often of great value for the quantitative detection of fluorescent molecules and biosimulation of living cells.

Detection of tumor cells with QDs is also one of the hotspots in the current studies. Scientists may have the transferrin and QDs covalently cross-linked, so that cervical cancer cells can be "swallowed" into the cell. Then, QDs connected with transferrin still have biological activity to achieve long-term observation of single-color fluorescence labeling. They use two kinds of QDs of different sizes in marking the mouse fibroblasts: one emits green fluorescence and the other emits red fluorescence. Meanwhile, the QDs emitting red fluorescence are specifically marked on the cell actin filament, while QDs emitting green light are combined with urea and acetic acid. Such QDs have high affinity with the nucleus, and simultaneously the red and green fluorescence in cells can be observed to achieve the two-color fluorescent labeling.

Also, other applications of QDs in biological detection have been frequently reported. For instance, QDs were linked with biotin, urea, acetic acid salts, and certain antibodies, and they successfully achieved specific cell structures. A variety of molecules can be used as guidance materials for QDs, including nucleic acids, lipids on the cell membrane, proteins closely connected with the carrier protein or carrier sugar, as well as some drugs, by which QDs can be guided to a specific cellular structure. Researchers are committed to the application of QDs in neurotransmitter studies (for the understanding of neural signal transduction). They marked the QDs on an important neurotransmitter, 5-hydroxytryptamine, and then observed how the transporter protein in the process of promoting neurotransmitters returned to the cell after the intermittent signal was transmitted through adjacent nerve cells. This can be applied to medical imaging technology.

In general, because of its unique marking characteristics, QD technology will certainly develop into a cutting-edge technology for future biomolecular detection, providing a more advanced approach for DNA testing (DNA chips), protein detection (protein chips), and the exploration of the mechanisms of protein–protein reactions (antigen–antibody, ligand–receptor, enzyme–substrate). Furthermore, this technology will greatly enhance the development of bioimaging technology and biopharmaceutical technology, leading to a huge step forward for disease diagnosis and treatment.

9.5.2 QDs for *In Vivo* Studies

Early detection of cancer is a major issue in the modern-day medical profession. Traditional detection methods required the tumors to reach a certain size before they could be discovered, and the best treatment is often adversely affected by time. The ideal situation is when just a small amount of cancer cells arise in a local area, and they can be detected and treated.

Different CdSe/ZnS core–shell-type QD surfaces modified with different peptides can be injected into mice for postbiopsy analysis. The results showed that different peptide-modified QDs can be specifically applied to the lungs and vascular system of a normal or a tumor-afflicted mouse. These results indicated that QDs can possibly be utilized in studies of disease diagnosis and drug delivery (Figure 9.7).

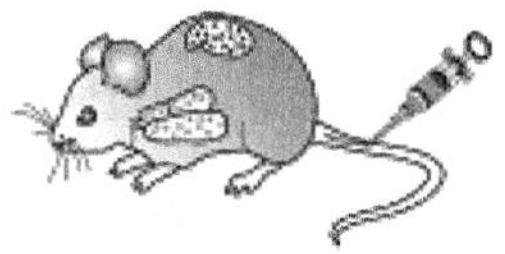

Figure 9.7 Schematic of QDs as target material.

Recently at the University of Cincinnati, Professor Donglu Shi and colleagues developed a multifunctional nano combination device [8,9]. Such nanoscale structure uses nanotubes as a substrate. After surface treatment, the outer surface of nanotubes can be coupled to connect QDs, which can be used to partially trace cells *in vivo*. Due to the strong luminance, QDs can be used for the imaging of the depth tissues (*in vivo* imaging) and characterization. After a special plasma coating, the nanotubes are loaded with an antibody that can indentify cancer cells on the outer surface to complete the so-called targeting role. The interior of the hollow nanotubes can be used to store anticancer drugs, which can be transported to the vicinity of cancer cells for controlled release to kill cancer cells, thus achieving a local treatment effect. This new method is far superior to conventional chemotherapy.

9.6 Research Progress of Nanomagnetic Materials in Hyperthermia

9.6.1 Background of Hyperthermia

After surgery, radiotherapy, and chemotherapy, the hyperthermic treatment of tumors has become a novel "green therapy" in treating cancer. Hyperthermia used as a treatment modality has a long history that can be dated back to 5,000 BC. It is said that breast tumors treated with heat were documented in the manuscripts of a Danish doctor named Edwin Smith. Legend has it that Hippocrates, the famous ancient Greek doctor, had used heat therapy in treating tumors and he had a proverb: "The disease which cannot be cured with iron can be cured with fire; the disease that fire cannot cure is incurable."

In 1884, Bruns reported that a case of advanced melanoma infected with erysipelas was accompanied by high fever of 40°C; a few days later, the tumor disappeared. This contributed to survival of 8 more years. Later, people began to use artificial bacterial infection or the injection of a chemical-induced heat source to cause the patients to have a high fever. The most renowned research was performed by Coley, who published a number of articles in 1893 that described his artificial methods of infection-induced fever of 38–42°C. As a result, 12 of the reported 38 cases of advanced cancer were cured, which caused a sensation.

Nineteenth-century German scholars have presented many reports and literature on hyperthermia. Heating technology was then rudimentary and even used inferior methods such as hot needles, burning surface lumps with a small hot iron, or soaking limbs with hot water and local infusion with hot water for heating, and so on. With the application of the electric knife since the nineteenth century, Westermark took the initiative to use radiofrequency coils as radiators for hyperthermia in cervical cancer. After World War II, the microwave technology developed rapidly. In the 1960s, more systematic studies were performed regarding the heating technology in electromagnetic hyperthermia. Despite a long history of hyperthermia, its development was limited as a result of underdeveloped science and technology and backward heating methods and equipment. Until the 1970s, with the rapid development of electronic technology and multidisciplinary involvement and coordination, modern hyperthermic oncology gradually formed its own disciplinary system on the basis of a large amount of basic and clinical research. Ultimately, it developed itself into an emerging discipline integrating oncology, biothermal methods, thermal physics, and electrical and mechanical disciplines with computer and other technologies.

In recent years, the United States, Russia, and other European countries have performed a wide range of research on tumor hyperthermia, showing that hyperthermia is indeed effective against cancer. Hyperthermia in China began in the late 1970s and is currently available in forms of microwave hyperthermia, focused ultrasound hyperthermia, radiofrequency hyperthermia, and the latest generation of endogenous field hyperthermia. Microwave heating can only be applied to superficial cancer treatment because of its shallow penetration depth. Focused high temperature of ultrasound hyperthermia may be 90°C or more. With a higher penetrating power, it can be used to treat deep tumors. However, the limited nature of ultrasound prevents it from either penetrating the gas-bearing tissues or passing through bones, and therefore it is not a candidate for the treatment of lung and esophagus cancers or liver cancer that is blocked by ribs. Moreover, ultrasound must also rely on water as the mediator, because it cannot reach the tumor without going through water on the surface of the focus. Therefore, the location of the disease plays a decisive role and the cost of this kind of therapy is high. Radiofrequency hyperthermia can also be used to heat deep tumors, but it has to be applied through local water cooling. Meanwhile, subcutaneous fat is prone to overheating and pain, resulting in rapid development and difficulties in operation. Endogenous field hyperthermia is the latest generation of the hyperthermia system that is a combination of the advantages of various types of hyperthermia at home and abroad, without the need of water cooling. Patients can be safe and comfortable in a supine position, and this therapy is applicable to the treatment of tumors in the chest, abdomen, pelvis, and other deep positions. Its significant effect has greatly enhanced the patient's quality of life.

Hyperthermia is the use of a variety of physical energy (e.g., microwave, radiofrequency, ultrasound) to produce thermal effects, so that the tissue temperature increases to the treatment temperature of 43°C and above to accelerate cell death. The role of hyperthermia in tumors includes the destruction of tumor blood vessels, tumor blood clots, and the inhibition of tumor angiogenesis. Statistics show the following: between 39°C and 40°C, the growth of most tumor cells is significantly inhibited; between 40°C and 41.5°C, tumor cell survival rate decreased rapidly; between 41.5°C and 43°C, tumor cells will die within a short period; and between 70°C and 120°C, the tumor cells burst instantaneously. Therefore, hyperthermia is a highly effective means of cancer treatment; it is safe, noninvasive, and is not associated with toxic side effects and infection. Currently, hyperthermia can be divided into two categories: whole body hyperthermia and local hyperthermia. The operation of whole body hyperthermia is less used because it is complicated and would cause strong systemic reactions, producing a certain degree of risk. Currently, local hyperthermia is the most applicable clinical tumor therapy; the most frequently used heat sources include IR, hot water bath, hot bath, ultrasonic waves, radiofrequency, microwave, and so on. Some experimental studies have shown that local hyperthermia not only can directly kill tumor cells but also can enhance the immune function. This possibly results from the high temperature applied locally that may lead to tumor degeneration and necrosis, and the absorption of necrotic tumor products will stimulate the body's immune function.

There is a need for comprehensive treatment strategies rather than a single treatment of cancer, whereas surgery, radiotherapy, and hyperthermia are all local treatments. Systemic treatment with drugs, for example chemotherapy with traditional medicine, can provide better efficacy. Cancer is a systemic disease. If a local tumor has grown to a volume of 1 cm^3,

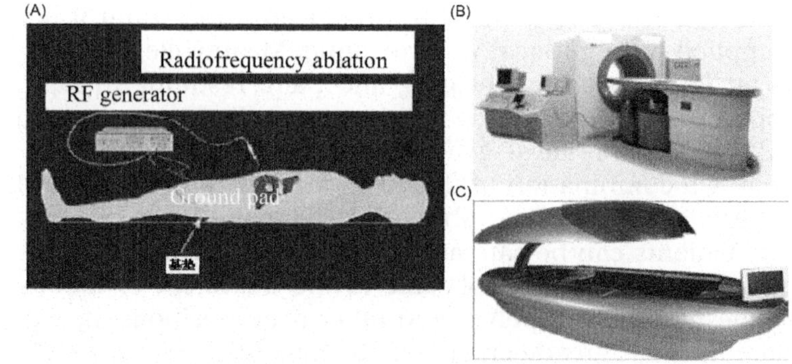

Figure 9.8 Clinical application of different hyperthermic methods: (A) RF ablation, (B) high-intensity focused ultrasound, and (C) whole body hyperthermia.

then other parts of the body may have small metastases that cannot be seen with the naked eye. Even with surgical excision, radiotherapy, or hyperthermia, there is still a need for coupling with systemic drugs before residual cancer cells *in vivo* can be wiped out, so that they cannot be left behind as the scourge of metastasis or have any recurrence in some parts of the body. Because hyperthermia is not a radical means and needs to be combined with systemic chemotherapy or traditional Chinese medicine treatment, efficacy could be enhanced by several times or even dozens of times. Here, the drug amounts are only one-third or one-half of the commonly used dose; therefore, the toxic side effects of the drugs are significantly reduced, whereas the treatment efficacy can be improved greatly. Patients with cancer are happy to accept such treatment because it can greatly improve their quality of life and allow new hope.

In this section, we highlight the use of nanomagnetic materials for hyperthermic treatment (Figure 9.8).

9.6.2 Magnetic Hyperthermia [10–13]

In magnetic hyperthermia, magnetic particles are transported to the treatment area; in an external alternating magnetic field, a hyperthermic effect is generated by the heat due to magnetic loss of the magnetic particles.

Magnetic hyperthermia has the following advantages. First is targeting within the tissues, and second is the bystander effect of heat treatment. A very good biocompatibility of materials can work to improve the efficacy of cancer chemotherapy or radiotherapy.

The main types of magnetic hyperthermia include embolism magnetic hyperthermia, liposome magnetic hyperthermia, intracellular hyperthermia, and whole body hyperthermia.

1. Embolism Magnetic Hyperthermia

   In embolization therapy, iodized oil or hot brine is injected via a catheter into tumor blood vessels of liver cancer patients. Magnetic hyperthermia plus embolization hyperthermia yields magnetic embolization hyperthermia. Embolism magnetic hyperthermia is an organic combination of embolization and hyperthermia; it can play the role of combined therapy to improve efficacy and also enhance the targeting of hyperthermia.
2. Liposome Magnetic Hyperthermia

   A liposome is a small body of closed vesicles, composed of a lipid bilayer with the aqueous phase inside. According to the structural properties of lipid membrane, liposomes can be divided into heat-sensitive liposomes, pH-sensitive liposomes, common liposomes, and reliposomes. As a targeted drug delivery system functioning with magnetic orientation, magnetic nanoliposomes have excellent drug-targeting capabilities and the unique role of local hyperthermia under the influence of an external magnetic field, showing attractive application prospects. Magnetoliposomes can combine hyperthermia with chemotherapy. Kubo achieved satisfactory efficacy by heating adriamycin containing 10 nm Fe_3O_4 and uracil for the treatment of the tumor osteosarcoma in hamsters.
3. Intracellular Hyperthermia

   A variety of modifications of the surface of magnetic nanoparticles may improve the efficiency and selectivity of magnetic nanoparticles in gaining access to the tumor cells. And the role of the alternating magnetic field *in vitro* may promote the magnetic nanoparticles within cells to generate heat and kill the tumor cells. This method is called intracellular magnetic hyperthermia.

   Studies show that cancer cells may absorb an amount of magnetic particles equivalent to 8 times to 400 times that of normal cells. Tumor cells containing ferromagnetic nanoparticles are susceptible to antimagnetic hyperthermia, so intracellular hyperthermia has excellent targeting features.
4. Whole Body Hyperthermia

   Babincova and colleagues injected dextran nanomagnetic particles into the patient's blood flow to perform whole body hyperthermia and effectively killed the cancer cells. This method can be used to treat malignant tumors with systemic blood metastasis or lymphoma-related diseases.

   The development of magnetic hyperthermia has opened new avenues for cancer therapy. Drug-loaded magnetic liposomes with targeted controlled release successfully solved the problems of cancer chemotherapy with severe systemic side effects and rapid elimination of drugs *in vivo*; magnetic microspheres with small particle size and their targeting characteristics provide a solution to the problem of embolization failure and ectopic embolization in tumor embolization. The bystander effect of heat treatment is expected to solve the problem of uniform heating of cancer cells; tumor cells with high efficiency in absorption of ferromagnetic nanoparticles make magnetic hyperthermia unique to other forms of hyperthermia in uniformity and efficiency in killing tumor cells.

9.6.3 Magnetic Materials for Hyperthermia

Magnetic materials for hyperthermia are required to have a high specific absorption rate (SAR). Used in the human body, their biological safety must be guaranteed. In addition, the product performance should maintain certain stability. At present, common magnetic materials for magnetic hyperthermia include the following three kinds: Fe_3O_4; γ-Fe_2O_3; and carbonized iron, iron powder, and glass–ceramic iron oxide. These magnetic materials can be synthesized using physical or chemical methods. Physical methods are the colloid mill, jetting, impact, and tear-ultrasonic methods, whereas chemical methods include coprecipitation, precipitation oxidation, sol–gel, and metal organic decomposition.

Magnetic materials for magnetic hyperthermia have SAR related to the particle size. The SAR is high when the particle diameter is approximately 10 nm; if the particle size is between 100 nm and 1 μm, then the SAR is low. To improve the SAR, surface modification is typically needed for the magnetic materials to be prepared in magnetic fluids. There are two common surfactants used for the surface modification: water-based magnetic fluids (dextran, polyvinyl alcohol, PEG) and oil-based magnetic fluids (oleic acid, polyester).

9.6.4 Thermogenesis Mechanism of Magnetic Materials for Magnetic Hyperthermia

In the magnetization process and the magnetization reversal process, part of the energy in magnetic materials would be irreversibly transformed into heat and the loss of energy is called magnetic loss. Therefore, for magnetic materials for magnetic hyperthermia, the heat is sourced from the magnetic loss. Magnetic loss W_m includes the eddy current loss W_e, hysteresis loss W_h, as well as other loss arising from the residual magnetic relaxation or magnetic after-effect W_r, that is, $W_m = W_e + W_h + W_r$. Under normal circumstances, magnetic losses in ferrites are mainly residual loss and hysteresis loss; in metallic magnetic materials, mainly eddy current loss and hysteresis loss occur.

For a magnetic conductor in an alternating magnetic field, electromagnetic induction may generate eddy currents. This causes magnetic field strength H and magnetic induction intensity B to have an uneven distribution of amplitude and phase inside the materials. Meanwhile, this will make the phase of B lag behind that of H, thus increasing part of the energy loss, known as eddy current loss. Experimental studies show that for some metallic magnetic materials, the measured magnetic loss is much greater than the sum of theoretical calculations of the eddy current loss and quasi-static losses. The difference between experiment and theory causing the additional loss is called abnormal loss. Abnormal loss comes as part of the micro-eddy current induced in the vicinity of the domain wall by electromagnetic induction in the domain wall movement; it also partially results from the deformation of domain walls or domain wall pinning. It is noteworthy that abnormal loss accounts for a large part of the total loss in a number of metallic magnetic materials (e.g., silicon steel sheet).

Hysteresis loss comes from the irreversible magnetization process existing in the magnetic material (the irreversible displacement of the domain wall and irreversible dynamic behavior of the magnetic domain). In the quasi-static magnetic case, the hysteresis loss is proportional to the area of the hysteresis loop. In medium and strong alternating magnetic fields, the hysteresis loss of some metal-based magnetic materials is suitable for empirical formula Steinmetz $W_h = \eta B_m^n f$, where f is the frequency and η and n are material-related constants. For example, for 3%Si–Fe alloy, $n \approx 1.6$ and $\eta \approx 1.2\ \mathrm{J/(m^3 \cdot T)}$.

Residual loss means all other losses apart from eddy current loss and hysteresis loss. It is caused by the magnetic relaxation processes in different mechanisms. In the low-frequency and weak magnetic fields, the residual loss is mainly magnetic after-effect loss and has nothing to do with the frequency. High-frequency residual losses include losses caused by size resonance, domain wall resonance, and the natural resonance. The residual loss dominates in ferrites.

Residual loss caused by magnetic after-effect is proportional to the frequency, domain wall displacement, and rotation damping of the magnetization vector. This loss is divided into two categories: Richter type and Jordan type. The former is related to temperature and frequency; the latter is slightly dependent on temperature and frequency. Richter-type loss is mainly caused by the induced anisotropy generated by the proliferation of impurities. Jordan-type losses are mainly caused by thermal fluctuations. Ferrite Richter loss is due to the proliferation of valence electrons between the ions.

In the high-frequency and ultrahigh-frequency areas at 10^4 Hz and above, the imaginary component of permeability on the ferrite spectrum related to magnetic loss μ'' may present several absorption peaks in different frequency regions. They correspond to the resonance loss, which is also a kind of relaxation loss. As the frequency increases, these absorption peaks will be caused by size resonance, domain wall resonance, natural resonance, and natural exchange resonance.

Studies have shown that magnetic particles approximately 10 nm in diameter will mainly have relaxation loss, showing a higher SAR; and hysteresis loss mainly occurs in magnetic particles of 100 nm–1 μm diameter, showing a lower SAR rate.

Two examples of the application of magnetic hyperthermia are described. Shinkai and colleagues used cationic liposomes (including Fe_3O_4 with a particle size of 10 nm) in the hyperthermia of rat glioma. Tumors in more than 80% of the rats disappeared completely. Brusentsov and associates used dextran-coated γ-Fe_2O_3 magnetic fluid (including γ-Fe_2O_3 with a diameter of 10 nm) in the hyperthermia of MX2 tumors in rats. Tumors in 33% of the rats showed complete recession, with the term of survival being extended by 150%.

Although the use of nanomaterials promoted the development of magnetic hyperthermia, there are still many problems in the application of magnetic hyperthermia. First is the need to improve the targeting of hyperthermia, the SAR of the materials, and the degree of biosecurity. It must be combined with other tumor treatment methods to achieve a satisfactory effect. Also, we must consider that the magnetic fluid should have the minimal residue in the body after treatment is completed. In general, nanomagnetic materials can be naturally excreted by the body. Furthermore, we can take advantage of the so-called magnetic field guidance technology to exclude the potential residual material. In addition, after the magnetic fluid enters the body, part of the fluid will inevitably find a way into normal tissue cells. We also need to conduct further studies on the impact of nanomagnetic materials on normal tissues, that is, their side effects.

References

[1] Y.B. Li (Ed.), Nano-Based Biomedical Materials, Chemical Industry Press, Beijing, 2004.
[2] Z.Y. Jiang, G.X. Cheng (Eds.), Nanobiotechnology, Wunan Publishing Co., Taiwan, 2004.
[3] X.Y. Yang, Magnetic nanoparticles in biomedical applications, Physics (bimonthly) 28 (2006) 692.
[4] W. Hou, Y.M. Shan, L.P. Wang, Progress in biomedical applications of quantum dots, Chin. J. Lab. Diagn. 8 (4) (2004) 436–438.
[5] M. Bruchez Jr., M. Moronne, P. Gin, S. Weiss, A.P. Alivisatos, Semiconductor nanocrystals as fluorescent biological labels, Science 281 (1998) 2013–2016.
[6] W. Chan, S. Nie, Quantum dot bioconjugates for ultrasensitive nonisotopic detection. Nonisotopic detection, Science 281 (1998) 2016–2018.
[7] L. Tian, The applications of quantum dots in life science, <http://snl.bjmu.edu.cn/course/reviews/bitech/Tian%20Liming.pdf>.
[8] Y. Guo, D. Shi, J. Lian, Z. Dong, W. Wang, H. Cho, et al., Quantum dot conjugated hydroxylapatite nanoparticles for *in vivo* imaging, Nanotechnology 19 (2008) 175102.
[9] Y. Guo, D. Shi, H. Cho, Z. Dong, A. Kulkarni, G.M. Pauletti, et al., *In vivo* imaging and drug storage by quantum-dot-conjugated carbon nanotubes, Adv. Funct. Mater. 18 (2008) 1–9.
[10] A. Jordan, R. Scholz, K. Maier-Hauff, M. Johannsen, P. Wust, J. Nadobny, et al., Presentation of a new magnetic field therapy system for the treatment of human solid tumors with magnetic fluid hyperthermia, J. Magn. Magn. Mater. 225 (2001) 118–126.
[11] N.A. Brusentsov, V.V. Gogosov, T.N. Brusentsova, A.V. Sergeev, N.Y. Jurchenko, A.A. Kuznetsov, et al., Evaluation of ferromagnetic fluids and suspensions for the site-specific radiofrequency-induced hyperthermia of MX11 sarcoma cells *in vitro*, J. Magn. Magn. Mater. 225 (2001) 113–117.
[12] C. Kun, M. Peng, G. Ning, Hyperthermia treatment of tumors with magnetic nanoparticles technology, Chin. Med. Equip. J. 27 (4) (2003) 276–278.
[13] R. Elghanian, J.J. Storhoff, R.C. Mucic, et al., Selecting colorimetric detection of polynucleotides based on the distance-dependent optical properties of gold nanoparticles, Science 277 (1997) 1078–1081.

10

NANOENERGY MATERIALS

CHAPTER OUTLINE

It has long been noted that there is a huge discrepancy between the finite fossil fuels on earth and the unlimited human needs. Traditional fossil energy is facing two main issues concerning its use: energy depletion and environmental pollution. Based on existing reserves and the rate of consumption, oil, coal, and other major energy sources will be depleted in the next few decades to hundreds of years. To make it worse, the use of traditional fossil energy sources has caused increasingly serious environmental pollution. Current global emissions of

Nanomaterials and Devices. DOI: http://dx.doi.org/10.1016/B978-1-4557-7754-9.00010-X

carbon dioxide each year have gone up to 2.1 million tons, leading to the deterioration of the environment; in the air, carbon dioxide and dust levels have seriously affected people's health and the natural environment for human survival. The use of fossil fuels is causing an ecological disaster on Earth, with global warming and acid rain posing serious threats to the survival of the Earth's flora and fauna. Therefore, years ago, researchers and activists called for the development of renewable energy beyond traditional fossil fuels.

Renewable energy sources are numerous, including wind, hydro-energies, as well as geothermal, tidal, and solar energies. Except for solar energy, other energy sources all have obvious regional characteristics.

Solar energy is abundant. It is estimated that energy generated from 40 min of solar radiation from the Earth can meet the 1-year energy requirements of people all around the world. As a clean energy, unlike oil, coal, and other fossil fuels, solar energy will cause no "greenhouse effect" or any environmental pollution. Easy use is another characteristic of solar energy. In comparison with hydro, wind, and other new energy sources, it is not subject to geographical restrictions and its use comes at a low cost. One important way to use solar energy is to develop solar cells that directly convert sunlight into electrical or chemical energy.

Environmental issues are closely related to energy issues. The types of available energy sources and their implementation have a positive or negative impact. Driven by the aforementioned issues, researchers hope to use nanotechnology to solve or alleviate the imminent energy crisis and find sustainable sources of energy for humanity.

Nanotechnology is based on atoms and molecules in manufacturing materials and products. This small to large manufacturing approach requires fewer materials and would cause less pollution. As a result of the miniaturization, nanotechnology products require fewer resources. The objectives of sustainable development are "low consumption, high efficiency," and lower costs, and these can be achieved. In the foreseeable future, large-scale machinery and equipment that are expensive and wasteful in regard to resources will be phased out with the aim of achieving "zero resource consumption." Therefore, promoting the use of nanotechnology is an energy-saving activity. Application technology in the development of nanoenergy resources is mainly focused on two key directions: "nanoenergy storage materials and technology" and "nanoenergy-saving technologies." Because nanoscale energy storage materials are characterized by high activity, large surface area (200–2,000 m^2/g), self-assembly (1–3 nm active catalyst), small sizes (10–30 nanostructure), and possible photoelectric effects, most developed countries are actively involved in the development of energy-related nanomaterials, attempting to provide nanofeatured energy density power storage systems that might be significantly superior to the current battery power. In the field of energy storage devices, competitive products being developed include micro-nanofuel cells with a 50-day standby time, high-performance storage appliances, and photochemical batteries two to three times more powerful than conventional lithium batteries. These products all aim at upgrading energy efficiency while reducing energy consumption. New energy-saving applications being developed include heat-exchange systems, energy-saving windows, and solar cells. These products would help the traditional energy-saving technologies reach a higher level.

Research and development of nanoenergy materials include the following main areas.

Nanomicro Fuel Cells

This entails the development of nanocatalytic electrodes (1–3 nm), nanocomposite proton-conducting membranes (1–5 nm channel), and the simulation computation and system miniaturization design. The development of nanocatalytic electrode structures will lead to improved fuel efficiency in fuel cells. Similarly, nanocomposite proton-conducting membranes can enhance fuel cell power density.

High-performance Storage Appliances

High-energy lithium batteries and nanostructured materials and devices have been developed for high-power energy storage (250 W·h/kg). Through the design of nanoelectrode materials and device fabrication technology, materials can have a higher capacity and current discharge rate. High-power (10–20 C) nanostructured electrode materials for high energy storage are based on the use of porous nanostructured (10–50 nm) materials with electrochemical activity such as high-energy electrode materials aiming to significantly increase power density to achieve high-power fast charge and discharge in a few minutes.

Nanothermal Fluid Technology

Nanothermal fluid test equipment, nanothermal fluid numerical simulation technology, nanofluid properties database, and microthermal-exchange system design technology have been established in an effort to guide the industry's investment in the development of new products and production technology.

Nanocrystal Application Technology

Preparation technology, such as porous nanostructured electrodes, the design technology in redox chromophores and solar dye molecules, and the development of nanocrystal energy-saving electrochromic windows and dye-sensitized solar cells (DSSCs), has been developed to guide the industry's investment in the mass production of technology development for new products.

10.1 Nanostorage Materials [1]

Nanomaterials are mainly based on hydrogen-based energy storage, and the present research on hydrogen storage materials is focused on the new composition of alloys and the change of microstructures.

Both physical and chemical hydrogen storage require expensive equipment, yet the carbon nanomaterials may bring about an efficient and clean hydrogen storage method. These materials, if used as the hydrogen storage materials in fuel cell vehicles, can effectively prevent air pollution or greenhouse gas emissions.

It has long been known that some solid materials (e.g., metal hydrides) can store a small amount of hydrogen (approximately 1–2% by weight) at room temperature. Some metal hydrides can store more hydrogen (5–7% by weight), but they usually require a temperature of 250°C or higher for hydrogen storage. Nanostorage materials are characterized by high activity, high surface area, and ultrafine particle size (10–30 nm structure). Using the characteristics of nanomaterials to provide an energy storage system of high energy density is the goal.

The larger specific surface area and available surface atoms of nanomaterials result in higher activity (200–2,000 m^2/g). Nano-based hydrogen storage materials will be able to provide more adsorption sites for molecular hydrogen. Meanwhile, the hydrogen molecules are more easily dissociated into hydrogen atoms, which, by diffusion, are able to move from the surface of the grain boundaries into the hydrogen storage materials.

For example, carbon nanotubes (CNTs) and nanofibers can also exhibit excellent performance in the absorption of hydrogen at room temperature, acting as a tiny hydrogen-absorbing "sponge." Materials of this kind have broad application potential and can be used to create hydrogen containers in fuel cell vehicles. To add fuel, a motor vehicle just needs to be driven to a gas station, and hydrogen can be injected into an empty vessel. Calculations show that as long as carbon materials have hydrogen storage of 6.5% by weight, fuel cell vehicles can have practical value (the distance between two stations is set as 500 km or 310 miles).

10.1.1 Features and Objectives of Hydrogen Energy

Hydrogen is one of the most common elements in nature, and there is no depletion of hydrogen resources. Hydrogen has high calorific value; with its combustion product of water, it is emission free in a real sense and is also free from pollution. There are many ways to utilize hydrogen, for example to generate heat via burning or to generate power electrochemically. Hydrogen in storage and in transport can be gas, liquid, solid, or in compound form.

The key technology to achieve hydrogen economy is the development of an inexpensive and highly efficient hydrogen production technology that enables safe and efficient hydrogen storage technologies. Therefore, the priority is to develop new, highly efficient, and safe hydrogen storage materials and hydrogen storage technologies.

Vehicle hydrogen storage systems are designed in accordance with two standards issued, respectively, by the IEA (International Energy Agency) and DOE (US Department of Energy). Here, the IEA standard requires the mass hydrogen storage capacity to be more than 5% and requires the volume capacity to be more than 50 kg $(H_2)/m^3$, whereas the DOE requires mass hydrogen storage capacity of more than 6.5% and volume capacity larger than 62 kg $(H_2)/m^3$.

10.1.2 Comparison of Different Hydrogen Storage Methods

The characteristics of various hydrogen storage methods are described in Table 10.1. Obviously, solid-state hydrogen storage has many advantages.

10.1.3 Technology Status of Hydrogen Storage Materials

There are three kinds of hydrogen storage materials: metal hydrides, coordination hydrides, and nanomaterials.

Table 10.1 Comparison of Different Hydrogen Storage Methods

Hydrogen Storage	Characteristics
Gaseous hydrogen	Low energy density
	Not safe
Liquefied hydrogen	High energy consumption, requiring high performance of the storage tank insulation
Solid-state hydrogen	High hydrogen storage volume
	No high pressure and thermal insulation vessel
	Safe, no risk of explosion
	High purity available to enhance the added value of hydrogen

Metal hydrides have the following hydrogen storage characteristics: reversible reactions, hydrogen atom storage structure, solid-state hydrogen storage, safety, reliability, and relatively high-volume density of hydrogen storage. At present, successful developments of metal hydrides as hydrogen storage materials are the RE La–Ni system, titanium–iron system, magnesium system, and titanium/zirconium system.

Hydrogen storage alloy in the RE La–Ni system has a representative, $LaNi_5$, which was first developed by the Philips Laboratories in the Netherlands. It is characterized by its easy activation, moderate and smooth equilibrium pressure, low equilibrium pressure in hydrogen absorption and desorption, good performance of anti-gas poisoning impurities, and is suitable for room temperature operation. By partial replacement of elements, $MmNi_{3.55}Co_{0.75}Mn_{0.47}Al_{0.3}$ (mixed rare earth, with major components of La, Ce, Pr, Nd) is widely used in nickel–hydrogen batteries.

The titanium–iron system has a representative, TiFe, which was first invented by the US-based Brookhaven National Laboratory. It features a low price, reversible hydrogen storage at room temperature, susceptibility to oxidation, difficult activation, and weak ability in anti-impurity gas poisoning; surface modification is required in the actual use of the alloy.

TiFe alloy is characterized by two hydride phases, namely β-phase ($TiFeH_{1.04}$) and γ-phase ($TiFeH_{1.95}$). The hydrogen absorption reactions of the TiFe alloy are:

$$2.13TiFeH_{0.10} + 1/2H_2 \rightarrow 2.13TiFeH_{1.04}$$

$$2.20TiFeH_{1.04} + 1/2H_2 \rightarrow 2.20TiFeH_{1.95}$$

The magnesium system has a representative, Mg_2Ni, which was first reported by the US Brookhaven National Laboratory. It is characterized by high hydrogen storage capacity, abundant resources, low cost, and high temperature of hydrogen release (250–300°C), but its hydrogen release dynamic performance is weak. It can go through mechanical alloying or can be added with TiFe and $CaCu_5$ for ball milling.

Titanium and zirconium materials are intermetallic compounds with the Laves phase structure. The atomic gap in these materials comes from the tetrahedral structure, with more spaces conducive to the adsorption of hydrogen atoms. Typical examples are $TiMn_{1.5}H_{2.5}$ and $Ti_{0.90}Zr_{0.1}Mn_{1.4}V_{0.2}Cr_{0.4}$. Characterized by good material activity, they are mainly used for hydrogen storage of hydrogen vehicles and the negative poles of batteries (Ovinic).

Coordination hydride hydrogen storage material is formed mainly by alkali metals (Li, Na, K) or alkaline earth metals (Mg, Ca) and group III elements (B, Al). It has high hydrogen storage capacity, and it is difficult to restore it to its hydride form (catalytic $LiAlH_4$ in $TiCl_3$ and $TiCl_4$ at 180°C, are able to obtain 5% of the reversible hydrogen storage capacity under a hydrogen pressure of 8 MPa).

The main properties of metal hydride ligands are indicated in Table 10.2.

Table 10.2 Main Properties of Metal Hydride Ligands

Hydride	Density	Hydrogen Content (Mass Fraction)	Decomposition Temperature
$LiBH_4$	0.66	185	–
$NaBH_4$	1.074	10.7	–
KBH_4	1.177	7.5	–
$Be(BH_4)_2$	–	20.8	123
$Mg(BH_4)_2$	–	149	320
$Ca(BH_4)_2$	–	11.6	260
$Zn(BH_4)_2$	–	8.5	50
$Al(BH_4)_3$	0.55	1,698	–
$Zr(BH_4)_4$	1.11	10.8	–
$Th(BH_4)_4$	2.59	5.5	204
$LiAlH_4$	0.92	10.6	190
$NaAlH_4$	1.28	7.5	190
$Mg(AlH_4)_2$	–	9.3	140

CNTs were discovered in 1991 by Professor Iijima of Japan's NEC Corporation, but research on the use of CNTs in hydrogen storage was made by the American scholar Dillon, who set a precedent in 1997 [2]. On coke containing 0.1–0.2 wt% single-wall carbon nanotubes (SWNTs), he used the temperature-programmed desorption method to measure the hydrogen adsorption capacity and then found that pure SWNTs have a hydrogen storage capacity of 5–10 wt% at room temperature and a hydrogen pressure of 300 Torr. He also pointed out that hydrogen at the adsorption sites with high temperature is physically adsorbed and the CNTs can have a hydrogen storage capacity 10 times that of activated carbon. The sample size was only 1 mg, and thus the results were obtained through extrapolation; therefore, these results are perhaps unconvincing. Nonetheless, the results are encouraging. As such, there has been much in-depth and extensive research conducted regarding CNTs with regard to their hydrogen storage properties [3].

CNT hydrogen storage can occur in a variety of ways. One is through adsorption of hydrogen storage of CNTs. Table 10.3 shows a comparison of the hydrogen storage capacity of a variety of CNTs in hydrogen absorption.

Another mode of hydrogen storage by CNTs is electrochemical hydrogen storage. Research in this area was first reported in 1999 by Nützenadel and associates. Multi-wall carbon nanotubes (MWNTs) after purification had a maximum

Table 10.3 Hydrogen Storage Capacities of CNTs and $LaNi_5$ for Comparison (Data Determined by IMR, RT, 10 MPa)

	Sample Weight (g)	Environmental Stress (MPa)	Hydrogen Storage Capacity (wt%)
Conventionally prepared MWNTs	0.7281	14.1	2.0
Pretreated MWNTs	0.1580	14.5	6.5
Conventionally prepared SWNTs	0.4801	10.0	2.4
Pretreated SWNTs	0.4927	10.0	4.2
$LaNi_3$	1.1760	12.4	1.4

discharge capacity of 1,157 mAh/g, which is equivalent to 4.1% of hydrogen storage capacity by weight. After 100 cycles of charge–discharge, 70% of its maximum capacity remained. The SWNTs had maximum discharge capacity of 503 mAh/g, which is equivalent to 1.84% of hydrogen storage capacity by weight. After 100 cycles of charge–discharge, 80% of its maximum capacity remained (Figures 10.1 and 10.2).

Hydrogen is the cleanest, most renewable energy and has gained great importance in developed countries over the past

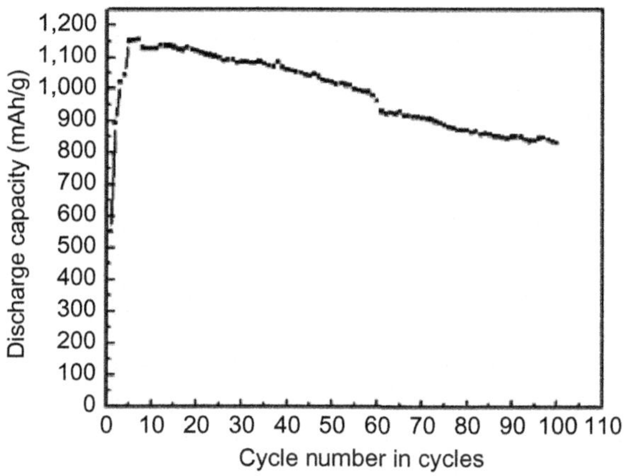

Figure 10.1 Cycle curve of MWNT electrode with 70% of the maximum capacity remaining after 100 cycles of charge–discharge. From Ref. [4].

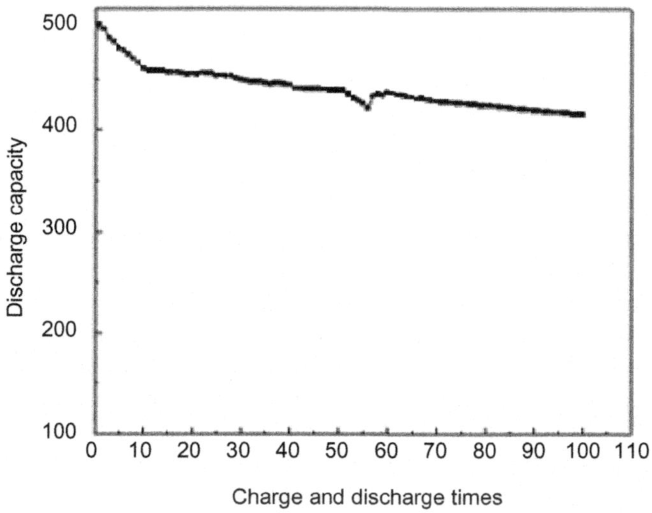

Figure 10.2 Cycle curve of SWNT electrode with 80% of the maximum capacity remaining after 100 cycles of charge–discharge. From Ref. [4].

decade. China, in recent years, has also invested heavily in research and development in this field. Pilot operations of hydrogen-driven vehicles have been performed in some developed countries, and China is in the process of introducing hydrogen-driven transportation. One of the barriers to the commercialization of hydrogen vehicles is the high cost of hydrogen storage. Taking into account safety and cost, both liquid hydrogen and high-pressure hydrogen gas are not the best choices for the commercialization of hydrogen vehicles. Most hydrogen storage alloy products have a large self-weight, and their life expectancy is also a problem, whereas magnesium-based alloys with a lower self-weight are difficult to store/release hydrogen at room temperature. In addition, reversible hydrogen storage of hydride ligands still needs further development and research. Carbon absorption materials for hydrogen storage have been taken seriously, but their basic research is still not sufficient. Whether it can be put into practical use is still a question. At present, hydrogen storage with nanomaterials has the following problems. There remains a sharp difference in the measured hydrogen storage capacities around the world (from 0.01 up to 67 wt%) without accurate methods for determination. Also, the hydrogen storage mechanism is still unknown. Therefore, the promise of hydrogen energy storage has a long way to go.

10.2 Fuel Cells [5]

10.2.1 Basic Concept

Fuel cells are one of the most important energy technologies in the twenty-first century and are a very promising alternative energy source. At present, the development of fuel cells used for vehicles and fuel cells for portable electronic devices has been classified as one of the top priorities in most developed countries. Fuel cells are a device in which an energy source can be directly converted; chemical energy in the fuel can combine with hydrogen and oxygen into water to produce electricity by using an electrochemical method rather than the burning mode. It is like installing a miniaturized generator in electronic products such as mobile phones, which require an endless fuel supply. In this sense, fuel cells differ from conventional secondary batteries. Therefore, fuel cells are actually power generators that are safe, clean, and characterized by long duration, long lifetime, portability, and cost-effectiveness. In addition, fuel cells can also be used in lighting devices, television sets, refrigerators and motors, and can also be widely used in military supplies. At present, the fuel cell is still in the development stage in the fields of materials and application. Any breakthrough in some key technologies will certainly result in a huge impact on the industry.

The basic principle of fuel cell power generation is based on a chemical reaction of hydrogen and oxygen that results in water, heat, and chemical energy, and the chemical energy can be directly converted into electricity. Specifically, the fuel cell is a "generator" using the reverse reaction of water electrolysis. It is composed of anode, cathode, and an electrolyte panel situated in the middle of the positive and negative electrodes. Initially, the electrolyte plate is formed by the use of electrolyte infiltration of the porous plate and is now being developed into a type that uses solids directly.

Under operating conditions, the negative terminal is supplied with fuel (hydrogen), while the positive terminal is supplied with oxidant (air). Hydrogen is broken down into H^+ ions and e^- electrons at the negative terminal. Hydrogen ions will enter the electrolyte, while electrons move along the external circuit to the positive terminal. Electricity load is connected to the external circuit. In the cathode, the oxygen in the air and the hydrogen ions in the electrolyte will result in the absorption of electrons entering the cathode to form water. This is precisely the reverse reaction of the water electrolysis process. By taking advantage of this principle, fuel cells in operation can continuously provide electricity to an external power source, hence the name "generator."

In fact, the fuel cell had been invented before the internal combustion engine was invented in the nineteenth century. In 1839, William Grove made the first primitive model of the fuel cell, which was based on the different redox potentials of H_2 and O_2 to obtain an available electromotive force. However, there was a lack of environmental awareness at that time, and with a large reserve of petroleum being discovered and explored at the same time, people chose to use the low-cost method of burning oil to generate energy. As a result, the development of fuel cells was overlooked. On entering the 1980s, the oil crisis started to emerge. In this global oil shortage crisis, people began to attach importance to the development of novel energy sources. Currently, great progress has been made in the development of renewable and clean energies, including solar energy, nuclear energy, wind energy, and tidal energy. Fuel cells can supply sustainable and stable electricity, and thus are widely studied. Current energy mainly comes from nuclear power, oil, natural gas, coal, and other minerals. Power generation is based on the principle of combustion or reaction of minerals to produce large amounts of heat, which is then converted into mechanical energy, and then into electrical energy. The use of internal combustion engines to generate electricity usually results in low efficiency and polluting emissions. High-efficiency fuel cells can generate electricity in a clean way (~83%) and may ensure a low pollution emission, so it can replace the conventional power generation methods of using internal combustion engines. So far, many developed countries have invested in fuel cell research and development.

Hydrogen fuel cell technology has currently been fully developed abroad and many products have been introduced, such as electric vehicles and hybrid cars, as shown in Figure 10.3. However, proton-exchange membrane (PEM) materials applied in methanol fuel cells are still far from satisfactory. Early electric vehicles were mainly dependent on lead–acid batteries as the main power source. Because of the long charging time and heavy weight of these batteries, electric vehicles have always been unable to compete with conventional internal combustion engine-driven ones. The most striking difference between the typical internal combustion engines and fuel cells is the lack of endurance. In less than 90 min, the fuel cell vehicles may run out of fuel. To make things worse, it is difficult for them to find premises for refueling. The key to the operation of the PEM is the thin platinum catalyst (expensive); it is plated on both sides of the membrane, accounting for 40% of the cost of the batteries. The biggest problem in the design of fuel cell vehicles is

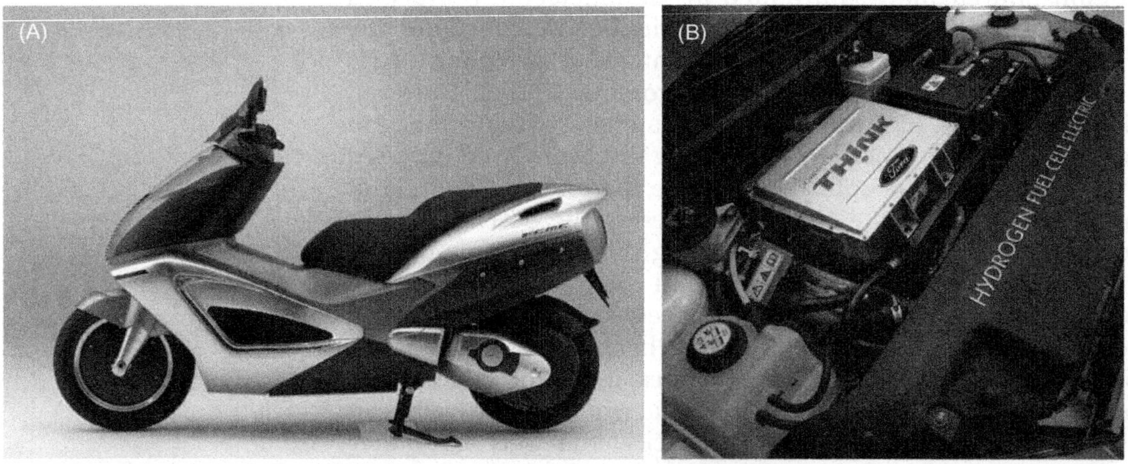

Figure 10.3 (A) Honda's fuel cell-powered vehicle and (B) Ford's hybrid power system.

how to make enough fuel available to meet the requirements of customers who drive a certain mileage. In general, a ride of approximately 640 km may take 5–7 kg of hydrogen, but the current prototype vehicles using fuel cells can only carry 2.5–3.5 kg. As a result, the design of fuel cell vehicles still has a long way to go.

10.2.2 Comparison of the Main Fuel Cells

There are many types of fuel cells. Depending on the electrolytes in use, fuel cells can be divided into the following categories: polymer electrolyte fuel cell (PEFC), solid polymer electrolyte fuel cell (SPEFC), proton-exchange membrane fuel cell (PEMFC), alkaline fuel cell (AFC), phosphoric acid fuel cell (PAFC), molten carbonate fuel cell (MCFC), and solid oxide fuel cell (SOFC). A comparison between these categories is indicated in Table 10.4. Currently, the PEMFC is one of the hotspots in fuel cell research, because it is operable at low temperatures and used with solid-state electrolytes. In addition, its application can also be found in industries relying on mobile energy, such as the automobile industry and 3C electronics industry. The core components of PEMFCs include membrane electrode assembly (MEA), gas diffusion layer (hydrogen system), and conductive bipolar plates. Here, MEA is formed by the catalytic electrode and proton conduction membrane, as shown in Figure 10.4. At present, it is urgent to overcome the many flaws in the applications of PEMFCs, such as improving the battery life and lowering the price. PEMFC in the fuel in use can be divided into two categories, hydrogen fuel system and methanol fuel system. In general, the PEMFC is based on hydrogen as fuel, and the system with methanol as fuel is called the direct methanol fuel cell (DMFC). Regarding the global fuel manufacturing giants or famous car makers, such as Ballard, Toyota, Ford, and Honda, PEMCF has matured in device technology development. At present, various fuel cell manufacturing giants are committed to the goal of developing reduced manufacturing costs of PEMCF.

DMFCs are characterized by large units of energy density, easy assembly, small volume, and ample fuel supplements, so they are suitable for application in 3C electronics, such as portable computers or mobile phones. Thus, European, American, and Japanese manufacturers, such as Smart Fuel, Gore Associates, Toshiba, NEC, and Panasonic, are focusing on funding research and development to develop DMFCs. However, studies found that a DMFC, if working with the same

Table 10.4 Comparison of the Major Fuel Cells

Type: By Temperature	Type: By Electrolyte	Fuel	Oxidant	Catalyst	Water Management	Power Generation Efficiency
Low-temperature fuel cell (60–200°C)	Alkaline fuel cell (AFC)	Pure hydrogen	Pure oxygen	Anode: Raney nickel, platinum/carbon Cathode: Raney nickel, platinum/carbon	Evaporation drainage	60–70%
	Proton-exchange membrane fuel cell (PEMFC)	Hydrogen, methanol	Oxygen, air	Anode: Pt/C Cathode: Pt/C	Evaporation drainage + Power drain	43–58%
Intermediate-temperature fuel cell (160–220°C)	Phosphoric acid fuel cell (PAFC)	Hydrogen		Anode: Pt/C, Pt–Ru/C Cathode: Pt/C	Evaporation drainage	37–42%
High-temperature fuel cell (600–1,000°C)	Molten carbonate fuel cell (MCFC)	Hydrogen, natural gas, coal gas, biogas		Anode: nickel–chromium alloy, nickel–aluminum alloy Cathode: Lithium-oriented–nickel oxide	Gaseous drainage	>50%
	Solid oxide fuel cell (SOFC)			Anode: Ni/YSZ Cathode: Manganese-doped lanthanum strontium		50–65%

accessories as hydrogen PEMFCs, may have very low power generation efficiency and short battery life. This results from the perfluorinated resin membrane more likely using a PEM in the hydrogen system. For example, Nafion, with a large swelling feature in methanol solutions, has poor barrier properties of methanol, resulting in the serious problem of the proliferation of methanol. Furthermore, it causes the platinum catalyst to be poisoned, which in turn leads to reduced life, or methanol may produce protons without being catalyzed by platinum, directly

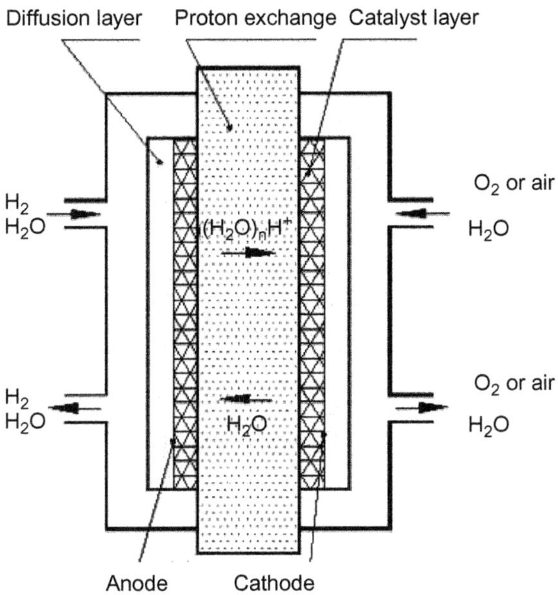

Figure 10.4 PEMFC: construction.

leading to a reaction with oxygen to reduce the efficiency of fuel use.

10.2.3 Proton-Exchange Membrane

PEM used in fuel cells is required to have the following basic properties: first, high proton conductivity; second, good mechanical strength; third, excellent chemical resistance; and fourth, good durability.

When operating at low temperatures (<80°C), PEMFCs and DMFCs work with the Nafion-based PEM manufactured by DuPont Co. When the operating temperature is higher than 120°C (120–200°C), high-temperature hydrocarbon polymers more commonly used, as reported in the literature, include poly (phenoxy phosphazene) (POP), sulfonated naphthalic polyimide, polybenzimidazole (PBI), alkylsulfonated polybenzimidazole (PBI-AS), sulfonated poly(arylene ether ether ketone) (PEEK-SO_3H), sulfonated poly(arylenesulfone) (PSU-SO_3H), and so on.

The PEM for the DMFCs is essentially the same as that of the hydrogen PEMFC. The difference is that methanol as a fuel needs special consideration regarding methanol permeability issues. Methanol permeability can cause the overall decline of the fuel cell discharge performance and a shortened life

Table 10.5 Proton-Exchange Membrane Materials of Foreign Producers

Company/University/Organization	Membranes
US PolyFuel Inc.	Acid-based polyether ether ketone
Toshiba	Acid-based polyether ether sulfone
Sony	OH-modified fullerene-based membranes
JSR Corporation, Japan	Polystyrene sulfonic acid-based compounds
Stuttgart University, Germany	Acid-based ionomer blends
Los Alamos National Laboratory (LANL)	Sulfonation sulfone polymer; PVDF-g-SPS
Ballard	Sulfonated-F-styrene
DuPont	Modified Nafion
Japan's Asahi Glass Co., Ltd. (Asahi Glass)	PFS/PTFE fibrils

expectancy. Thus, concerning the PEM for direct methanol, we must take into account the methanol permeability problem in the material selection and design.

There are currently many PEM producers. These producers and the materials used for PEM are listed in Table 10.5. As noted, in the hydrogen PEM fuel system, PEM must have good proton conductivity, mechanical strength, and durability. In 1960, DuPont developed Nafion, which was the first PEM broadly in line with these requirements. At present, some other companies also produce fluorine polymer proton-conducting membranes, but they are mostly modeled on the Nafion polymer structure, as shown in Figure 10.5. However, Nafion cannot be directly used in the DMFC system, because the perfluorinated PEM has extremely high penetration of methanol that may reduce the discharge life and performance of fuel cells. The high methanol permeability is rooted in a large channel formed by a large group of ions, and the formation of this large channel is associated with its molecular structure.

As indicated in Figure 10.5, perfluorinated PEM materials are usually designed by using carbon–fluorine resin as a polymer main chain to increase the membrane's chemical stability, thermal stability, and mechanical strength. In addition, the side chain-induced sulfonic fluorocarbon short chain has a proton conduction function. Nafion has high proton conductivity because the short-chain fluorocarbon sulfonic acid group has good mobility and can form acid-based clusters, which, in the case of water absorption, will soften the main chain of

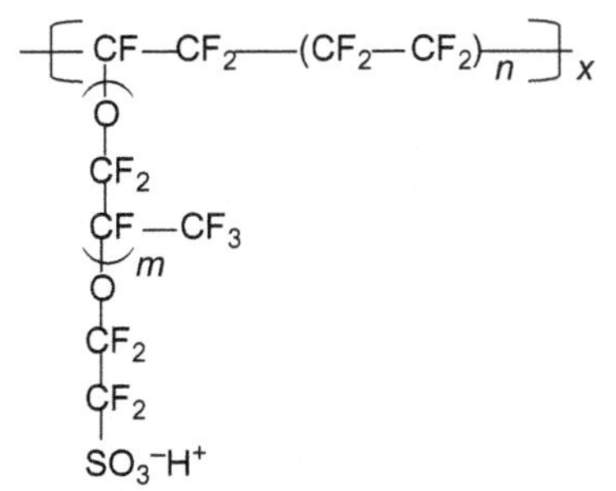

Figure 10.5 Chemical structure of perfluorinated proton-exchange membrane material—Nafion.

fluorocarbon polymer and increase the proton conduction pathway. However, because of the increased proton pathway, in the methanol fuel cell system the methanol molecules travel through the large channel at an increased rate. A perfluorinated resin with high methanol permeability would cause a shortening of life expectancy as a whole, dramatically reducing the power generation efficiency and effectiveness. To increase the efficiency of methanol fuel cells in power generation and battery life, the current research and development trend in methanol fuel cells is to reduce the methanol poisoning of the platinum catalyst or the methanol permeability of the PEM. To reduce the methanol permeability of the PEM, the PEM must be improved. This can be achieved in two ways. One way is the original perfluorinated proton-exchange membrane. For example, Nafion is modified by way of adding solid acid or inorganic oxides to reduce the number of channels in the perfluorinated membranes for methanol permeability, or by the use of different polymers doped with other polymers to limit the swelling of Nafion. Alternatively, the inorganic porous material can be impregnated with Nafion to reduce the swelling of Nafion, or the Nafion PEM is added with a good methanol barrier membrane on both sides. Another method is to use nonfluoride, high-performance engineering plastics containing a benzene ring, such as polyamide, PES, PEEK, or polymers containing more benzene. These materials are inexpensive and characterized by good chemical stability, mechanical strength, and thermal stability. Many researchers are trying to use the nonfluorine polymer sulfonic acid treatment to obtain a sulfonation polymer that can be used in PEM. The generated sulfonic acid enables the sulfonic acid polymer to have proton conductivity, which is related to sulfonation. For the same polymer, the level of sulfonation is proportional to the level of proton conductivity. However, as sulfonation is increased, water absorption of materials will be increased as well. When sulfonation is greater than a specific value, the membrane immersed in water will absorb a large amount of water, causing the swelling of polymer or even dissolution. In the preparation of a PEM using sulfonic acid polymer, the requirement for proton conductivity to achieve the level of Nafion will lead to an increase of sulfonation. However, while enhancing the proton conductivity, membrane dissolution will occur subject to the increased moisture in the membrane. This makes the sulfonic membrane distinct from Nafion in terms of proton conductivity. Therefore, in the case of low moisture content, the postsulfonic acid polymer cannot have high proton conductivity. To improve this, research has shifted toward the direct synthesis of nonfluorinated sulfonic acid polymer systems. The results showed that, by way of direct synthesis of sulfonic acid-based polyimide, the resulting PEM can have conductivity slightly lower than Nafion. Nonetheless, the methanol permeability was less than one-tenth that of Nafion. In addition, sulfonic acid-based polyamide is also lower than Nafion in moisture content. It is critical for the proton conductive membrane to be able to maintain high proton conductivity at low moisture content. This is because when the DMFC anode has contact with air, proton conductive membranes with low moisture content may have a water evaporation rate lower than that of proton conductive membranes with high moisture content. Therefore, the water dissipation can be reduced to increase the usage time of methanol fuel.

10.2.4 Nanofuel Cells

The nanofuel cell is one of the fuel cells based on the use of nanomaterials and nanotechnology. The small fuel cells (< 500 W) are currently undergoing research and development. It involves using 2-nm to 3-nm electrochemical catalysis and technology of organic–inorganic hybrid nanomaterials (e.g., PEM), leading to the development of integrated fuel cell technology. Low cost and high efficiency have become the most critical bottlenecks for its practical use.

Conventional fuel cells generally use hydrogen fuel. Compared with hydrogen, methanol as a liquid has high density and can be more safely stored, so the methanol fuel system has considerable potential for development. In addition, propane gas is currently applied in some of the nanofuel cells being developed. This material can be easily purchased.

The US-based NanoDynamics, Inc. is now developing a 50-W SOFC. The prototype is only the size of a loaf of bread, with 20% of its components being manufactured with nanomaterials. As such, a battery can store 3,000 W·h of electricity using just 5 pounds of propane. Under the same conditions, the traditional SOFC can only store one-third to one-half the electricity of the present fuel cell. This 50-W SOFC with nanomaterials is a prototype of fuel cells that was originally designed for battlefield soldiers. Such a battery can supersede the traditional batteries weighing 35 pounds.

In addition to using nanomaterials, NanoDynamics also renovated the various parts of these battlefield batteries dedicated to the soldiers, including the membrane, electrode, and battery catalyst, to enable them to become more portable. At the same time, the storage capacity of the fuel cell is improved. This should be higher in the requirements for battery storage capacity.

Fuel cell energy comes from the chemical energy released from the generation of water in the reaction of hydrogen and oxygen. Platinum in the fuel cells is the catalyst in the catalytic reaction. However, subject to the side effects occurring in the vicinity of the cathode, platinum would gradually lose catalytic oxidation ability. This is one of the challenges in the practical application of fuel cells. Studies have shown that platinum electrodes for fuel cells, if plated on gold nanoparticles with a diameter of approximately 2–3 nm, will increase the platinum oxidation potential, making the metal less susceptible to oxidation. The platinum electrode can be protected from damage in this way. The studies also showed that the catalytic ability of platinum will not decline when the surface is blocked by gold nanoparticles, that is, platinum plated with nanogold still has catalytic reaction capabilities.

10.3 Dye-Sensitized Nanocrystalline Solar Cells

10.3.1 Status of Solar Cells

In 1954, Bell Labs in the United States produced the world's first practical solar cell with efficiency of 4%, which was applied to the US satellite *Vanguard-1* in 1958. Later, solar cells gradually entered into the ground application of electricity from space. A home photovoltaic system of 4 kW on the roof is able to meet the average family's electricity needs, with the annual reduced CO_2 emissions equivalent to the annual emissions of a family car. Along with the continuous improvement in the material, structure, process, and other aspects, the present price of solar cells is less than 1% of that in the 1970s. It is expected that in 10 years, solar energy will be able to compete with thermal power in terms of cost of power generation in the United States, Japan, and Europe. At present, solar energy has an average annual growth rate of 35%, presenting the fastest growth in the energy technology industry.

10.3.2 Types of Solar Cell

Solar cells can be divided into the inorganic type or the organic type, according to the materials used. Inorganic materials commonly used in solar cells are semiconductor silicon (monocrystalline, polycrystalline, amorphous, complex, etc.) and compound semiconductors (GaAs, $CuInSe_2$, CdTe, InP, etc.). Silicon solar cells (monocrystalline silicon, polycrystalline silicon, amorphous silicon) have relatively higher photoelectric conversion efficiency, but the cost is also high and the preparation process is complex. In addition, the batteries made from multi-compound materials, such as inorganic salts, including gallium arsenide, cadmium sulfide, and copper indium selenium, contain cadmium and other toxic elements, as well as indium, selenium, and other rare elements, and therefore they are inferior in security and price.

Materials commonly used in organic solar cells are organic semiconductors (zinc phthalocyanine, polyaniline, poly *p*-phenylene vinylene, etc.). There are also other photochemical solar cells using nano-TiO_2. Solar cells made from functional polymer materials are presently still in the early stage of research and development and are characterized by low conversion efficiency and short service life.

10.3.2.1 Inorganic Solar Cells

Semiconductors can make use of different barriers, such as p–n junction, Schottky barrier, and heterojunction, to have a photovoltaic effect. When the solar cells are exposed to sunlight, the interaction of light and semiconductors can produce photocarrier and the resulting electron–hole pairs separated from the poles depend on the formation of the semiconductor barrier, while the positive and negative electric charges are collected by the upper and lower electrodes respectively. Formed by the charge accumulation, the electric current will flow to the load through the metal wire. Inorganic solar cells, especially thin-film batteries, are currently predominant on the market. The following is a brief introduction to the principal types on the market.

10.3.2.1.1 Silicon Wafer Solar Cells

Wafer mostly refers to monocrystalline silicon wafer, derived from the ordinary system of silicon salad. It is the most commonly used semiconductor material and can be divided into 4-, 5-, 6-, and 8-in. sizes by diameter. Recently, a 12 in. or greater specification has been developed. The larger the wafer, the more devices can be integrated on the same piece. At the same time, enlarging the size can reduce the cost but requires higher technology for materials and production. In 1954, Bell Labs claimed the successful development of the first silicon wafer solar cell, which showed luminous efficiency of 6%. With the subsequent development of integrated circuits, this type of solar cell has been predominant on the market; its market share has never fallen below 80%. If we only consider the market power of more than 1 kW, then this figure can almost reach 100%. This can be broadly attributed to three factors: first, cost and price; second, the module efficiency; and third, the scale and capacity utilization.

As the technology advances, remarkable improvements have been made in terms of the wafer thickness, cutting technology, wafer size, as well as wafer price. Since the 1960s, such cell power generation has led to a decline of approximately 50 times the unit per watts cost. The current price is approximately US $2.5 to $3/watt. Over the past 10 years, thin-film solar cells have seen a substantial decline in manufacturing costs, showing a trend even faster than that of the silicon wafer; however, so far, their price has generally remained at approximately 50% higher than the silicon wafer.

The optical efficiency in laboratories for the silicon wafer single-cell system now stands at 25%, which is very close to the theoretical value of 29%. Since the 1970s, the commercial products have also advanced greatly in optical efficiency, amounting to approximately 12% in recent years. This technical achievement, relatively speaking, is beyond the reach of the majority of thin-film technologies.

Production costs are often deeply affected by production scale, and solar cells are no exception. In comparison to the thin-film type and silicon wafer type, in general the former is approximately 10 times that of the current production capacity of the latter. Therefore, fixed costs can be substantially divided. Followed by capacity utilization, the current manufacturers of the silicon wafer type, due to the continuous significant market growth in recent years, present an average capacity utilization of approximately 80%, while the thin-film manufacturers only present an average capacity utilization of approximately 40%. This makes the silicon wafer type more competitive when it comes to production costs, thus contributing to its larger market share.

10.3.2.1.2 Amorphous Silicon Solar Cells

Amorphous silicon solar cells are the most well-developed thin-film solar cell. The structure usually has the p-i-n (or n-i-p) type of duality, where p-layer and n-layer are mainly used for establishing an internal electric field (i-layer) comprising amorphous silicon. Because amorphous silicon has a high absorption capacity, the i-layer usually has a thickness of 0.2–0.5 μm. Its absorption frequency ranges between 1.1 and 1.7 eV, which is different from that of the silicon wafer, which has an absorption frequency of 1.1 eV. Unlike the crystal, the structural homogeneity of amorphous material is relatively low. Electrons and holes are conducted inside the material; therefore, in the case of long distance conduction, there may be a high composite probability of the two. To avoid this phenomenon, the i-layer should not be too thick or too thin, because the latter problem can easily lead to inadequate absorption. To overcome this predicament, a multilayer structured stack is often used in the design of amorphous silicon solar cells to achieve a balance between the optical absorption and photoelectric efficiency.

The inherent inadequacy of such photovoltaic cells is a short-term but significant decline in the performance after light exposure, also known as the SWE effect, with the rate of decline between 155% and 35%. It is caused by some of the unsaturated silicon atoms in the material, which may show structural changes due to light exposure. The aforementioned multilayer stack approach is a candidate for overcoming the SWE effect.

The manufacture of amorphous silicon photovoltaic cells is based on plasma-enhanced chemical vapor deposition (PECVD), which can be used to produce silicon thin film. Substrate can be made of the flexible and inexpensive material in larger sizes, for example stainless steel or plastic materials. The process is the roll-to-roll method. The evaporation process is slow, and the high-quality conductive glass layer is very costly, so its total manufacturing cost is only slightly less than that of crystal solar cells. As for the multilayered stack model, improved battery efficiency also comes with an increased cost. In summary, although it does not have advantages in terms of cost, this type of solar battery output has shown rapid growth in the past 3 years; in 2003, there was an annual growth of 113%. It is expected that this trend will continue.

To reduce manufacturing costs, it has been developed using VHF plasma in the manufacture of membrane, enabling the process speed to be increased fivefold. Meanwhile, SnO_2 may supersede ZnO as the conductive glass material to reduce the TCO costs. Once this process is successfully developed, it would greatly enhance the competitiveness of amorphous silicon solar cells. The greatest weakness is its low photoelectric conversion efficiency. At present, the efficiency is approximately 13.5% in laboratories only and 4–8% in commercial modules. It seems the scope for future improvement may be quite limited.

10.3.2.1.3 Copper Indium Gallium Diselenide Solar Cells

There are two kinds of copper indium gallium diselenide solar cells. One is referred to as CIS, containing copper, indium, and selenide, and the other is CIGS, containing copper, indium, gallium, and selenide. It is popular because of its higher efficiency and low cost of photoelectric materials. The optical efficiency of CIGS solar cells made in laboratories can be up to approximately 19%; when it comes to the module, the maximum value can still reach approximately 13%. Depending on the differences in the percentage of indium and gallium, CIGS may have a light absorption range that varies from 1.02 to 1.68 eV. This feature can be used for multistack modules to further enhance the organizational effectiveness of the battery. In addition, due to the high absorption efficiency ($\alpha > 10^5\ cm^{-1}$), the required thickness of photoelectric materials will not be larger than 1 μm, and 99% or more of the photons can be absorbed. It is generally estimated that the semiconductor materials required for the mass production of this kind of cell may be US$0.03/W only.

CIGS photovoltaic cells are different from amorphous silicon solar cells in structure, mainly due to a buffer layer that exists between the optical layer and the conductive glass. The layer material is usually chromium sulfide (CdS). The carrier can also be made of a flexible material, so the roll-to-roll mode is available for the process. At present, the commercial process is being developed by Shell Solar and includes a series of vacuum procedures, which has led to very high investment in hardware and manufacturing costs. An estimate showed that investment in this production process is approximately US$33 for 1 m^2. Simultaneous evaporation is commonly used in laboratory production techniques, but its amplification is not easy and may not be viable commercially. Another company, ISET, has been actively involved in the development of nonvacuum techniques trying to use nanotechnology similar to the ink process in preparation of the layered results. The company reported that initial success has been achieved, but whether it can develop itself into a commercial manufacturing process is unknown. In addition, NREL in the United States has successfully developed a three-stage process that was a great success in the laboratory in obtaining solar cells with an optical efficiency of 19.2%. However, the complexity of the production process and the high costs are detrimental to any further progress being made.

In summary, despite advantages such as high photovoltaic efficiency and low material cost, CIGS also has many difficulties. For example, three problems must be solved. The first is complex process and high investment costs. Although many organizations are working to improve the process, it is rather difficult to reduce costs in this way. Second is supply of key raw materials. CIGS material involves the use of the indium element, which is a potential problem because the natural reserves of indium are very limited. Researchers abroad have made some calculations based on a 10% battery efficiency, indicating a wide application of CIGS photovoltaic electricity as a major energy resource will use all the natural indium within only a few years. Third is the potential toxicity of the CdS buffer layer. The toxicity of cadmium (Cd) has been a concern. The fact that improper leakage of CdS in the cell will cause environmental harm makes people uneasy. Therefore, most countries in Europe have stopped studying this type of photovoltaic cell.

10.3.2.1.4 Cadmium Telluride Thin-Film Solar Cells

Cadmium telluride thin-film solar cells (CdTe) have the longest history of the thin-film photoelectric cells, so they have been the focus of more research. In 1982, Kodak took the first step forward in the manufacture of cadmium telluride thin-film solar cells with optical efficiency of more than 10%. At present, the highest electro-optical efficiency achieved in the laboratory is 16.5%, which was completed by the NREL laboratory in the United States. The goal is to make further incremental changes in the established battery configuration, and such changes are used in some of the materials.

The typical structure of CdTe photovoltaic cells displays the formation of the main body of an approximately 2-μm p-type CdTe layer and the n-type CdS with thickness of only 0.1 μm. Photon absorption occurs mainly in the CdTe layer, with efficiency of the absorption coefficient greater than $10^5\ cm^{-1}$. Therefore, a thickness of only a few microns is available to absorb more than 90% of the photons. The upper edge of the CdS layer first links to TCO and then connects to the substrate, while the upper edge of CdTe is linked to the back panel to form a photovoltaic cell structure. It is known that in making CdTe solar cells, regardless of the battery structure, a cadmium chloride-activated semiconductor layer is required. The preparation methods include wet or dry steam. The dry steam method is used by the industrial sector more often.

As for the preparation of the membrane used for CdTe photovoltaic cells, there are a number of available processes, among which are a few methods that are suitable for mass production. The known methods are sputtering, CVD, atomic layer epitaxy, screen printing, galvanic deposition, chemical spraying, close-packed sublimation, modified close-packed sublimation, and sublimation–condensation. Each method has its advantages and disadvantages. Current deposition is the least expensive one, and thus becomes the main approach used in industry. It consumes the least tellurium.

Although the CdTe solar cells possess many of these procompetitive factors, its market share is still low. The reason why CdTe solar cells cannot be promoted to the mainstream market may be summarized as follows. Both the module and the substrate material are too costly. The overall CdTe solar cell material accounts for 53% of the total cost; here, semiconductor materials only account for approximately 5.5%. The natural reserves of tellurium are limited. If humans rely on it for substantial and comprehensive photovoltaic power generation, then its total amount cannot meet the demand. Cadmium is toxic. People cannot use such photovoltaic cells without worrying about the toxicity.

10.3.2.1.5 Silicon Thin-Film Solar Cells

The development of silicon thin-film solar cells started in the 1970s, but there was no big breakthrough until the 1980s. The thickness of the silicon crystal layer should be only 5–50 mm, so the acceptable substrates can be sub-silicon materials, glass, ceramic, or graphite. In addition to the substantially reduced amount of silicon material used, this type of photovoltaic cell is not as strictly required as silicon wafer solar cells in the purity of silicon material due to the short distance of conduction between electrons and holes, so material costs can further decline. Silicon material does not have as high an absorption efficiency as other semiconductor photovoltaic cell materials in development; in optical cells of this type, the film thickness of the silicon layer is not as thin as silicon-based solar cells, which can reach approximately 300 microns of silicon layer thickness. Thus, in design, the concept of *optical hysteresis* should be introduced to improve the light absorption rate. This point is different from other thin-film photovoltaic cells.

The preparation methods for this type of photovoltaic cells include liquid phase epitaxy, many types of CVD, including low pressure and atmospheric pressure chemical vapor deposition (LPCVD, APCVD), PECVD, ion-assisted chemical vapor deposition, as well as hot-wire chemical vapor deposition. Unfortunately, none of these methods can be used in industry. Nonetheless, it is generally believed that APCVD can be adopted in mass production. For the aforementioned vapor deposition method, its operating temperature ranges from 300 to 1,200°, depending on the differences in substrate materials.

This type of photovoltaic cell has reached 21% of optical efficiency at the maximum laboratory level. In the market, the product is only available in Astropower. With a graphite substrate, the efficiency is approximately 13.4%. Because the graphite material is expensive, the current studies concentrate on the use of glass or high-temperature substrate, or the transfer of the semi-finished products of the monocrystalline silicon layer to the glass substrate to supersede the graphite. Japan's Mitsubishi Corporation has successfully used this method in the preparation of components of 100 cm^2, with optical efficiency of up to 16%. Overall, this type of photovoltaic system is still in the verification period of feasibility. Whether the laboratory preparation technology can be developed into a mass production process to bring about economic benefits is another focus of attention.

10.3.2.2 Organic Solar Cells

Organic solar cells are made with organic semiconductors, which follow the working principle that organic semiconductors may produce electrons and holes that are bound in excitons while separation of the electron and hole occurs in the interface (the junction of electrodes and conductive polymers).

In recent years, the major advance in organic solar cells has been the first generation of plastic solar cells developed with plastic nanotechnology by scientists at the University of California (Berkeley) in 2002. These cells can be installed in various portable devices and wearable electronic devices to provide 0.7 V voltage.

Organic solar cells are characterized by low price, easy shaping, and performance control by chemical modification.

In organic solar cells, the most important type is the dye-sensitized nanocrystalline solar cell, which is actually an organic and inorganic complex.

10.3.3 Dye-Sensitized Nanocrystalline Solar Cells [6–8]

DSSCs are developed mainly on the basis of mimicking photosynthesis. It is a new type of solar cell, with its main advantages being abundance of raw materials, low cost, relatively simple technology, and more advantages in large-scale industrial production. Furthermore, all raw materials and production processes are nontoxic and nonpolluting, and some of the materials can be fully recovered. This is of great significance to the protection of the human environment. Dye-sensitized nanocrystalline solar cells may have a cost that is merely one-fifth to one-tenth that of silicon solar cells and a service life of up to 15–20 years. Thus, DSSCs represent a promising clean solar energy conversion device. Research on DSSCs will help alleviate the energy crisis in the world today with significant practical potential.

10.3.3.1 The History of Dye-Sensitized Nanocrystalline Solar Cells

In 1991, O'Regan and Gratzel [9] at Ecole Polytechnique Fédérale de Lausanne, Switzerland, reported a new type of highly efficient solar cell with dye-sensitized nanocrystalline TiO_2 films as light anode, symbolizing the emergence of the world's first nanosolar cell. DSSCs can achieve photoelectric conversion efficiency more than 7% at a lower cost, providing a new way to use solar power.

In 1997, the DSSC's photoelectric conversion efficiency reached 10–11%, short-circuit current reached up to 18 mA/cm^2, and open circuit voltage reached 720 mV.

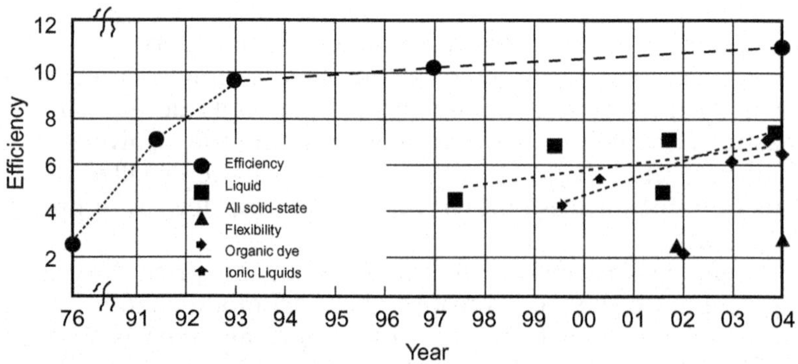

Figure 10.6 Schedule for the overall efficiency of the various DSSCs.

In 1998, the use of solid organic hole transport materials to supersede the solid-state DSSCs with liquid electrolytes was successful. Its monochromatic photoelectric conversion efficiency of up to 33% attracted worldwide attention.

At present, the photoelectric conversion efficiency of DSSCs can be maintained at 10%, life expectancy is 15–20 years, and its manufacturing cost is only one-fifth to one-tenth that of silicon solar cells [10].

DSSCs comprise an organic–inorganic composite system. The advantage of DSSCs is that they can be made into products of transparency, so they have a wide range of applications. They can be used in a variety of lighting conditions and have high light utilization efficiency. In addition, they are not sensitive to light and shadows and can work in a wide temperature range (Figure 10.6).

10.3.3.2 Cell Structure

Dye-sensitized nanocrystalline solar cells (DSSCs; also known as Grätzel-type photoelectrochemical solar cells) mainly include glass substrates coated with a transparent conductive film, the dye-sensitized semiconductor materials, such as electrode pair and electrolyte several parts (Figure 10.7). In Figure 10.7, the anode in the electrode pair is dye-sensitized semiconductor film, such as TiO_2 film (5–20 μm, 1–4 mg/cm^2); the cathode is platinum-plated conductive glass and the electrolyte is I_3^-/I^-.

Titanium dioxide thin-films electrolyte Pt mirror

10.3.3.3 Working Principle

The working principle of DSSCs is shown in Figure 10.8. Semiconductor TiO_2 has a band gap of 3.1 eV, so absorption

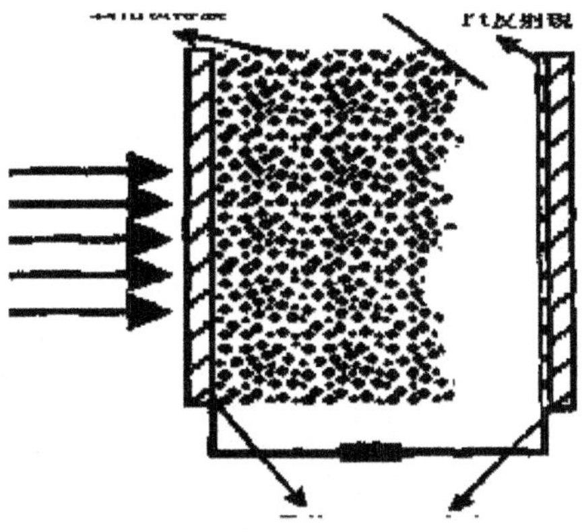

Figure 10.7 Cell structure.

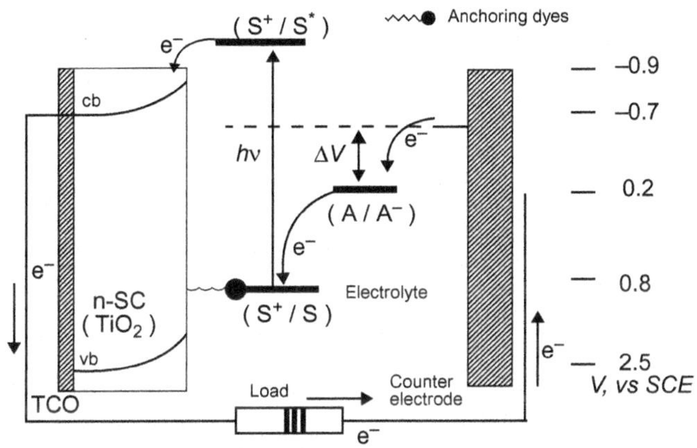

Figure 10.8 Diagram of working principle.

occurs in the ultraviolet region. After surface adsorption to the dye sensitization, the good response of dye to visible light can help to expand its absorption into the visible light region. DSSCs are actually a heterojunction photoelectric conversion device. The sunlight on the cell will be absorbed by electrons in dye molecules, which can then jump to the excited state. Excited state electrons will be quickly injected into the TiO_2 conduction band via TiO_2 film, quickly reaching the contact surface between the membrane and conductive glass. It will be collected in the conductive substrate and flow toward the electrode through the external circuit.

The principle in Figure 10.8 can be expressed by the following reaction equation, with the same symbols as shown in the figure:

$$S + h\nu \rightarrow S^*$$
$$S + h\nu \rightarrow S^*$$
$$S^* \rightarrow S^+ + e^- \rightarrow CB(TiO_2)$$
$$S^* \rightarrow S^+ + e^- \rightarrow CB(TiO_2)$$
$$S^+ + A^- \rightarrow S + A$$
$$S^+ + A^- \rightarrow S + A$$
$$A + e^-(CE) \rightarrow A^-$$
$$A + e^-(CE) \rightarrow A^-$$

10.3.3.4 Parameters for Performance Evaluation

The performance indicators of DSSCs are mainly the incident photon-to-current conversion efficiency (IPCE) and the total conversion efficiency (ratio of output power and input power) η_{total}.

The incident monochromatic light photoelectric conversion efficiency (IPCE) is defined as

$$IPCE = \frac{(1.25 \times 10^3) \cdot J_p}{\lambda \cdot \Phi_p} \tag{10.1}$$

Here, J_p is photocurrent density in $\mu A/cm^2$, λ is the optical wavelength in nm, and Φ_p is the flux by W/m^2.

Alternatively, IPCE can be expressed as:

$$IPCE = LHE(\lambda)\phi_{inj}\eta_c \tag{10.2}$$

Here, LHE(λ) is the light absorption efficiency, $LHE(\lambda) = 1 - 10^{-\Gamma\delta(\lambda)}$; Γ is the number of moles of dye covering the surface of membrane per unit of square centimeter, $\delta(\lambda)$ is cross-sectional area of dye absorption, ϕ_{inj} is the efficiency of electron injection, $\phi_{inj} = k_{inj}/(\tau^{-1} + k_{inj})$, k_{inj} indicates rate constant of electron injection, τ is the lifetime of the excited state, and η_c is the efficiency of the electrode in charge collection.

There are three main factors that may affect the photoelectric conversion efficiency of the cells: (1) lighting efficiency, mainly dependent on the optical absorption of an organic photosensitive dye; (2) electronic injection, mainly dependent on the energy level matching between the organic photosensitive materials and nanocrystalline semiconductor materials; and (3) collection efficiency, mainly determined by the proliferation of electronic properties in the film.

The total conversion efficiency (ratio of output power and input power) is calculated as follows:

$$\eta_{global} = i_{ph} V_{oc}(ff)/I_s \tag{10.3}$$

Here, i_{ph} is short-circuit photocurrent or, in other words, the current at short circuit (i.e., resistance is zero, only to connect the electrode pair and ammeter), which is the largest photocurrent that can be generated by photovoltaic cells. V_{oc} is open circuit photovoltage, which is the voltage at open circuit (infinite resistance, only to connect the reference electrode and voltmeter). It is the maximum voltage that can be generated by photovoltaic cells. Theoretically, the open circuit voltage V_{oc} is equal to the Fermi energy level of the semiconductor TiO_2 in light minus the Nernst reversible redox potential of electron pairs in electrolyte; *ff* is fill factor. It is defined as the ratio between the product of current (I_{opt}) and voltage (V_{opt}) against the product of the battery short-circuit current and open circuit voltage when the battery has a maximum output power (P_{opt}):

$$ff = \frac{P_{opt}}{I_{sc} \cdot V_{oc}} = \frac{I_{opt} \cdot V_{opt}}{I_{sc} \cdot V_{oc}}, \tag{10.4}$$

I_s is the incident light intensity.

10.3.3.5 Research Progress

There has been progress in the research on dye-sensitized nanocrystalline solar cells in the aspects of sensitizer, nanosemiconductor materials, and electrolytes.

10.3.3.5.1 Sensitizer

Dye-sensitized nanocrystalline solar cells require the sensitizer to be:

1. able to absorb sunlight as much as possible;
2. closely adsorbed on the surface of the nanocrystalline network electrode;
3. able to match the corresponding nanocrystalline energy band;
4. long enough in the excited state;
5. stable in the long term.

Dye-sensitized nanocrystalline solar cell sensitizers are mainly divided into the following four series: bipyridyl metal complex series, phthalocyanine series, porphyrin series, and pure organic dye series.

At present, bipyridyl metal complex is still the best and the most widely used dye sensitizer. Bipyridyl metal complex series have typical representatives N3 and black dye. Figure 10.9 indicates the formulae of N3 and black dye.

Figure 10.10 is a comparison of N3 and black dye regarding the incident monochromatic IPCE.

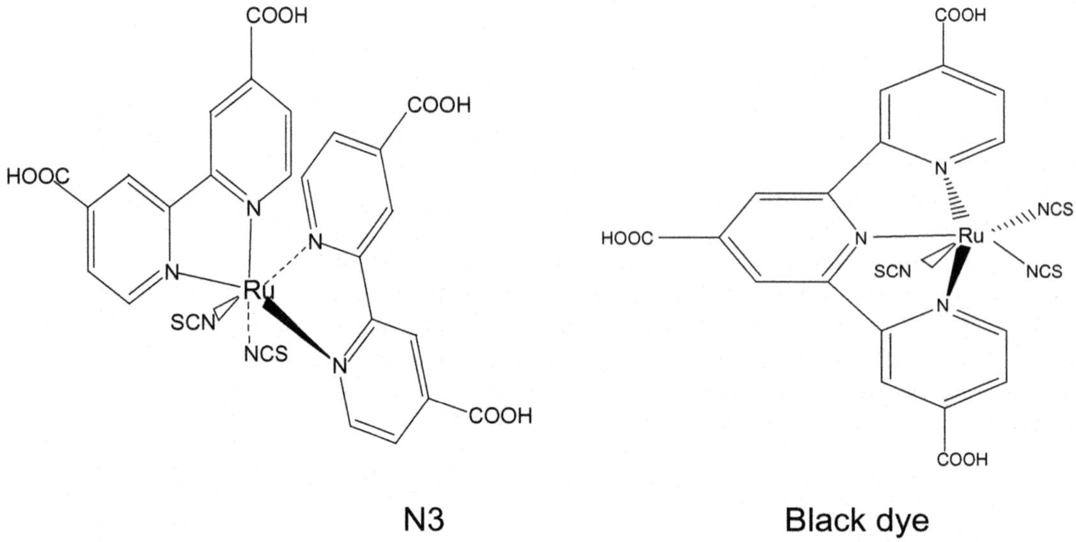

Figure 10.9 N3 and black dye molecular structures.

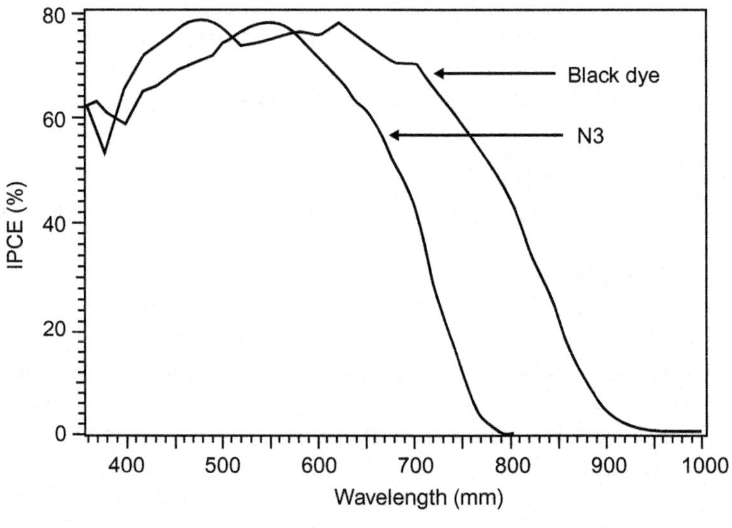

Figure 10.10 A comparison of N3 and the black dye as regards incident monochromatic IPCE.

The pure organic dye family may be either semi-cyanine dye or coumarin derivatives. Their performance is not discussed here.

10.3.3.5.2 Nanosemiconductor materials

Oxides of metal sulfides, metal selenium compounds, perovskite, and titanium, tin, zinc, tungsten, zirconium, hafnium, strontium, iron, and cerium can all be used as semiconductor materials in DSSCs. For example, the optical and chemical properties of Nb_2O_5 DSSCs were reported in 1999 [11], and those of the nanocrystalline In_2O_3 thin-film electrode were reported in 2000 [12].

The use of nanoscale semiconductor materials such as TiO_2, ZnO, and SnO_2 as photovoltaic solar cells is extremely popular in research worldwide; the nontoxic nano-TiO_2 of light stability is the most commonly used material for study of photovoltaic solar energy conversion cells.

As mentioned, the two major crises humankind is facing in the twenty-first century are energy and pollution. TiO_2 will play a significant role in resolving these two issues. In Chapter 9, for example, we introduced the photocatalytic application of TiO_2. Here, we mainly focus on the application of solar cells.

As indicated in Table 10.6, TiO_2 is a high-refractive index semiconductor with a high dielectric constant that is conducive to the absorption of UV light; therefore, it can be used as micro-electromechanical material, photocatalyst, and in micro-light decomposition, as well as anti-fog and self-cleaning coatings and for light-grating. TiO_2 nanoporous membrane has the advantages of having high porosity and a large surface area. When applied in DSSCs, it can absorb more dye molecules.

Table 10.6 TiO_2 Features

Physical Properties	Rutile	Anatase
Molecular weight (g/mol)	79.866	79.866
Density (cm^3/g)	4.25	3.89
Lattice constant *a* (nm)	0.4594	0.3785
Lattice constant *c* (nm)	0.2962	0.9514
Crystal structure	Square	Square
Number of formula units in the cell	2	4
Band gap (eV)	3.0	3.2
(corresponding to UV light wavelength, nm)	~410	~385
Hardness (Mohs)	6–7	5.5–6
Refractive index		
Air	2.7	2.55
Water	2.1	1.9
Oil	1.85	1.7
Dielectric constant (powder state)	114	48
Specific heat (kJ/°C kg)	0.7	0.7
Melting point(°C)	1,855	Transformed into rutile

In addition, the mutual reflections between the grains within the films are able to enhance the absorption of sunlight. Thus, in addition to ensuring high photoelectric conversion quantum efficiency, dye-sensitized nanocrystalline TiO_2 semiconductor electrodes can further guarantee the high efficiency of light capture. The nanometer TiO_2 electrode is the key to the performance of solar cells, which is directly related to the efficiency of solar cells.

Methods of preparation of nano-TiO_2 thin-film electrode materials mainly include the sol–gel method, hydrothermal reaction, sputtering, alkoxide solution, sputtering deposition, plasma spraying, and screen-printing methods.

10.3.3.5.3 Electrolyte

Liquid electrolyte is found to have some obvious shortcomings. It tends to lead to dye-sensitized desorption. It is a volatile solvent in reaction with sensitizing dyes and leads to dye degradation. It has a complex sealing process, with long-term placement caused by electrolyte leakage. Sealant reaction with the electrolyte still exists in cell, electrode, with light corrosion, and can easily become dye-sensitized in desorption. It has a very slow carrier mobility rate and instability in high-intensity illumination. It also has the presence of other oxidation–reduction reactions. At present, researchers have prepared all solid-state nanosolar cells using solid-state organic hole transport materials to replace the liquid electrolyte, allowing for gratifying progress. For example, in 1996, Masamitsu and colleagues prepared complete solid-state solar cells using solid polymer electrolyte and obtained high ionic conductivity of the electrolyte using a special preparation method. A continuous photocurrent was achieved with 0.49% of the photoelectric conversion efficiency. In 1998, Grätzel and associates used 2,2′, 7,7′-4 (*N*, *N*-2 pairs of *p*-methoxyphenyl-amino)-9,9′-spiro-2-fluorene (OMeTAD, shown in Figure 10.11) as a hole transport material. The result was a cell with 0.74% of the photoelectric conversion efficiency and color efficiency as high as 33%. This has sparked worldwide attention and enabled nanosolar cells to make a huge step forward in the solid state [13].

10.3.3.6 Main Problems

1. Dyes

   Dye photosensitizer is key to the battery. It requires not only a wide range of absorption of sunlight, combined with good adsorption properties of semiconductors, but also

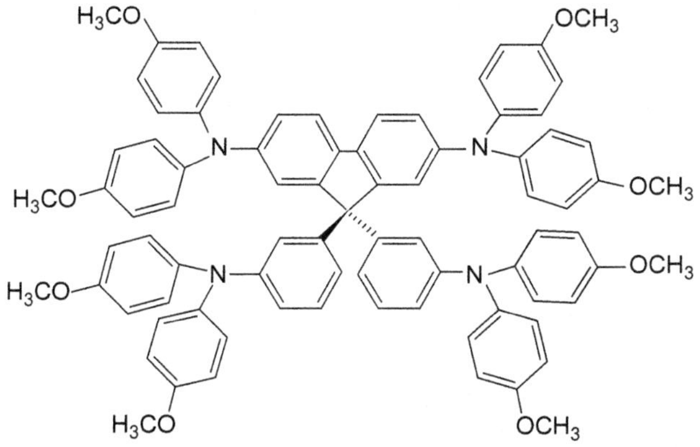

Figure 10.11 OMeTAD molecular structure.

suitable oxidation–reduction potential (excited dye to match with the semiconductor conduction band potential enables electrons to be effectively injected into the semiconductors), long lifetime of excited states, good photoluminescence and stability that can withstand at least 108 excited oxidation–reduction occurrences, ease of synthesis, and low cost. Now, N3 is a widely recognized dye that shows better performance, but its preparation process is more complicated and therefore more expensive. At present, the search for low-cost dyes exhibiting good performance has become a popular topic of research. Development of new dyes to supersede organic ruthenium metal (referred to as N3), the currently accepted best dye, is also very popular. Organic dye chemistry is a field undergoing considerable academic and industrial development. Many people believe that after a certain stage of theoretical preparation and a series of experiments, there should be opportunities to develop organic dyes with an absorption capacity better than N3. If this becomes true, then it may reduce the cost of dyes and be free from the requirement of using the precious metal ruthenium.

2. Nanomaterials

   The wider the band gap of semiconductor, the higher its light stability. From this perspective, if a material with a gap wider than that of titanium dioxide is used, then better DSSC product durability can be expected. At present, many scientists are engaged in research in this area, but no satisfactory results have been reported.

3. Electrolyte and Substrate Materials

   Liquid electrolyte can easily experience leakage during working conditions, which is a serious problem for the battery package, and there is also major difficulty in commercializing DSSCs. To achieve commercial objectives, the solution electrolyte should be gradually replaced by solid electrolyte, aiming to improve stability and provide longer service life.

4. Series and Parallel Connection of Cells

   One of the international focuses of research is nano-TiO_2 single-liquid junction solar cells that are arranged in series to improve the open circuit voltage.

10.3.3.7 Flexible DSSC Cells

In recent years, a flexible DSSC has attracted the attention of researchers. It has polymer materials (e.g., flexible ITO) that are used to replace the traditional glass substrate, so that the weight of the battery can be reduced and the battery can even be bent, thus expanding the scope of application of the cells. Satoshi Uchida at Tohoku University, Japan [14], used TiO_2 powder hot-synthesized with TiO_2 precursor sol and water as a coating for ITO/PET and sintered it for 5 min with 28 GHz microwave to prepare a TiO_2 porous electrode. Its battery photoelectric conversion efficiency increased to 2.16%. Resistivity of ITO conductive film shows rapid growth in high-temperature conditions, and PET polymer substrate is subject to heat-resistant temperatures below 150°. Therefore, the critical step is to achieve low-temperature preparation of TiO_2 thin films on flexible ITO/PET conductive substrate.

References

[1] C. Nützenadel, A. Züttel, D. Chartouni, L. Schlapbach, Electrochemical storage of hydrogen in nanotube materials, Electrochem. Solid-State Lett. 2 (1) (1999) 30–32.

[2] A.C. Dillon, K.M. Jones, T.A. Bekkedahl, et al., Storage of hydrogen in single-walled carbon nanotubes, Nature 386 (27) (1997) 377–379.

[3] C. Liu, Y.Y. Fan, H.M. Cheng, et al., Hydrogen storage in single-walled carbon nanotubes at room temperature, Science 286 (5) (1999) 1127–1129.

[4] P. Zhao, Introduction to Hydrogen Storage Materials, <http://www.docin.com/p-677641189.html>.

[5] H.L. Wu, Z.J. Ma, Research Status of Proton Exchange Membranes for Fuel Cell. <http://140.114.46.151/Laboratory Results/Laboratory Results.html>.

[6] S.G. Sun, Y.Q. Xu, L. Shi, X.J. Peng, L.C. Sun, Dye-sensitized nanocrystalline photovoltaic cells, Contemp. Chem. Indus. 33 (1) (2004) 47–50.

[7] M. Liang, Z.L. Tao, J. Chen, Sensitizers of dye-sensitized solar cell, Chemistry 12 (2005).

[8] J.H. Wu, S.C. Hao, J.M. Lin, Y. Huang, Progress in the study of dye-sensitized nanocrystalline TiO_2 solar cell, J. Huaqiao Univ. (Natural Science) 24 (2003) 336.

[9] B. O'Regan, M. Gratzel, A low-cost, high efficiency solar cell based on dye-sensitized colloidal TiO_2 films, Nature 353 (6346) (1991) 737.
[10] M. Grätzel, Photoelectrochemical cells, Nature 414 (2001) 338.
[11] P. Guo, M.A. Aegenter, RU(II) sensitized Nb_2O_5 solar cell made by the sol-gel process, Thin Solid Film 351 (1999) 290.
[12] S.K. Poznyak, A. Kulak, Characterization and photoelectrochemical properties of nanocrystalline In_2O_3 film electrodes, Electrochimica Acta 45 (2000) 1595.
[13] U. Bach, D. Lupo, P. Comte, et al., Solid-state dye-sensitized mesoporous TiO_2 solar cells with high photon-to-electron conversion efficiencies, Nature 395 (1998) 583–585.
[14] U. Satoshi, T. Miho, Flexible dye-sensitized solar cells by 28 GHz microwave irradiation, J Photochem. Photobiol. A Chem. 164 (2004) 93–96.
[15] A. Hagfeldt, M. Grätzel, Molecular photovoltaics, Acc. Chem. Res. 33 (2000) 269–277.
[16] D. Zhang, H. Hu, L. Li, D. Shi, Low-temperature preparation of amorphous-shell/nanocrystalline-core nanostructured TiO_2 electrodes for flexible dye-sensitized solar cells, J. Nanomater. (2008) 1–4 [ID 271631].

11

NANOCOMPOSITES

CHAPTER OUTLINE

Nanomaterials and Devices. DOI: http://dx.doi.org/10.1016/B978-1-4557-7754-9.00011-1

11.1 Concept and History

The history of humanity is closely related to the development of materials. Historians view the use of tools and toolmaking materials as a sign of the progress of civilization and divide the history of human development accordingly into the Stone Age, Pottery Age, Bronze Age, and Iron Age.

Ever since the beginning of human civilization, mankind has continued to develop new materials in advancing production and science. As for materials, any single material may come along with certain advantages and obvious shortcomings. Sometimes, it is difficult to improve the properties of materials and avoid their shortcomings. In the past three decades, the rapid development of science and technology has placed an increasing demand on the performance of materials. Materials made of single elements are therefore no longer capable of meeting this demand. Studies proved that two or more materials combined in some manner can be made into new materials (*composite materials*), with the advantages of the original components retained in many cases. These new materials are made to overcome or compensate for the shortcomings of the original elements and, in some cases, with new properties.

Composite materials first emerged in 5,000 BC, when people in the Middle East learned to mix asphalt with reeds for shipbuilding. In 3,000 BC, the ancient Indians were already able to produce composite boards with shellac resin. In China, a composite adhesive was used in the construction of the Forbidden City. Composite material made of thatch and soil in building houses is another early example. However, as a discipline and a new kind of industry, composite materials did not emerge until the 1940s. In 1940–1960, glass fibers, the so-called reinforced plastics or glass fiber-reinforced plastic, were introduced together with the emergence of boron fibers and carbon fiber-reinforced plastics. These are known as the first generation of composite materials. The period of 1960–1980 witnessed the advent of fiber-reinforced plastics, Kevlar fiber-reinforced plastics, SiC, and metal fiber-reinforced plastic Al_2O_3, which belong to the second generation of composite materials. This period underwent the development of advanced composite materials. In the period of 1980–2000, the advanced composite materials were well developed. This was called the third generation of composite materials, which have undergone rapid development in aviation and aerospace fields, and wide applications are found in various fields. At the same time, fiber-reinforced metal and nanomaterials were known to people as *equivalent dispersion composite materials.*

The definitions of composite materials vary. Some define a composite material as "a material composed of two or more kinds of single materials and it displays a number of new properties." This interpretation is easy to understand, but scientifically it is imperfect and imprecise. The most sound definition was given by the ISO (International Organization for Standardization): "composite materials (or composites for short) are engineered materials made from two or more constituent materials with significantly different physical or chemical properties which remain separate and distinct on a macroscopic level within the finished structure."

Nanocomposites [1] are one of the new composite materials formed by nanometer-sized materials that are dispersed in a 3-D substrate. They can provide obvious quantum size effects (e.g., the blue shift of the optical absorption edge) and an increased third-order nonlinear optical coefficient, which are expected to have applications in optical bistability, fast response, phase conjugation, optical waveguides, and other optoelectronic devices.

In contrast to the nanophase materials from single-component nanocrystalline materials, nanocomposite materials can be either intragranular nanocomposite (i.e., nanocrystalline phase comes into the larger phase grains) or nanocomposite in the grain boundary (i.e., nanograin distribution in the grain boundary over the larger grains of the parent phase); the latter applies to ceramics, metal, and polymer materials.

11.2 Surface Modification of Nanomaterials [2,3] and Their Applications

In the preparation process of nanocomposite materials, nanoparticles are very easy to join together because of their large specific surface area and high surface energy; however, nanoparticles have poor affinity with the substrate with a relatively low surface energy. The two are not compatible in mutual mixing, leading to interface voids, that is, the phenomenon of phase separation. Only after a satisfying solution is found for the agglomeration of nanoparticles in the material can the special effects of nanoparticles be well applied in the materials. Ultimately, the material performances in the mechanical, optical, thermal, and other aspects will be greatly improved. To ensure the nanosized existence of nanoparticles in the material, nanoparticle surface modification has become a critical goal of nanopowder research surface modification of nanotechnology and is an important part of surface engineering.

11.2.1 Nanosurface Engineering [4]

Surface engineering is an integrated design as a system of the surface and substrate of a material; it is system engineering involving the use of a variety of surface technologies that enable the surface of the material to have the desired performance not present in the material itself.

Nanosurface engineering technology is built on the basis of surface scientific theory. It is an advanced technology derived from the combination of modern physical and chemical methods and conventional surface technologies, involving all areas of surfaces and interfaces, as well as a variety of metals and nonmetal, ceramic, composite materials. Nanosurface engineering includes surface coating technology, surface thin film technology, and surface modification technologies. At the same time, modern surface technology also involves the preparation of coatings and thin film materials, processes and equipment, process and quality control, surface analysis, surface properties and evaluation, as well as testing technology, testing methods and standards, and the process of environmental injury and its mechanism.

With the development of nanotechnology, coupled with the growth of design and manufacture of microelectromechanical systems, manufacturing technology has moved from the submicron level to the nanoscale and even atomic or molecular level. Electromechanical systems like nanorobots, nanotweezers, and nanomotors may involve considerable surface science and technology issues. Meanwhile, with the size reduction and the appearance of surface effects, traditional surface design and processing methods are no longer adaptable. There are a growing number of cases that require materials to be serviced under special conditions, such as ultrahigh temperature/low temperature, high pressure, high vacuum, strong oxidation, and reduction or the corrosive environment, and in the presence of radiation, sound absorption, signal shielding, and bearing point load. As nanomaterials are presented with characteristics different from those of macromaterial in the aspects of power, electricity, sound, light, heat, and magnetism, the surface of nanotechnology is particularly important to traditional materials, and thus nanosurface engineering has emerged.

Nanosurface engineering is a systematic type of engineering that equips materials with the surface of nanotechnology via specific processing. The nanostructure process enables the material surface to be strengthened or modified or gives new properties to the surface to produce smart surfaces, nanointellectual surfaces, and devices with nanoscale surfaces.

Compared with traditional surface engineering, nanosurface engineering is characterized by factors that depend on the substrate properties being weakened, freedom for surface treatment, modification, and functional properties to expand, and the more prominent role of the surface processing technology and higher value-added products. Attributed to its nanosurface engineering characteristics and the properties that may be acquired, it has a broader range of applications, virtually covering every aspect of industrial and civilian use in high-tech areas, such as aerospace, aviation, marine, computer technology, electronics, information, transportation, oil, chemistry, construction, irrigation, and machinery industries, as well as all other areas of people's lives. Statistics show that industries directly related to surface technology have accounted for 7% of national economic output. With the development of nanomaterials and nanotechnology, the use of the related research achievements in the development of surfaces in nanotechnology and engineering is an important direction of the development of surface technology and also an important part of nanotechnology today.

11.2.2 Mechanism of Surface Modification of Nanoparticles

Surface modification of nanoparticles refers to the reaction between the surface of nanoparticles and surface modifier to improve the solubility of the nanoparticle and enhance the interfacial compatibility of nanoparticles in the medium, so that nanoparticles can be easily dispersed in organic compounds or water. The molecular structure of a surface modifier must have the characteristics of the substrate group, which easily reacts with the surface of nanoparticles. This characteristics group can be obtained from the design of the molecular structure of the surface modifier. According to the reaction mode between nanoparticles and the surface modifier, the modified mechanism can be divided into coated modification and coupling modification.

11.2.2.1 Coating Modification

Coating modification specifically refers to the surface-coated treatment of nanoparticles with inorganic or organic compounds (water-soluble or oil-soluble polymers, fatty acid soap, etc.), aiming to weaken or block the agglomeration of nanoparticles. Coated materials can generate a steric hindrance repulsion, which makes it quite difficult to obtain agglomeration of particles to achieve the purpose of modification. The mechanism used in coating can be coated adsorption, attachment, simple chemical reaction, or deposition of the coating. The preparation of nano-TiO_2 may introduce the hydroxypropyl cellulose modifier, whereby modifier molecules adsorbed on the TiO_2 particles may show a stereo-hindrance effect that can effectively prevent the particles from further accumulating growth, thus improving the dispersion and uniformity of the hydrated TiO_2 particles. At the same time, a particle surface with adsorption of these molecules can thoroughly "shadow" the nonbridging carbonyl and the absorbed water between particles to reduce the surface tension and become less prone to aggregation. Polyvinyl alcohol (PVA) may be added in the preparation of nanometal oxides. PVA contains a large amount of free, strong polar hydroxyl groups, which may form a chelate bond with metal ions in aqueous solution, which is closely coated around the metal ions. A finite structure can be formed with PVA chains to limit the size of the synthesis of nanoparticles, achieving the purpose of modification. In the preparation of nanosilver particles, polyvinylpyrrolidone (PVP) may be doped to have coordination between PVP molecules and surface atoms of nanosilver particles through the N and O atoms, leaving behind a long chain of CH stretching around to prevent the mutual agglomeration of nanosilver particles between them. With this method, silver powder with an average particle size of 25 nm can be prepared with good dispersion and distribution in a uniform particle size.

11.2.2.2 Coupling Modification

Coupling modification refers to the chemical coupling reactions that may occur on the surface of nanoparticles. In addition to the interactions of van der Waals forces, hydrogen bonds, or coordination bonds, there is also a combination of covalent bonds or ionic bonds between the two components. Via treatment with a coupling agent, the surface of nanoparticles can produce very good compatibility with organic matter. Coupling agent molecules must have two kinds of groups: one has a chemical reaction with the surface of inorganic nanoparticles or the precursors to prepare the nanoparticles, and the other (organic functional groups) can be reactive or compatible with the organic substrate, such as dioxin acetate titanate (2-*n*-octyl coke phosphate), vinyl triethoxysilane, and so on. Coupling modification can be easier to operate and have more choices of coupling agents, so the method is more often applied in nanocomposite materials. The preparation of polymethyl methacrylate–silica nanocomposite material requires methylpropionyl oxypropyltrimethoxysilane to act as coupling agents, with its carbon–carbon double bond copolymerized with the polymethyl methacrylate (PMMA). At the same time, the trimethoxysilane group is hydrolyzed with tetraethoxysilane to produce a silica bond, so that a composite system can be uniformly dispersed and remain stable.

11.2.3 Surface Modifiers of Nanoparticles

11.2.3.1 Inorganic Compounds for the Surface Modification of Nanoparticles

Usually, Al_2O_3, SiO_2, and ZnO are used as modifiers for the surface modification of nano-TiO_2. After proper treatment, anatase TiO_2 will exhibit strong UV absorption capabilities and can be safely applied to cosmetics, paper, paint, and other fields. The fluoride-modified α-Al_2O_3 can generate alumina powder that is uniformly dispersed, with an average particle size less than 50 nm. In the Surface Engineering Research Center for military equipment maintenance in China, the nickel-coating method is used for the treatment of nano-Al_2O_3, SiC, and diamond powder surfaces to improve the conductivity of nanoparticles. It has been proven that this method can increase the amount of nanopowder deposition in the nickel-based composite coating and enhance the uniformity of nanopowder coating.

11.2.3.2 Surface Modification with Nanoparticles

Surface modification with nanoparticles is actually the use of a compound between nanoparticles to improve some aspect of the performance of the nanoparticles being processed. For example, the sol–gel method can be used for preparation of composite nanoparticles. Specifically, the sol–gel method is first used to make *n*-butyltitanate particles into nano-TiO_2, which is then made into a transparent solution. Before this solution changes into a gel, it is added with another precursor, by which the sol–gel method can be used to form nanoparticles. An example is using ammonium solution as WO_3 tungsten precursors. After the formation of gel in the mixture, a compound of nanoparticles WO_3/TiO_2 can be formed through heat treatment. The new layer of nanoparticle film formed on the surface of nanoparticles will act to stabilize the inner nanoparticles and give a new performance to the particles. By using the gel method, Fe_2O_3 nanoparticles coated with a layer of SiO_2 film can significantly improve the dispersity of modified nanoparticles in PMMA solution.

11.2.3.3 Surface Modification with Organic Compounds

Organic compounds are a major modifier of nanoparticles that can bring about unique changes in properties. CdS nanoparticles are capable of producing weak red light. Modification with a small amount of alkylamine can significantly enhance the fluorescence of CdS nanoparticles and produce a blue shift. However, the high concentration of alkyl amines may quench the CdS fluorescence; mercaptan also functions in the same way as alkyl amines. TiO_2 nanoparticles coated with stearic acid will have a very low degree of crystallinity, and the obtained TiO_2 nanoparticles have an average particle size of 5–6 nm. Meanwhile, the absorption spectra of TiO_2 nanoparticles will be generated with a clear blue shift and light-induced luminescence may occur at room temperature. When resinate RS-2 and fatty acid salt RS-1 are used in the surface modification of nanocalcium carbonate, the modifier coated on the surface of calcium carbonate will increase their oleophilic properties, and its dispersion in nonpolar medium can be improved accordingly. Because nano-ZnO is dispersed in nonaqueous media, the surfactants sodium dodecyl benzene sulfonate and sodium laurate can be added as a surface modifier, which will be adsorbed on the nano-ZnO to ensure a stable and homogeneous dispersion. Stearic acid-modified CdO nanoparticles may have an absorption blue shift with strong luminescence. In the sol–gel preparation process of SiO_2 nanoparticles, when the polymer in the nonhydrophobic chain (polyethylene glycol) is used as surfactant for particle modification, sol clusters will form a rod-like network structure; with hydrophobic–hydrophilic polymers (fatty alcohol polyoxyethylene ether) as surfactant for modification, sol clusters will form a rod-like network structure; with polyoxyethylene–polyoxypropylene ether–polyoxyethylene triblock copolymers as surfactant for modification, a dendritic network structure will be formed. Increase of the hydrophobic part (polyoxypropylene ether) helps form the network structure and keeps a consistent nanoparticle size.

11.2.3.4 Surface Modification with Polymers

Surface modification of nanoparticles with polymers is the use of a polymer network to acquire the stability of nanoparticles. After a polymer network is treated with carboxylic acid salts (zinc, separated), sulfonates (zinc, cadmium, copper, and their multicomposite metal ions, etc.), and hydrogen sulfide gas-flow processing to generate the sulfide nanoparticles, the obtained particles may have an average size of only a few nanometers. Under the 3-D protection of polymer networks, the stability of nanoparticles can be increased to achieve micro-control of the special nature of nanoparticles. Because of its excellent optical properties and easy processability, polymers can be a good supportive agent in the formation of nanoparticles. In the synthetic process of ZnO nanoparticles, PVP demonstrated protection of nanoparticles. This can be helpful in the synthesis while improving the properties of nanoparticles. MoO_2 nanoparticles synthesized in γ-radiation aqueous solution, by adding PVA as a surfactant, can be significantly better than the surfactant sodium dodecyl sulfate with respect to purity and thermal stability.

11.2.4 Implementation of Nanoparticle Modification

Means for the implementation of modifying nanoparticles can be divided into mechanical dispersion, ultrasonic dispersion, and high-energy processing methods. Mechanical force dispersion mainly uses mechanical effects by external shearing force or impact force, and the special surface structure of nanoparticles is prone to chemical reactions, so that nanoparticles may undergo chemical changes with the surrounding medium (e.g., the surrounding solid, liquid, or gas) to form a layer of branched-chain organic compounds or a protection layer on its surface to make it easier for the dispersion of nanoparticles. Fe_3O_4 powder and micron polyvinyl chloride (PVC), when dispersed in high-energy ball milling, are able to form α-Fe_3O_4/PVC nanocomposites, where α-Fe_3O_4 may have a particle size of 10 nm. Ultrasound is widely used in chemistry, playing an important role in the synthesis of compounds, polymer degradation, and the dispersion of particulate matter. When nano-$CrSi_2$ particles (with an average diameter of 10 nm) are added to tetrahydrofuran solution containing acrylonitrile–styrene copolymer via ultrasonic dispersion, nanocrystals may become available for coating the polymer materials. The high–energy approach indicates the use of ultraviolet, infrared, corona discharge, and the plasma radiation method to perform the surface modification of nanoparticles. In addition, methyl methacrylate can be grafted onto nano-MgO with the help of UV radiation. Such surface modification can greatly improve the dispersion of nanopowder in high-density polyethylene (PE).

Plasma is a system comprising a large number of charged particles (ions, electrons) from ionized gas and the neutral particles in excited states (atoms, molecules). The system contains the same total number of positively and negatively charged particles. Over the past 20 years, plasma technology has made fruitful achievements in the fields of chemical synthesis, new materials development, fine chemicals, and surface treatment. In recent years, there have been more reports on studies and research on using plasma technology for surface modification of nanoparticles for biomedical purposes [5]. Plasma technology is relatively low cost, easy to operate, and useful to control material modification by optimizing process conditions. In theory, it can achieve the surface modification of the materials of any nature in any shape without changing the physical properties of the nanomaterial.

There are two main types of reactions between plasma and the material surface, namely plasma polymerization and plasma surface treatment. By way of plasma polymerization, organic monomer can be converted into a plasma state to produce various active species (free radicals) for surface polymerization. Plasma surface treatment involves the use of energy particles and reactive species in the plasma of nonconvergent gas (e.g., argon, nitrogen, oxygen) to react with the surface of the material to be processed, resulting in the surface producing a particular functional group (e.g., $-OH$, $-N_2H$). These two types of reactions can change the surface composition and structure, aiming to achieve the purpose of material surface modification.

With the development of medical research, a variety of transplant technologies for therapeutic, diagnostic, and orthopedic surgeries were developed, and the implants had to satisfy many requirements, such as biocompatibility, permeability, anti-aging and nontoxic properties, and so on. Many polymer materials usually have satisfactory mechanical properties such as bulk properties, but they cannot be used as implants because their surface properties often fail to meet these requirements. Over the years, many domestic and foreign researchers have been trying to introduce a variety of physical, chemical, or biological means to perform surface modification on biological materials. Plasma technology has been proven to be a very effective approach.

11.2.5 Application of Modified Nanoparticles

11.2.5.1 Application in Plastics

Due to the small size effect of nanoparticles, as well as large surface area and strong interfacial bonding, nanomaterials can be effective in increasing strength and toughness, as well as in improving the aging resistance of plastics. When SiO_2 (with a particle size of 14 nm) treated with dimethyl silane has a volume fraction that is 4% of that of PE, the casting method can be used to prepare SiO_2/PE composite materials, which exhibited a tensile strength approximately double that of the substrate. Polypropylene filled with CH-IA-treated nano-$CaCO_3$ powder will show marked improvement in composite material toughness and impact resistance. By surface modification of high-energy radiation, the obtained SiO_2-filled polypropylene composites will also have increased modulus and strength, whereas their toughness can also be improved significantly.

11.2.5.2 Application in Composite Fire-Retardant Materials

After traditional inorganic flame retardants are changed into nano-based materials, with nano-Sb_2O_3 as the carrier, a type of highly efficient flame retardant can be made by way of the surface modification, with its oxygen index being several times that of common flame retardants. In addition, the nano-Sb_2O_3 and polyolefin can be perfectly matched with plastic. It has features of good thermal stability, nontoxicity, and durable resistance against fire.

11.2.5.3 Application in Composite Catalysts

Nanoparticles are small, so the surface/volume ratio is large, which makes the bonding state and the electronic state of the surface different from that inside the particles; the surface smoothness may deteriorate and form uneven atomic levels. Meanwhile, this will increase the contact surface for chemical reactions and produce incomplete atomic coordination, leading to an increase in active surface sites. All these properties are the basic conditions that cause it to be a type of catalyst. Copper chromite is a good catalyst that can promote ammonium perchlorate decomposition. However, it is likely to contain agglomeration of previously prepared copper chromite and perchloric acid ultrafine particles. Ammonium perchlorate crystals used for coated nanocopper chromite to form composite particles provide a better solution to this problem.

11.2.5.4 Application in the Field of Lubrication

The application of nanomaterials in lubrication is a new area of research. Because nanomaterials have large surface/volume ratio, high proliferation, low sintering, and melting point reduction, the new lubricating materials, which are prepared with nanomaterials and applied in the friction system, will have a lubricating function that is different from that of traditional load additives. In addition to the formation of a layer of film that is easy to cut on the friction surface to reduce the friction coefficient, the new lubricating materials can help fill and restore the surface, thus enhancing the resistance of the system.

11.2.5.5 Applications in Composite Coating

The unique role of nanomaterials will have a far-reaching impact on paint applications. The combination of nanomaterials with traditional paint in manufacturing nanocomposite coatings is an important direction for paint development. The excellence of new composite coatings is increasingly reflected in a variety of nanocomposite coatings that have been successfully developed. After surface modification, nanomaterials can be produced with hydrophobic and oleophobic features. Nissan and Toyota have applied nanomaterials with self-cleaning and anti-fog properties as the mirror surface coating for their automobiles.

11.2.5.6 Application in Rubber

After rigid nanoparticles are added to the rubber-toughening system, the special effects of nanoparticles will provide the rubber with reinforcement, barrier, and processing properties. Rubber and the modified SiO_2 nanoparticles in composite materials may be dispersed very evenly. In addition, the dispersed phase can be controlled with respect to its chemical composition and structure, size and distribution, and surface characteristics. The prepared nanocomposite material has extremely high tensile and tear strength, as well as excellent heat and dynamic lag/static compression performance. In the most optimum conditions, its overall performance can significantly surpass that of rubber nanocomposites reinforced with carbon black and silica. By using this technology, the mixing process can also be partly omitted.

11.3 Core–Shell Structure Composite Nanomaterials [6–9]

As researchers' understanding of nanomaterials expands, the continuous research also undergoes a gradual transition from the simple preparation of nanoparticles to the design and controlled synthesis of nanocomposite materials that have specific functional properties. Composition of materials in the nanometer scale is a popular research topic. Nanocomposites with core–shell structures are a typical form of composite material of this kind. Over the past 10 years, making core–shell structures in materials has stimulated the interest of scientists. Core–shell material is formed with the core and the external shell having exceptional composition, size, and structure, and it can provide better performance than that of conventional materials. Core–shell structured nanomaterial has an excellent range of adjustable features. To prepare different types of core and shell, new materials have become the focus of studies in many countries.

11.3.1 Characteristics of Core–Shell Composite Structures

Core–shell nanoparticle composite structures are an ordered assembly of nanoscale structures. This new structure is formed by using a particular nanomaterial to coat another kind of nanomaterial via chemical bonds or other interactions. It is a higher level of composite nanostructure. This structure provides many new properties beyond the reach of single nanoparticles, which are widely appreciated for wider application prospects than single nanoparticles. The design and controllable synthesis of composite nanomaterial with core–shell structures has been an increasingly important forefront of material science in recent years. As functional materials with a new structure, such materials are favored by researchers because they have many unique properties, such as single dispersion, core–shell operability, stability, control, self-assembly, and the capabilities involved in light, electricity, magnetic, catalytic, chemical, and biological responses. Thus, by providing a reasonable design of experimental conditions, regulation can be largely achieved for many properties of composite nanomaterials.

Core–shell nanomaterials can be divided into different categories by applying different criteria. Depending on whether a chemical reaction may occur between core–shell particles, it can be divided into physical coating and chemical coating types. Depending on the different core–shell components, it can be divided into three categories: organic–inorganic, inorganic–organic, and inorganic–inorganic types. These core–shell structures are designed for specific purposes. The relatively stable nature of the shell is used to protect the core particle from the occurrence of physical and chemical changes; another potential objective of the shell is to improve the surface activity of the core particles, as well as stability, dispersion, and so on. Through surface coating, core particles can have magnetic, optical, and catalytic properties that are unique to shell particles.

11.3.2 Composite Method

With the in-depth research supported by improved experimental means, core–shell structure composites can be prepared using a variety of methods such as surface modification, polymerization, deposition film formation of the most commonly used *in situ* composite, self-assembly technology, electroless plating, and so on. These methods not only can lead to shell thickness and uniformity control but also can prepare multishell structured composite materials. A few commonly used methods are given below.

11.3.2.1 Polymerization Chemical Reaction

Polymerization chemical reaction usually refers to the polymerization reaction of organic monomers in solution containing the particles to be coated with the formed polymer deposited on the particle surface to form a coating layer. It includes monomer adsorption polymerization and emulsion polymerization.

In monomer polymerization, a core with high catalytic activity is usually used as coating particles, such as α-Fe_2O_3, CeO_2, CuO, and SiO_2. Strong interactions may exist between monomers and the coated particles, which can be directly adsorbed to the surface of inorganic particles, and then trigger monomer polymerization to complete the coating. The key to the use of monomer-coated particles is that polymerization must take place at the particle surface. Experiments show that the thickness of the coating can be adjusted by changing the time of contact reaction between the core and organic matter. This method is simple to operate and applicable for a wide range of conditions.

Additionally, a low-molecular-weight surface-active agent is capable of forming double-layer micelles at the particle surface. Based on this, monomers can be inclusive in the micelles to cause polymerization to achieve surface modification of particles. With this method, a thin polymer coating layer (2–10 nm) can be formed on the organic or inorganic particle surface, especially for particles with an irregular surface shape. It can maintain a certain thickness in thin-layer coatings along the contours of the surface of the particles.

11.3.2.2 Biological Macromolecular Method

Biological macromolecules are another candidate for nanoparticle coatings. Their main purpose is to enable ordinary particles with special genes and response capabilities that may be found in certain proteins or organisms and can be widely applied to clinical analysis, immunological tests, and the study of biological characteristics. There are several techniques that can be used to help biological macromolecules adhere to the surface of solid particles, such as nonvalence-bond adsorption, valence-bond adsorption, sol–gel capture, and electrostatic self-assembly. One of the most commonly used methods is valence-bond adsorption. It can be used for the coating of various types of proteins and antibodies against solid particles (e.g., polystyrene, polyaniline). Nonetheless, such coating usually is not solid enough and may be accompanied by loss of activity. This is especially true for smaller biological particles. The sol–gel method can achieve a variety of complex forms of coating that may be difficult for a general coating method; for some complex biological systems, the coating can be completed via infiltration of sol–gel solution with the requirements of nondestruction of its structure and function.

11.3.2.3 Surface Deposition and Surface Chemical Reaction Method

In particles coated using surface deposition techniques, it is required that the coating and coated particles are both dispersed in aqueous solution. By conditioning pH or heating, coated materials may undergo precipitation or hydrolysis, followed by deposition on the core material to form a core–shell structure, or specific functional groups can be used to complete the direct reaction on the surface for coating. This method can be used to prepare inorganic coating layers, such as SiO_2, basic yttrium carbonate, TiO_2, ZrO_2, and so on. Studies of the TiO_2 surface coated with SiO_2 showed that a large number of particles may be agglomerated with the deposition of SiO_2. Meanwhile, the sol–gel combined hydrogen reduction method can be used to prepare Fe/SiO_2 core–shell nanoparticles. Based on this, the acetylene pyrolysis deposition method is introduced to prepare core–shell structured carbon-coated Fe/SiO_2 particles; iron nanoparticles can be evenly coated in silica and carbon shell to further improve the thermal stability, as shown in Figure 11.1.

11.3.2.4 Controlled Deposition of Inorganic Colloidal Particles on the Core Particle Surface

Controlled deposition of inorganic colloidal particles coated on the core particle surface is generally a result of the surface static interaction between inorganic nanoparticles and large particles. For example, under solvent-controlled deposition, a polystyrene (PS) colloidal surface can be coated with a layer of CdTe nanocrystals. When modified with mercapto-glycine, CdTe nanocrystals will have a negatively charged surface, –COO, which via electrostatic action will be adsorbed to the surface of PS particles positively charged with $-NH_3$ to form a single-layer coating. The coating can then be completed by the following condensation of nanocrystals. A certain thickness of sedimentary layer is available by controlling the rate of cohesion. The thickness and shape of the inorganic coating layer are determined by the initial concentration of reactants, aging time, and temperature.

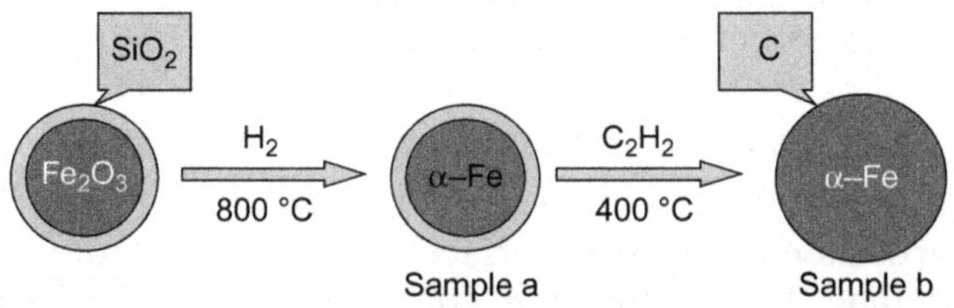

Figure 11.1 Diagram of carbon-coated Fe/SiO_2 nanoparticles. From Ref. [6].

11.3.2.5 Ultrasonic Chemical Method

Sonochemical methods have been considered to be quite effective in the preparation of new materials. The mechanism is largely due to the ultrasonic cavitation effect, for example the formation of liquid in microbubbles and raising and collapse by implosion. In present experiments, numerous nanocoated materials have been chemically synthesized with ultrasound, including iron oxide nanoparticles coated on carbon balls, gold nanoparticles deposited on SiO_2 microballs, Eu_2O_3 and Tb_2O_3 coated on SiO_2, Al_2O_3, and ZrO_2, oxides of the transition metals Fe, Co, and Ni deposited on SiO_2 or Al_2O_3 microballs, Eu_2O_3 coated on TiO_2 nanoballs, SiO_2 coated on ZnS particles, and so on. A series of experiments showed that ultrasound may increase the chemical interaction between the surfaces of the coating material and the particles being coated. This is conducive to the formation of chemical bonds.

11.3.2.6 Self-assembly

Self-assembly is an effective method in the preparation of core–shell structured composite nanoparticles. In self-assembly, nanoparticles are made first, and are then taken as a template for surface coating. Finally, core–shell structured composite nanoparticles are constructed. For example, a spherical biodegradable polymer is taken as a template. The first step is the use of the polymer electrolyte to be modified to make the smooth surface charged with static electricity, and then to attach gold nanoparticles and silica particles to the surface, followed by repeated centrifugation, washing, and removal of the particles not being adsorbed. These steps can be repeated to achieve uniform and compact multicoated gold and silica nanoparticles. The dissolved template can generate a material with special optical properties. In addition, the vapor deposition method and chemical plating are both often used in the preparation of core/shell-type materials. For instance, prepared by chemical vapor deposition, SiO_x-coated FeCoNi nanowires will have excellent soft magnetic properties and good thermal stability and can be used in nanodevices for high-density magnetic recording. In addition, the chemical plating method has also been cited in reports regarding its application to carbon nanotubes with Ag coating. Due to the low reactivity of carbon nanotubes, adequate surface oxidation, sensitization, and activation treatment are required before conducting chemical silvering to attain a smooth uniform coating. Meanwhile, reactions should be performed at the lowest possible rate.

11.3.3 Mechanism of Formation of Core–Shell Structures

The mechanism of formation of the core–shell structure through the coating of organic or inorganic particles is described in the following subsection.

11.3.3.1 Mechanism of Chemical Bonding

In research on SiO_2-coated TiO_2, we find that the two are combined through the formation of the Ti_2O_2Si bond. This is because inorganic oxide nanoparticles in water, such as SiO_2 and TiO_2, may be hydrated with water molecules to produce hydroxides such as Si–OH on the surface of silica particles. These groups are susceptible to chemical reaction with the hydroxyl groups on other inorganic particle surfaces, or some of the functional groups of the polymer chain (such as –COOH, –OH), thus enabling the formation of chemical bonds. In the reaction system, the introduction of the coupling agent also allows the formation of chemical bonds between the coating material and the coated material. For example, in studies on the preparation of Au coated with SiO_2, Au nanoparticles cannot exist stably in solution, and there is no affinity between Au and SiO_2, making it impossible to apply the coating directly; therefore, citric acid is used and adsorbed on the surface of Au nanoparticles to prevent their agglomeration, followed by doping with 3-aminopropyl methyl siloxane coupling agent and sodium silicate. Thus, the process of coating Au nanoparticles with SiO_2 can be completed with the help of chemical bonds.

11.3.3.2 Mechanism of Coulomb Electrostatic Force

According to this mechanism, coating agent accompanies a charge that is opposite to that on the substrate surface. Affected by Coulomb attraction, this charge enables particles of the coating agent to be adsorbed on the surface of the particles being coated.

11.3.3.3 Mechanism of Adsorption Layer Media

Inorganic particles can undergo surface treatment to form a layer of organic adsorbed layer. With this kind of particle as core, the adsorption layer may act as the medium to improve the compatibility between inorganic particles and organic matter. Thus, the organic monomer can be aggregated to obtain complex encapsulated particles. For example, using citric acid

Y_2O_3/Eu for surface modification can make the surface adsorbed with an organic layer. By the polymerization of styrene in the subsequent process, we can obtain PS-coated Y_2O_3/Eu composite particles.

Coating of the particle surface, inorganic or organic, generally involves the use of the mechanism mentioned previously, and some coating processes may involve the combination of two or more mechanisms. For example, electrostatic adsorption and chemical bonding can be combined to enable Au nanoparticles to be coated on the surface of PS particles. First, the positive poly(ethylene imine)-static charge is adsorbed on the negatively charged PS latex particle surface, followed by the bonding of Au nanoparticles with amino groups of polyethylene imine, which is coated with the PS particle surface. Next, the PS particles are used as crystal seeds in the reaction with NH_4OH and $HAuCl_4$ to improve the coverage of Au particles on the PS particle surface.

11.3.4 Changes in Material Properties

Coated nanomaterials have physical and chemical properties that differ from those of noncoated material and show some special optical, electrical, magnetic, or biological properties.

11.3.4.1 Changes in Optical Properties

Semiconductor nanocrystals can be used in biological fluorescence labeling and optoelectronic devices, but they require a very high fluorescence quantum efficiency and strong stability of light degradation. Coating semiconductor nanocrystalline particles with semiconductor material with a broader band gap is an effective way to improve these properties. In this way, shell modification can greatly enhance the fluorescence quantum yield of the core and the stability of the material, with the band gap energy being adjustable in certain bands. For example, at room temperature, CdS/Cd $(OH)_2$ may have a fluorescence quantum yield of 50%, which is much higher than that of CdS itself.

11.3.4.2 Increase in the Stability of Particles

A nanoparticle is characterized by its small size and large specific surface area. This brings about properties such as active nature, large surface energy, and ease of aggregation. The most widespread application of surface coating is to improve the chemical stability of the material being coated. Some metal nanoparticles are easily oxidized in the air, or even result in spontaneous combustion. The solution is to coat such nanoparticles with one or more layers of inert compounds that are able to isolate them from the external environment. Magnetic nanoparticles, such as Fe_2O_3, are widely used in magnetic fluid but are easy to aggregate and susceptible to acid corrosion. Coating its surface with a layer of inert material (e.g., SiO_2) can improve its stability and prevent acid corrosion. For example, after the surface of iron nanoparticles is coated with SiO_2 and Al_2O_3, its stability in air is greatly improved. As the volume of the coating increases, the stability becomes enhanced. As shown in Figure 11.2, iron particles with a diameter of 30–40 nm, through the coating, are able to provide a steady presence in the air and can effectively prevent oxidation from occurring.

11.3.4.3 Catalyst Stability and Changes in Catalytic Activity

TiO_2 is widely used as a white coating, catalyst, and catalyst carrier. As a catalyst, TiO_2 with high surface area is thermally unstable and can easily aggregate, leading to the reduction of the specific surface area. A common practice to improve thermal stability of TiO_2 is to coat particles with a high surface area with TiO_2. It was reported that TiO_2 catalyst on SiO_2-coated surfaces can be stabilized to 1,058 K, with an increase of two orders of magnitude in 12-propanol dehydrogenation reactivity.

11.3.4.4 Changes in Magnetic

The magnetic core can be coated with nonmagnetic, antiferromagnetic, or ferromagnetic/ferrimagnetic shells. Nonmagnetic

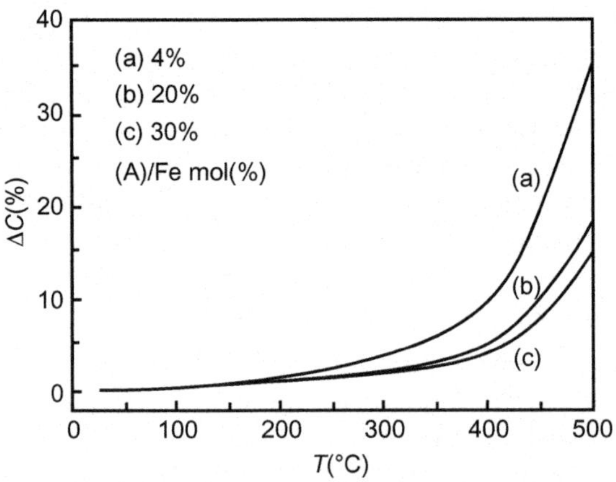

Figure 11.2 *T–G* curves of Fe/Al_2O_3 nanoparticles in different coating quantities. From Ref. [7].

coating is mainly used to increase the magnetic stability of the core or to achieve surface function for biomedicinal purposes. Ferromagnetic core coated with antiferromagnetic may lead to exchange bias (hysteresis loop shift along the field direction) and improved thermal stability. For the core and shell made of strong magnetic (ferromagnetic or ferrimagnetic) materials, close contact between the core and shell may lead to an effective exchange coupling, making the magnetism adjustable. It is reported that FePt core coated with an MFe_2O_4 (M = Fe, Co) layer with a thickness of 0.5–3 nm, under the action of an external electric field, will have a smoother magnetic hysteresis curve, with the coercivity depending on the volume ratio of the hard and soft magnetic phases. Magnetic properties can be adjusted by changing the chemical composition of the coating and thickness. The soft magnetic properties of the magnetic core made with iron and their alloys coated with SiO_2 and Al_2O_3 can be significantly improved. Meanwhile, the resistivity of the material can also be greatly improved, thereby reducing eddy current losses. In addition, the high-frequency soft magnetic properties will be significantly enhanced accordingly, and the real part of magnetic conductivity can be maintained constant even up to 1 GHz, as shown in Figure 11.3.

11.3.5 Applications of Core–Shell Composite Nanomaterials

As an ordered composite structure, core–shell composite nanomaterials have many properties that single nanomaterials

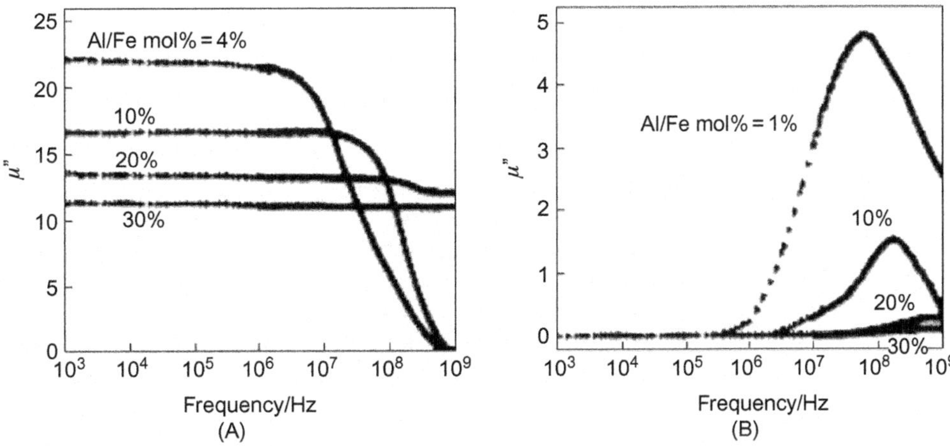

Figure 11.3 Spectral curve of Fe/Al_2O_3 magnetic nanoparticles coated in different quantities. From Ref. [7].

do not have, and they have many new applications as well. The differences in the categories of nanomaterials that constitute a core–shell may lead to various target properties and applications. For example, the two kinds of semiconductor nanoparticles with different energy bands can be combined as indicated, and narrow band gap semiconductor particles can be used in the sensitization of semiconductor nanoparticles with a wide band gap. The result of such a match can improve material properties in applications like photovoltaic conversion, nonlinear optical properties, electric color conversion, solar cells, and high-density information storage devices. In recent years, some breakthroughs have been made in the preparation of water-soluble semiconductor nanomaterials. Compared with single-core semiconductor nanometer quantum dots, the core–shell structured compound semiconductor nanoparticles have shown superior optical properties; for example, core–shell nanoparticles may have their fluorescence quantum yield significantly increased and their stability greatly enhanced. In recent years, computer performance and computational tools have been and are being improved constantly. Researchers have been able to perform calculations based on the first principle of the core–shell structured compound semiconductor nanoparticles. These theoretical studies have greatly enhanced the understanding of the properties of core–shell structured compound semiconductor nanoparticles [10,11].

Core–shell nanocomposite structures composed of biocompatible polymers can protect the biological activity of bioenzymes, DNA, and other bioactive substances as carriers for controlled release of these substances to achieve targeted therapy. Nanoparticles embedded by using block copolymers or end-group functionalized polymers can have many new properties and reactivity on the surface, which means that the surface of the nanoparticles is modified. As for magnetic Fe_3O_4 nanocomposite structured particles coated with metal Au, because Fe_3O_4 is superparamagnetic, nano-Au can be used as a biologically active agent in clinical tests. Nanoparticles of this composite structure are expected to have superparamagnetic features plus the advantages of easy separation and easy modification of gold surface, making it easier to use. Regarding catalysis, the core–shell structured catalysts can be controlled to achieve a catalytic reaction. This structure can also protect the core material from chemical erosion from the external environment and prevent agglomeration of nanoparticles. By removing the core material, we can get a hollow nanocage, which can be used for nanoparticle synthesis reactors and separators.

References

[1] G. Guo, Y. Yu, Z. Luo, X.J. Chen, Y.C Liu, Progress in polymer-based nanocomposites research, Guizhou Sci. 20 (2) (2002) 22–28.
[2] Y. Liu, L.H. He, R. Song, Research progress in nano-ZnO surface modification, Chem. Online 11 (2007) 823–828.
[3] Y. Wang, H. Xu, W. Qiang, H. Gu, D. Shi, Asymmetric composite nanoparticles with anisotropic surface functionalities, J. Nanomater. (2009) 1–5, ID 620269.
[4] B.S. Xu, Z.W. Ou, S.N. Ma, Nano-surface engineering, Chin. Mech. Eng. 11 (6) (2000) 707–712.
[5] Y. Guo, H. Cho, D. Shi, J. Lian, Y. Song, J. Abot, et al., Effects of plasma surface modification on interfacial behaviors and mechanical properties of carbon nanotube-Al_2O_3 nanocomposites, Appl. Phys. Lett. 91 (2007) 261903.
[6] W. Liu, W. Zhong, Y.W. Du, Research progress in core/shell composite nanomaterials, Mater. Rev. 21 (3) (2007) 59–62.
[7] W. Liu, W. Zhong, H.Y. Jiang, N.J. Tang, X.L. Wu, Y.W. Du, Highly stable alumina-coated iron nanocomposites synthesized by wet chemistry method, Surf. Coating Technol. 200 (16–17) (2006) 5170.
[8] S.Y. Gao, Y.H. Qiu, C.Y. Peng, D.H. Sun, L.N. Sun, H.J. Zhang, Preparation of inorganic—inorganic core-shell composite materials and the applications, Henan Normal Univ. (Natural Science) 33 (3) (2005) 62–66.
[9] T.H. Xie, S. Liu, Research progress in core-shell nano-titanium dioxide composite materials, Inorganic Chem. Ind. 39 (7) (2007) 8–11.
[10] J. Li, L.W. Wang, First principle study of core/shell structure quantum dots, Appl. Phys. Lett. 84 (18) (2004) 3648–3650.
[11] A.C. Bartnik, F.W. Wise, A. Kigel, E. Lifshitz, Electronic structure of PbSe/PbS core-shell quantum dots, Phys. Rev. B 75 (2007) 245424.

12

DNA NANOTECHNOLOGY

CHAPTER OUTLINE

12.1 Basics of DNA

12.1.1 Unique Structure of DNA

DNA is a biomacromolecule that plays a very important role in storing the genetic information of organisms. It is formed by four neucleotides, each with a unique side change base: adenine (A), guanine (G), cytosine (C), and thymine (T). The commonly envisioned picture of DNA consists of a helical structure of two individual DNA strands brought together by a series of nucleotides. Between the two single strands of DNA, nucleotides will form hydrogen bonded pairs of bases at corresponding positions. These nucleotides are mutually complementary: A combined with T, C combined with G. Through this base-pairing, the single strands twist together to form the famous *double helix*.

Nanomaterials and Devices. DOI: http://dx.doi.org/10.1016/B978-1-4557-7754-9.00012-3

The DNA sequence (the series of bases along the polymeric chain) contains all the genetic information of organisms that is ultimately responsible for the organism's growth, reproducibility, adaptability, and function. Specific fragments of DNA are known as genes or coding regions that guide cellular protein synthesis. Other areas of note on the DNA molecule are the noncoding region (with no specific role or an undiscovered role) and some regulation and control areas. The human DNA sequence is as long as 3 billion nucleotides, consisting of approximately 30,000 genes, with a noncoding region accounting for more than 75% of all nucleotides. The shortest known sequences belong to viruses, which number in the range of a few thousand nucleotides. By comparison, bacterial DNA commonly contains approximately a few million nucleotides. These short sequences exhibit a lower percentage of noncoding DNA, which is sometimes zero for most efficient use.

Since the discovery of the DNA double-helix structure by Watson and Crick, the biochemical function of DNA has been extensively studied. Molecularly, the structure is modestly simple, consisting of two single chains that are wound into a double helix directed by base-pairing. By various treatment methods, the double helix can easily be separated into the two single chains. Moreover, two individual DNA strands containing matched base pairs (sometimes referred to as complementary chains) can be brought together to form a helical structure. This attribute of DNA is useful in the study of biological and chemical properties of DNA molecules and make the material useful in the fields of nanotechnology and nanoelectronics.

12.1.2 DNA Conductivity [1,2]

The electrical nature of DNA was proposed as early as the 1960s. While the fundamental electrical properties of DNA can help assist in the understanding of the biological properties and processes this material has in nature, others have expressed interest in the material for the electronics field. Presently, 90-nm scale processes are widely used in commercial integrated circuit manufacturing, with smaller size scales of 65 and 45 nm now also arriving on the market. Although there is a push for smaller scale materials to increase the performance of integrated microchip technology, current manufacturing techniques have been developed using classical physics. Quantum effects would gradually become evident as the size scales continue to shrink, resulting in unpredictable and often undesirable materials features. Therefore, as the sizes of electronic devices are pushed to 10 nm or less, a paradigm shift is required with completely different materials and technology. DNA may play an important role in this field, which is discussed in detail later.

In the double-helix structure, the base pairs are parallel to each other and arranged axially due to strong intramolecular π-electron interactions in the bases (sometimes referred to as the π-stack). The average distance between the base pairs is 0.34 nm (Figure 12.1), similar to the spacing between the planes of carbon atoms in graphite. Due to the interactions of π-electrons between lamellar regions of graphite, electron conductivity is achievable in the lamellar direction. DNA base pairs have electrons in a similar orientation, with base pair spacing of identical length compared with that of graphite. Electron transfer between the base pairs of DNA molecules is dependent on the direction of electron mobility, wherein longitudinal electron movement along the DNA axis gives the material conductive properties. In fact, some conductive organic molecules and polymers have a similar atomic structure; however, these materials typically exhibit higher degrees of intramolecular $\pi-\pi$ interactions and are more crystalline in nature than DNA. Disordered systems lacking periodicity tend to result in localization of electrons that impede electron mobility and, hence, are not conductive; this is called Anderson localization. That stated, DNA sequences are not completely disordered systems, so the electrical properties can be between those of a conductor and those of an insulator.

Interest in DNA conductivity is not only driven by fundamental knowledge of biological processes but also driven by the potential of DNA in electronic applications. The current integrated circuit (IC) manufacturing industry has reached feature sizes of 90 nm with respect to the process of the main volume (circuit wire width) and is continuing to push toward smaller sizes of 65 and 45 nm. Nevertheless, we know that quantum effects will become more evident as the size continues to be reduced. At this scale, traditional processes and materials will fail

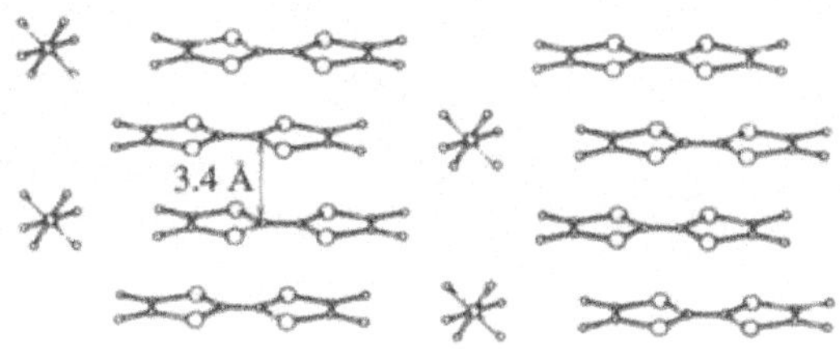

Figure 12.1 Axial stacking arrangement of DNA molecules. From Ref. [1].

to produce usable technology utilizing extremely small scales. Therefore, electronic devices with fewer features require completely different materials and technology.

To determine whether DNA can be used as nanoscale wires in future ICs, direct electrical measurements of single DNA molecules are required to establish possible viability. To accomplish this, the DNA molecule can be directly tethered between the two electrodes as shown in Figure 12.2. To perform this experiment successfully, two aspects of the experiment require careful attention. The first critical aspect is the preparation of electrodes, wherein smaller electrode spacing can provide more reliable measurements, with a minimum value of only a few nanometers (Figure 12.2). The next step is to reliably link the DNA molecules to the electrode while maintaining good contact between them.

There have been numerous experimental studies of the electrical properties of DNA, with reports of insulating, semiconducting, conducting, or even superconducting properties. Such variation is not surprising because the experiments were performed using different DNA sequences, preparation, and measurement methods and conditions. As stated, an electron microscope experimental setup can be used to study a DNA chain between two metal electrodes for direct measurement of the current–voltage relationship. This experimental setup can be problematic, unfortunately. Contact quality between DNA and the electrode will seriously affect the measurement. If a poor DNA–electrode interface is obtained, then the results are unreliable. Alternatively, if a short circuit is present between the

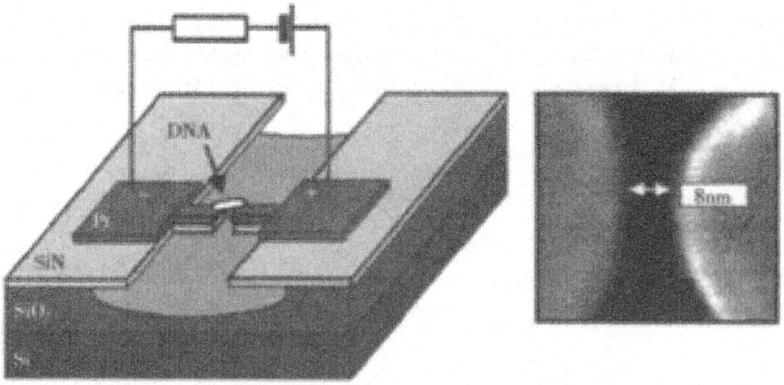

(A) (B)

Figure 12.2 (A) Schematic of the DNA electrical measuring device, with DNA molecules to be placed in between a pair of electrodes, and (B) photographs of the electrodes under a scanning tunneling microscope, showing electrode spacing of only 8 nm. From Ref. [2].

electrodes, then the results of measurements may show DNA conductivity, even if the DNA is in fact insulating. In this regard, current experimental techniques still need further advancements.

Theoretical studies are an attractive route to study DNA conductivity and are most commonly pursued using two methods: first principles calculations and model calculations. First principles calculations rely solely on the basic laws of physics, namely of quantum mechanics. Specifically, the first principles calculation takes into account all the atoms (more accurately the outer electron shell) and all types of interaction for the entire system to solve the Schrödinger equation. From introductory quantum physics, one should remember that an exact solution is not directly available but can be approximated using the Hartree–Folk (HF) method or density functional theory to seek the approximate solution. The advantage of this approach is that it is straightforward, starting from the basic laws of quantum mechanics without the addition of adjustable parameters. The main drawback, however, is that the complexity of the calculation increases exponentially with an increasing number of atoms. This often results in the modeling of smaller systems. In the context of DNA, this method is usually only used to calculate properties of very short sequences because larger sequences are computationally expensive. The addition of solution and counter ions, making the simulation more realistic, further compounds this issue. As such, it would be impractical to use first principle calculations to study the conduction of electric charges in longer DNA (more than a dozen nucleotides).

Model calculation is also known as the "effective model" approach. Regarding DNA conductivity, one approach is to use experimental data or first principles calculations to simplify a model DNA system for the establishment of a relatively simple model. For example, first principles calculations can obtain all electronic energy levels and wave functions for the small segment of a DNA molecule. At room temperature, electrons at the lower molecular orbitals are very difficult to move, so we only need to consider the energy of electrons in the highest occupied molecular orbital (HOMO) and the energy level and potential occupation at the lowest unoccupied molecular orbital (LUMO). Further approximations can then be made. Each nucleotide is considered an individual unit (or an imaginary "atom," if you will), with each unit possessing two energy levels: the LUMO and HOMO mixed via the adjacent LUMO and HOMO orbitals, and the respective ionization potentials of four different nucleotides. Thus, the calculation can be governed by the charge on the potential energy as a stop at each nucleotide as well as the kinetic energy as it moves to the adjacent nucleotides. With this approximation, DNA molecular structure has been greatly simplified and thus can be utilized for calculations of longer DNA sequences. Of course, the results from these calculations must be taken with a grain of salt, given that the level of simplification may not be entirely accurate in a physical system.

These methods have advantages and disadvantages, with the results being heavily influenced by lack of current computational power or oversimplification. Ultimately, the merit of computational results depends on how well the calculations reproduce experimental data without user bias.

12.1.3 Simplest Equivalent Model of DNA Conduction [3]

The simplest theoretical treatment defines a DNA molecule as a one-dimensional periodic structure. With this approximation, concepts in solid-state electronic band theory can be used to theorize the conductivity of DNA. Depending on the implementation of theory, DNA is either a semiconductor or an insulator.

The method introduced in the following part is one of the effective tight-binding models; the hole (in the DNA, the hole is the charge for electrical conductivity) has a Hamiltonian that can be approximated as follows:

$$H = \sum_i (t_{i,i+1} c_i^+ c_{i+1} + h.c.) + \sum_i \varepsilon_i c_i^\dagger c_i \tag{12.1}$$

Physicists are very familiar with the mathematical form of this *Hamiltonian*, which is, to them, the Anderson localization model used in the study of disordered systems. Here, $c_i^+(c_i)$ is the operator for generation (elimination) of an ***i***-point electron (hole). ε_i is a random diagonal block that follows a certain distribution in the $[-W/2, W/2]$ interval, and W is the so-called disorder degree. N is the number of grid points and t_{ij} is the transition matrix element. If ε_i is disordered and if t_{ij} is a constant, then we refer to it as diagonal disorder; if ε_i is ordered and t_{ij} is disordered, then it is referred to as nondiagonal disorder; if ε_i is disordered and t_{ij} is also disordered, then it is called total disorder.

Equation (12.1) can be rewritten in the form of a tight-binding Schrödinger equation:

$$-t(\psi_{i+1} + \psi_{i-1}) + \varepsilon_i \psi_i = E\psi_i \tag{12.2}$$

This equation is then rewritten in matrix form:

$$\begin{pmatrix} \varepsilon_1 & -t & 0 & 0 & \cdots & 0 & 0 \\ -t & \varepsilon_2 & -t & 0 & \cdots & 0 & 0 \\ 0 & -t & \varepsilon_3 & -t & \cdots & 0 & 0 \\ \vdots & \vdots & \vdots & \vdots & \ddots & \vdots & \vdots \\ 0 & 0 & 0 & 0 & \cdots & \varepsilon_{N-1} & -t \\ 0 & 0 & 0 & 0 & \cdots & -t & \varepsilon_N \end{pmatrix} \begin{pmatrix} \psi_1 \\ \psi_2 \\ \psi_3 \\ \vdots \\ \psi_{N-1} \\ \psi_N \end{pmatrix} = E \begin{pmatrix} \psi_1 \\ \psi_2 \\ \psi_3 \\ \vdots \\ \psi_{N-1} \\ \psi_N \end{pmatrix} \tag{12.3}$$

Through diagonalization of this Hamiltonian, we can get the intrinsic energy and wave function of the system.

For DNA, potential energy ε_i is not entirely generated by random numbers like disordered systems but depends on the arrangement of the nucleotide sequences of DNA. In Eq. (12.1), the first item $(t_{i,i+1})$ is the kinetic energy of the hole for a transition from the lattice point i to lattice point $i+1$. In a strict definition, this term is concerned with the nucleotide species at the two points; however, to simplify the model, all current transition items are first treated as equal, that is $t_{i,i+1} = t_{\mathrm{DNA}}$. The second term (ε_i) is the potential energy for the hole to remain stationary at the lattice point i, related to nucleotide types at that point: $\varepsilon_i = \varepsilon_{\mathrm{A}}, \varepsilon_{\mathrm{T}}, \varepsilon_{\mathrm{C}}, \varepsilon_{\mathrm{G}}$.

In this model, according to the results of first principles calculation, ε_i is 8.24, 9.14, 8.87, or 7.75 eV for A, T, C, and G nucleotides respectively, and represents the potential energy at the lattice point i. It is important to note that these values are calculated on "gaseous" nucleotides, that is, isolated nucleotides in vacuum conditions. In biological systems, the potential energy values will become lower. As for the kinetic energy t_{DNA}, its value lies in the range of 0.03–0.4 eV. Because of the influence of the chemical and biological environments, however, it will result in larger t_{DNA} values. With these considerations, the energy parameter ε_i actually has the aforementioned value, but t_{DNA} is in the range of 0.4–1.0 eV.

As mentioned, the potential energy sequence of DNA is different from the periodic structure, yet it is not a completely disordered system. Accordingly, we need to build a model that is quasi-periodic. Albuquerque and associates used the RS model to generate a four-element (G, C, A, T) quasi-periodic sequence that is based on the general rules as G→GC; C→GA; A→TC; T→TA. Starting from G (the first generation), the initial generations are G, GC, GCGA, GCGAGCTC, GCGAGCTC GCGATAGA.., making F_i the number of elements of the R-S sequence of generation i, and then we have $F_{i+1} = 2F_i (i \geq 1)$. Therefore, the number of elements in the first few generations follows the sequence of 1, 2, 4, 8, 16,. . ., and for the 12th generation the number of lattice points (elements) will be 2,048.

12.1.4 Advantages of DNA Molecular Devices

DNA is considered one of the best candidates for the next generation of nanoelectronics given its variety of electrical properties. The electrical properties of DNA are directly related to the nucleotide sequence, and therefore sequence design can lead to the desired conduction properties. Electrical conductivity is critical for circuit design.

Although scientists have been able to produce nanoelectronic devices in a research laboratory setting, adaptation of these methods to large-scale production is impossible in most instances. For example, the use of an STEM tip to move atoms or atomic groups in a particular arrangement is most definitely not suitable for mass production. Due to the development of biotechnology, the mass production of DNA with desired specific nucleotide sequences is obtainable. Moreover, the ability to fold DNA molecules into a variety of different shapes is easily accomplishable with many known techniques. As such, there is great potential for the use of DNA in the electronics industry, provided that there is complete understanding of the electrical properties of DNA and that they translate well to manufacturing electronic devices.

Irrespective of the electrical properties of DNA, the charged phosphate backbone is a useful template for other materials. One such example is the creation of metallic nanowire using DNA as a template. Metal ions in solution can be adsorbed on the DNA molecule, allowing for easy assembly of metallic nanowires. Such processes make DNA "conductive," regardless of the particular DNA sequence. Note that this method likely results in the loss of biological function of DNA. The creation of Ag nanowires using DNA templates is detailed here. Note that this method can be used, in theory, for most metallic materials.

First, DNA molecules are deposited in the substrate and are immersed in the solution containing Ag^+. The negatively charged phosphate backbone of DNA electrostatically interacts with the Ag^+ in solution. A reducing agent is used to create zero-valent Ag nanowires, templated in a parallel arrangement in the DNA backbone. Electrical measurements have shown that metallic nanowires made using this method have good conductivity. Using this method, metalized DNA molecules can be used to construct components of next-generation ICs.

12.2 DNA Nanotechnology [4–7]

In the DNA replication process, the simplicity and near-constant matching of base pairs coupled with the diversity of genetic information along with the conformational specificity are all desired principles in the design of nanoscale materials. The mutual noncovalent bonding between single-strand DNA molecules via the complementary bases leads to self-assembly of DNA and the further production of a functional assembly of aggregates. Nanotechnology based on the physical and chemical properties of DNA can be referred to as DNA nanotechnology.

DNA used for gene chips has long been the focus of research. Apart from gene chips, DNA molecules are applied in two main areas of nanotechnology. In one area, DNA is used as a template to assemble nanoparticles. Nanoparticles may be incorporated in a DNA template in many different ways. One of the early and main applications of nanoparticle assembly in this regard is the aggregation of Au nanoparticles functionalized with DNA, which was pioneered by the team of Chad Mirkin. Taking advantage of the base-pairing seen in DNA, complementary DNA strands can assemble Au nanoparticles together, changing the system's optical properties and allowing for applications in medical diagnostics and sensing. Likewise, manipulation of the sequence and structure of DNA can yield complex patterns and assemblies of nanoparticles, such as in the preparation of molecular wires.

12.2.1 DNA for the Assembly of Nanoparticles

In certain applications, the assembly and arrangement of nanoparticles are crucial. Assembly methods for nanoparticles can be divided into two categories, namely the colloidal crystal method and the template method. Among them, the template method is the most intensively studied and therefore most developed. The concepts of biomimicry and bio-inspired molecular recognition are becoming a very promising method for assembling nanoparticles. Furthermore, biomolecular templating methods may be more accurate in the control of generating particles with the desired structure, size, and shape to influence properties. Using DNA as a template for sophisticated spatial arrangements, research on DNA-functionalized Au nanoparticles was the first area of aggressive research. As stated, pioneering work in this area was reported by the group of Chad Mirkin at Northwestern University [8]. In their work, a thiol was introduced to the ends of two noncomplementary oligonucleotides and absorbed onto Au nanoparticles using the known affinity of thiols with metallic surfaces. A specifically designed DNA connector was introduced into the system containing a double-strand area and two single-chain areas that are complementary with the two oligonucleotides on the surface of the Au nanoparticles. The addition of this DNA connector causes accumulation and slow sedimentation into a DNA gel. The aggregates contain a large number of DNA molecules with reasonable structural order that can be identified by transmission electron microscopy (TEM). TEM images show a closely assembled monolayer and two-dimensional colloidal aggregates formed by the Au nanoparticles.

Nanoparticle assembly using DNA has considerable advantages over traditional approaches, because DNA can be used as templates for very unique assembly arrangements. As shown, an appropriate choice of a DNA sequence can lead to the precise assembly of nanoparticles through complementary base-pairing. Additionally, because the driving force for assembly is dependent, in part, on the length of the oligonucleotide, this method is suitable for the assembly of aggregates of different sizes.

12.2.2 Driving Force for Self-Assembly of DNA Templates

The earliest reports on applications of DNA as the template for successful self-assembly of nanoparticles were performed using quantum dot materials. Using low concentrations of plasmid DNA mixed with Cd^{2+} solution, the negatively charged phosphate backbone would electrostatically interact with the metal cation. On exposure to H_2S gas, CdS quantum dots were obtained. This early report clearly shows that DNA molecules can play a role in self-assembly and nanoparticle synthesis through electrostatic interactions with the DNA backbone. Using similar methodologies, other self-assembled nanoparticle structures can be made.

When electrostatic interactions are used to create nanomaterials, the self-assembled nanostructures are often in the form of one-dimensional metallic nanowires. For example, electrostatic deposition of Ag^+ onto DNA, followed by reduction of Ag^+ to Ag, successfully created silver nanowires with widths of 100 nm and lengths up to 12 μm. Such metallic nanostructures cannot be made via conventional synthesis techniques. The resulting DNA-templated Ag nanowire exhibited promising electrical properties, suggesting that such materials have promise in future nanoelectronic devices. Using similar methods, Pd, Pt, and Au nanowires have also been prepared.

For inorganic nanomaterials, if the driving force of self-assembly is from biotic–abiotic interactions or base-pair recognition between oligonucleotides, then two-dimensional or three-dimensional networks of assembly systems of nanoparticles can be obtained. For example, a DNA-based reversible assembly process was used to make a macroscopic network of Au nanomaterials [4]. Starting with a gold nanoparticle sol, two thiol-containing nucleotides were added to the Au, wherein the corresponding bases were noncomplementary to each other and thus were noninteracting. By adding a connector DNA exhibiting the complementary bases of both DNA-functionalized Au nanoparticles, the Au nanoparticle would become attached during the sol, changing from a red to a purple color. As the DNA-functionalized Au nanoparticles aggregated beyond the point of colloidal stability, a magenta precipitate would settle out of solution. TEM images (Figure 12.3) show a network structure of Au nanoparticles connected with DNA. Through the use of higher temperatures, this process was reversible if the temperature was greater than the melting temperature (T_m) of DNA. The input of additional energy would disrupt the hydrogen bonding between complementary bases, resulting in dissolution of the composite biological material while generating individual Au particles that were redistributed in solution. Such control of

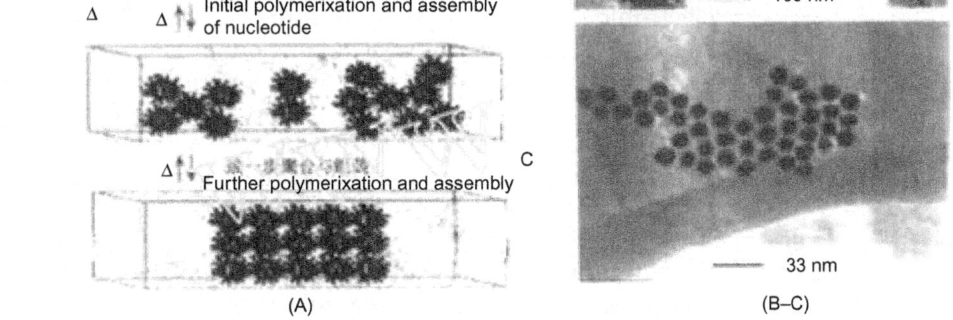

Figure 12.3 Diagram of the mechanism of DNA-based assembly of gold nanoparticles and TEM photographs of the assembled particles.

this self-assembly process could have interesting applications in electronic and optical properties.

12.2.3 DNA as a Template to Prepare Molecular Wire

A molecular wire can be defined as a bridge between molecular devices or between molecular devices and a macroscopic transducer. Effective molecular wires are a key component in making future molecular circuits and must meet the following criteria: reasonable conductivity, a length with low polydispersity, connection points that can be tethered at different components in a system, oxidation–reduction chemistry available at the endpoints, and insulation from the surroundings to prevent any transfer of electrons.

There are many different types of molecular wires, with current research focused on polyphenylacetylene (PPE) and its derivatives, carbon chain-type molecular wires, porphyrin molecular wires, and DNA molecular wires. An obvious common theme of these materials is the delocalization of a π conjugation system for the high mobility of electrons. Many issues currently exist in the field of molecular wires, including chemical synthesis, characterization, performance testing, and appropriate theoretical calculations. Fortunately, new breakthroughs have been made in these areas in recent years.

The structure and properties of DNA are tunable through modifications of the base sequences. Such tunable properties include overall size, molecular rigidity, and double-strand or single-strand confirmations of the overall conformation of DNA. Such modifications can result in a regular two-dimensional or three-dimensional nanoscale network. Such versatility makes DNA very useful in biotechnology and possibly for future nanoelectronic devices. One such example is the use of Au nanoparticles assembled into two-dimensional DNA templates for electronic applications, setting the foundations of a DNA-based molecular memory device. The fabrication of DNA onto a silicon chip can lead to potential applications in semiconductor devices as well. Jiang Xiao-Hua and colleagues performed electric field-induced absorption by use of different types of DNA on the surface of highly regular pyrolytic graphite (HOPG) and obtained the nano-network structures of different morphologies with the goal of preparing a DNA-based molecular wire. Monson and Woolley [7] deposited copper onto surface-absorbed λ-DNA, which created similar structured nanowires. To accomplish this,

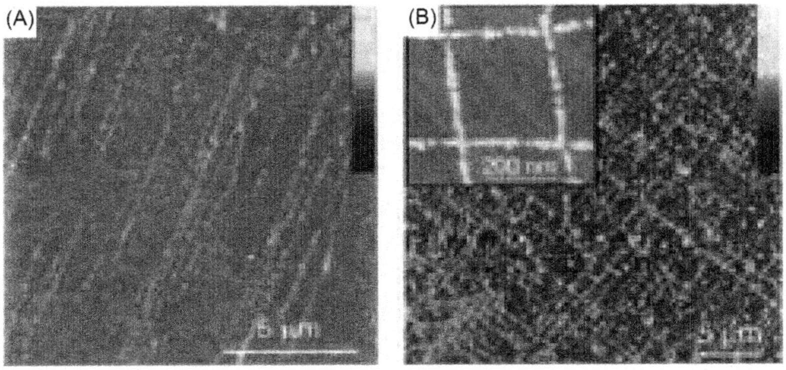

Figure 12.4 AFM image of palladium nanowires on DNA template: (A) one-dimensional parallel palladium nanowires and (B) two-dimensional cross palladium nanowires. Height gauge is (A) 15 nm and (B) 30 nm. From Ref. [9].

λ-DNA is adsorbed on the surface of silicon substrate and treated with a solution of $Cu(NO_3)_2$. Via electrostatic interactions previously described, Cu nanowires were formed after reduction with ascorbic acid. Atomic force microscopy observations show that additional copper and/or ascorbic acid results in a denser metal layer on the DNA surface. These copper nanowires could serve as a connector used in the nanometer integrated circuits and have laid a foundation for the production of functional electronic devices, such as electronic switches and dipoles. Various other metals can be produced in this fashion, including Ag, Pt, and Pd, while DNA is capable of intercalating Zn^{2+}, Ni^{2+}, and Co^{2+} ions, leading to another route to conductive DNA nanowires.

Many applications in nanoscience often require highly ordered nanowire or nanotube arrays. Creating such structures in a controlled and repeatable fashion has been challenging thus far, particularly regarding nanoelectronic applications. Using DNA, such hurdles may be overcome. By aligning DNA molecules on a mica surface using mechanical stresses in the form of compressed air, DNA can template nanowire arrays in an ordered orientation, as shown in Figure 12.4.

12.3 DNA Molecular Motors

12.3.1 Drexler Conjecture

K. Eric Drexler is a renowned space technology expert at Massachusetts Institute of Technology. In 1986, he published the book *The Engines of Creation*. In the book he says that the continuous progress of one's ability in controlling substances will result in the birth of molecular-sized mechanical components that can be assembled into micro-machines that are much smaller than cells, so that humankind can have a direct effect on biological mechanisms. He also points out that the organism is but a combination of molecular machinery derived from natural evolution. Human-made "cell repair machines" manipulated by nanoscale computers may do a better job than the forces of nature. It can fix atoms one-by-one, correct DNA errors, and perform maintenance of all the ingredients of the individual cells. Based on this, he declared in his book, "With the help of such man-made machines as 'cell repair machines,' aging can be terminated, and humankind could eventually overcome death."

Are nanoscale machines realistic and feasible? Can humans develop the technology to build nanomachinery? Would such nanomachines be miniaturized copies of current machines or run with totally different mechanisms?

To answer these questions, we should first clearly define machinery. The machines that are mentioned here are mechanical devices. Large mechanical devices, ranging from aircraft and submarines to toasters and microwave ovens, are very familiar. Mechanical devices have many definitions, but in this analysis a mechanical device is "a device designed for the implementation of mechanical motion." Accordingly, we know that a mechanical force must be used; its mode of operation is determined by its manufacturer. Note that here we do not deliberately draw a line between machinery and machine. In fact, the machine is a special kind of mechanical device; it is a "special mechanical device that can run a specific requirement of mechanical motion with a specific energy input." The concept of a molecular motor is such a machine.

According to this definition of machinery, nanoscale machinery is found in living cells where functional molecular devices, such as proteins or RNA molecules, perform a specific task of mechanical motion. So, the answer to whether nanoscale machinery exists has already been answered by nature.

In cells, there are some molecule devices that are similar to human-made machinery, such as the rotary motor on bacterial cell membranes, which looks like an electric motor. There is also molecular machinery that is slightly similar to some of our human-made machines. One example is the combination of RNA and protein, namely the ribosome is the "machinery" to make proteins in an assembly line fashion. Some molecular machinery, such as topoisomerase (a protein with double-strand DNA that can be entangled together with helicase), has no similarity with mechanical devices in the macro-world. The machinery in the cells is involved in efficient synthesis of macromolecules, including the manufacturing process of molecular self-assembly. This process is much more complicated than the mechanical assembly we are familiar with on the macroscale; therefore, some are inspired by such processes to create models of economic and organizational processes based on nature's complexity.

12.3.2 Molecular Motors [10–13]

The development of modern medical equipment is an integral part of the development of medicine. Medical devices based on nanoscale components will open a new territory of nanomedicine. At present, much research effort is focused on molecular motors. Molecular motors can be defined as molecular machinery with a composite structure of larger molecules with mobility at a molecular level that also uses the smallest entity that can be used as mechanical components. They are driven by way of an external stimulus (such as the use of chemical, electrochemical, photochemical, and other methods to change the environment) to bring significant change to the molecular structure, configuration, or conformation, and it must be ensured that this change is controlled and modulated, akin to human-made machinery.

Current molecular motors comprise biological macromolecules and a nanoassembly that uses chemical energy to perform mechanical work. Natural molecular motors, such as kinesin, RNA polymerase, and myosin, are involved in a series of important life activities in organisms, such as cytoplasmic transport, DNA replication, cell division, and muscle contraction. Molecular motors fall into two general categories: linear progression and the rotating type. Linear molecular motors are biological molecules that can convert chemical energy into mechanical energy and move in a linear fashion. This class includes myosin, kinesin, DNA helicase, and RNA polymerase (and many others), among which muscle myosin is more often studied. With actin representing a linear track, the process of movement is coupled with ATP hydrolysis. The kinesin takes tubulin as the track to move along the microtubules, and thus completes the different transfer functions inside and outside cells. To elucidate the mechanism for kinesin movement, a "hand-over-hand" model was proposed: kinesin's two heads are alternately combined with the microtube, moving on foot along the microtubule. In the course of their "walk," certain conformational changes will occur to convert chemical energy into mechanical energy, achieving linear movement of molecules themselves. At present, between ATP hydrolysis and myosin or kinesin, the chemical and mechanical coupling relationships of their mechanical movement are still unclear. Recent studies have found that they share the same central core structure, and ATP energy is converted into protein conformational movements under similar changes. DNA helicase is a linear molecular motor, with DNA molecules as a "track," and it is coupled with the energy released by ATP hydrolysis. In the release of ADP and phosphate at the same time, the DNA double-strand is separated into two complementary single strands. RNA polymerase in the DNA transcription process will rapidly move along the DNA template, with the consumed energy coming from the nucleotide polymerization and RNA folding reactions.

Rotary molecular motors can convert chemical energy directly into mechanical energy via the hydrolysis reaction. Their structure can be more appropriately called a "motor," because their motion is similar to that seen in human-made motors. The typical rotary molecular motor is the F1-ATP enzyme. ATP enzymes consist of two parts: one is integrated in the mitochondrial membrane, known as F0, and the other part is outside the membrane, known as F1. The subunits a, b, and c in the F0-ATP enzyme constitute the channel flowing through the membrane proton. When the proton flows through F0, the torque will be generated to facilitate the rotation of the g subunit in the F1-ATP enzyme. The clockwise and counterclockwise rotations of subunit g are associated with the ATP synthesis and hydrolysis respectively. The F1-ATP enzyme has a diameter of less than 12 nm, and it can produce forces greater than 100 pN, with a no-load speed of up to 17 rev/s. The F1-ATP enzyme, in combination with the nanomicroelectromechanical system, has become a new kind of nanomechanical device.

12.3.3 Basic Principle of Molecular Motors

The concept of molecular motors was first proposed by the Polish scientist M. Smoluchowski. In the beginning, the feasibility of molecular motors was widely questioned. Later, its feasibility was proven in the studies by Feynman and Magnasco. More molecular motors have been found to exist in biological systems, and the design of artificial molecular motors is also of immense interest to the scientific community.

The concept of the molecular motor has a very broad definition, such as the Brownian motor, molecular motors, atomic machines, ion pumps, and others. Also included are the quantum interval charge pump and the spin pump.

As the name would suggest, the Brownian motor is related to Brownian motion. At the nanoscale and microscale, collisions constantly occur between particles; therefore, particle motion would not follow a fixed path (Figure 12.5) and would undergo random motion. Similarly, a particle's own energy is not fixed, but rather is a randomly distributed range. Random collisions are the physical mechanism causing such an energy distribution. Any random collisions will constitute the driving force found in micro-world movement. Meanwhile, the collision at a certain moment during the process of the particle motion is also likely to have slightly higher energy than the average energy. In addition to the collision, the particles found between the interaction are called viscous force, which may result in the dissipation of particle energy. Figure 12.5 shows the so-called Einstein fluctuation–dissipation theorem.

Random collisions in the microscopic world reflect the nature of the particle movement, sometimes referred to as spontaneous fluctuations. As long as the temperature is sufficient, there is a fluctuation force. Therefore, the "on-table" particles of ⊓-type (Figure 12.6) are likely to fall "under-table" through spontaneous acts of pushing, and it might be squeezed from "under-table" up to "on-table." This can explain why there is a difference between the microscopic world and macroscopic world.

This random fluctuation is also called "white noise." This kind of force is not directional, so there is no effective power

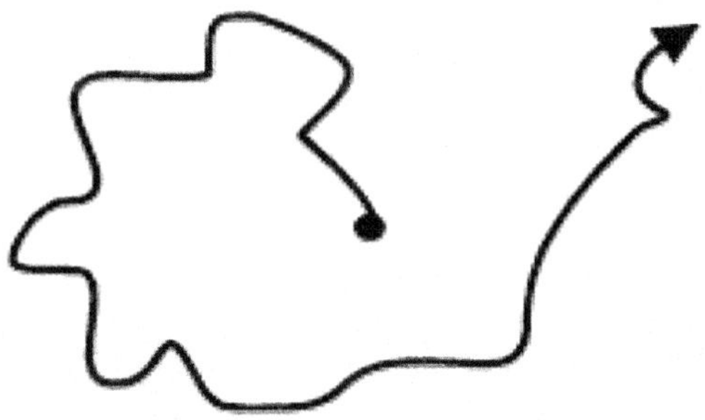

Figure 12.5 Trajectory of Brownian motion. From Ref. [11].

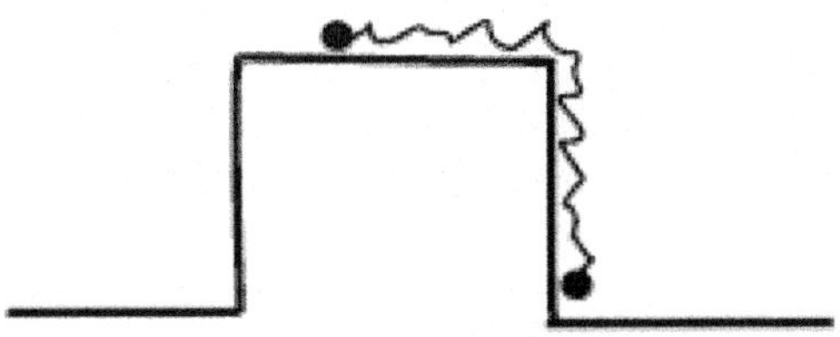

Figure 12.6 ⊓-type potential energy, • for Brownian particles. From Ref. [11].

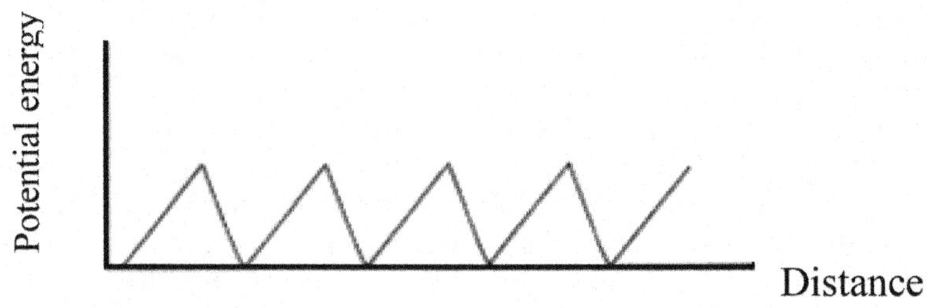

Figure 12.7 Sawtooth model. From Ref. [11].

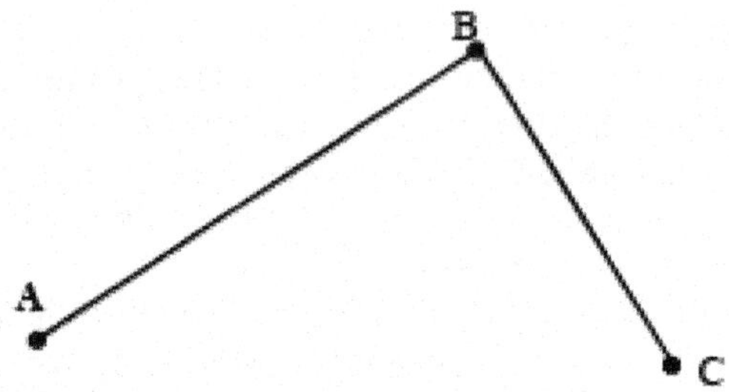

Figure 12.8 Part of sawtooth potential energy. From Ref. [11].

in it. The molecular motor provides the movement at molecular orientation in the absence of an external macroscopic force, which can translate the energy provided by the outside world, for example chemical energy into power to achieve directional movement.

Here, we use a simple model to explain why molecular motors can generate directional movement in the microscopic world, a key basic principle of molecular motors. Feynman has proposed a ratchet potential model (Figure 12.7) to illustrate the feasibility of molecular motors. Here, we analyze the details of its principles.

We already know that on the microscale, when particles are pressed, their energy will increase. On movement, the length of the path is not a concern, for example $\overline{AB}$ or $\overline{BC}$ (Figure 12.8); however, only the energy level of point B is of interest. Therefore, in the absence of external forces and under the condition of random fluctuation, there is a chance that the particle distribution at A is equivalent to that found at C, which is referred to as the "detailed balance" principle.

To make more particles distributed at point C than at point A, more work is needed. Rather than providing a fixed external force (macroforce), we can use an external fluctuating force with an average of zero external forces over time; however, unlike thermal noise, it is unrelated. Because the particle has a force on $\overline{AB}$ that is the same on $\overline{BC}$, the average effect of the external fluctuation forces on the right and left are offset by each other, so there will be no net effect. Therefore, it does not undermine the delicate balance. As mentioned, once the force on $\overline{AB}$ is not equal to that on $\overline{BC}$, thermal noise will not be able to produce such an imbalance to disrupt the system; however, after the external fluctuation force is combined with the forces on $\overline{AB}$ and $\overline{BC}$, the forces on the right and left are no longer in balance and will produce a net force. This results in net flow for promotion of the particle.

12.3.4 DNA Molecular Motors

12.3.4.1 DNA Applications in Molecular Devices

In this section, molecular devices refer to devices at the molecular scale or molecular level, for example molecular wires, molecular switches, molecular rectifiers, molecular memory, molecular motors, molecular sensors, molecular logic, molecular computers, and so on. In recent years, molecular devices have been increasingly studied regarding the function of a single molecule. Another trend lies in the use of organic materials to supersede inorganic materials for enhancing the flexibility of molecules. Clearly, among the various materials, DNA is undoubtedly the most appropriate choice. Therefore, DNA has been given more and more attention in the study of molecular devices. For example, DNA molecules have a natural length, which has been seen as an ideal molecular wire and as the materials to build a molecular wire. Second, DNA molecules based on single-molecule transistors, sensors, molecular motors, and biochips have also been favored by more and more scientific researchers. Here, we focus on DNA molecular motors.

12.3.4.2 DNA Molecular Motors

As mentioned, molecular motors have an interesting potential application. Micromolecular motors can be used as building blocks for nanorobots or nanodevices and components in power sources. The DNA-based molecular motor is one hotspot of the research performed on molecular motors. There have been reports on various methods of using DNA molecular assembly of DNA molecular motors [5–17]. Here, we provide a brief introduction of the work performed by Li and Tan, with the details published in *Nano Letters* (2002) [14].

The key to create a molecular "motor" is the realization of the "automatic" elements. Li and Tan used synthetic single-strand DNA molecules to assemble a DNA molecular motor.

Such a DNA molecular motor works based on the changes of DNA molecules in two different configurations: intramolecular tetraplex (TE) configuration and intermolecular duplex (DU) configuration (Figure 12.9). Interconversion of these two configurations is achieved through DNA hybridization interactions with the strand exchange reaction, so that DNA molecules can be flexible, like an inchworm. Typically, they use the DNA molecule, which contains 17 units and is naturally present in a fourfold chain form. Its two ends are closer, equivalent to the "contracted" state of molecular motors. When this "contracted" DNA molecule finds the target pairing DNA chain α, its fourfold chain will naturally stretch to form a double strand of DNA with α (process I in Figure 12.9), which is equivalent to a "stretched" state of the molecular motor. Then, with the emergence of another DNA chain, β, as a result of the chain substitution reaction, the original chain in the double-strand DNA will be replaced by β (process II in Figure 12.9), where α and β will form a longer, more stable αβ chain. The original chain is released (Figure 12.9 in the process of III), which will reform a fourfold chain structure (process IV in Figure 12.9). Processes I–IV form the complete working course of this molecular motors.

How does one quantify the movement of such DNA molecular motors? We can allow the DNA molecular motor to be connected with an organic fluorophore (abbreviated as F) and a fluorescence quencher (abbreviated as Q) on both of its sides. This movement of the DNA molecular motor can be directly monitored by the fluorescence signal (Figure 12.9, right).

The DNA molecular motor described is a typical one, featuring a simple structure, stable performance, and easy operation, and it can work in solution or on the surface of nanoparticles.

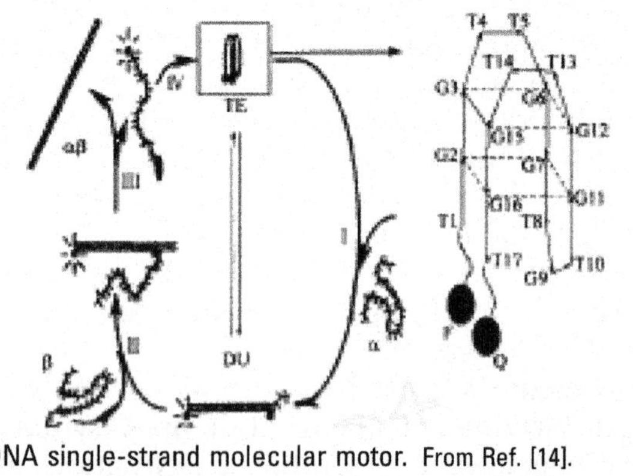

Figure 12.9 DNA single-strand molecular motor. From Ref. [14].

References

[1] Q.T. Shi, DNA electrical transport properties, Physics (bi-monthly) 28 (2006) 654.

[2] G.L. Li, J. Hu, M.G. Li, DNA molecule manipulation and molecular electronics, Micronanoelectron. Technol. 3 (2002) 12.

[3] Z.Z. Guo, Entanglement in one-dimensional Anderson model with long-range correlated-disorder, Chin. Phys. Lett. 25 (2008) 1079–1082.

[4] C.L. Zhu, Y.P. Liu, W.H. Huang, Z.Y. Chen, DNA template self-assembly of nano-particles and the possible applications in nano-electronic devices, Physics 32 (8) (2003) 515–519.

[5] H. Jiang, H.X. Li, X.J. Yang, N.Q. Zhao, Y.C. Liu, W.P. Hu, et al., Advances in DNA determination of the electrical properties and DNA molecular devices, Prog. Nat. Sci. 15 (7) (2005) 769–775.

[6] X.H. Jiang, W.Q. Liu, J.J. Chen, X.Q. Lin, Research and application of DNA nanotechnology, Prog. Chem. 19 (4) (2007) 608–613.

[7] C.F. Monson, A.T. Woolley, DNA-templated construction of copper nanowires, Nano Lett. 3 (2003) 359–363.

[8] C.A. Mirkin, R.L. Letsinger, R.C. Mucic, et al., A DNA based method for rat ionally assembling nanoparticles into macroscopic materials, Nature 382 (1996) 607–609.

[9] Z.X. Deng, C.D. Mao, DNA-templated fabrication of 1D parallel and 2D crossed metallic nanowire arrays, Nano Lett. 3 (2003) 1545–1548.

[10] Z.X. Deng, C.D. Mao, Molecular lithography with DNA nanostructures, Angew. Chem. Int. Ed. 43 (2004) 4068–4070.

[11] D.Y. Yang, Molecular motors and ion pumps, Physics (bi-monthly) 27 (3) (2005) 487–491.

[12] Y.G. Shu, Z.C. Ouyang, Biological molecular motors, Physics 36 (10) (2007) 735.

[13] S.A. Endow, H. Higuchi, A mutant of the motor protein kinesin that moves in both directions on microtubules, Nature 406 (2000) 913–916.

[14] J.J. Li, W.A. Tan, A single DNA molecule nanometer, Nano Lett. 2 (4) (2002) 315–318.

[15] R. Vale, The molecular motor toolbox for intracellular transport, Cell 112 (4) (2003) 467–480.

[16] Y. Chen, M. Wang, C. Mao, An autonomous DNA nanomotor powered by a DNA enzyme, Angew. Chem. Int. Ed. 43 (2004) 3554–3557.

[17] W.B. Sherman, N.C. Seeman, A precisely controlled DNA biped walking device, Nano Lett. 4 (7) (2004) 1203–1207.

INDEX

Note: Page numbers followed by "*f*" and "*t*" refers to figures and tables, respectively.

D

E

F

G

N

CPSIA information can be obtained at www.ICGtesting.com
Printed in the USA
LVOW02*2323290615

444357LV00001B/2/P

9 781455 777549